149
Advances in Biochemical Engineering/Biotechnology

Aims and Scope

This book series reviews current trends in modern biotechnology and biochemical engineering. Its aim is to cover all aspects of these interdisciplinary disciplines, where knowledge, methods and expertise are required from chemistry, biochemistry, microbiology, molecular biology, chemical engineering and computer science.

Volumes are organized topically and provide a comprehensive discussion of developments in the field over the past 3–5 years. The series also discusses new discoveries and applications. Special volumes are dedicated to selected topics which focus on new biotechnological products and new processes for their synthesis and purification.

In general, volumes are edited by well-known guest editors. The series editor and publisher will, however, always be pleased to receive suggestions and supplementary information. Manuscripts are accepted in English.

In references, Advances in Biochemical Engineering/Biotechnology is abbreviated as *Adv. Biochem. Engin./Biotechnol.* and cited as a journal.

More information about this series at http://www.springer.com/series/10

Rainer Krull · Thomas Bley
Editors

Filaments in Bioprocesses

With contributions by

Dominique Anne-Archard · Wellington Balmant
Marcin Bizukojc · Thomas Bley · Jana Blotenberg
G. Corkidi · Markus Fiedler · Luc Fillaudeau
Agenor Furigo · E. Galindo · Krist V. Gernaey
Timo Hagemann · Kim Hansen · A. Holguín-Salas
Rudibert King · Nadia Krieger · Rainer Krull
Stanislaw Ledakowicz · Felix Lenk · Vera Meyer
David Alexander Mitchell · Judith Moench-Tegeder
Tien Cuong Nguyen · Benjamin Nitsche
Daniela Quintanilla · J.A. Rocha-Valadéz
L. Serrano-Carreón · Susanne Steudler
Maura Harumi Sugai-Guérios · Li Tian
Robert Walisko · Thomas Wucherpfennig

Springer

Editors
Rainer Krull
TU Braunschweig
Braunschweig
Germany

Thomas Bley
TU Dresden
Dresden
Germany

ISSN 0724-6145 ISSN 1616-8542 (electronic)
Advances in Biochemical Engineering/Biotechnology
ISBN 978-3-319-20510-6 ISBN 978-3-319-20511-3 (eBook)
DOI 10.1007/978-3-319-20511-3

Library of Congress Control Number: 2015942612

Springer Cham Heidelberg New York Dordrecht London

Printed on acid-free paper

Springer International Publishing AG Switzerland is part of Springer Science+Business Media
(www.springer.com)

Preface

Filamentous cells and organisms are of major biotechnological importance. The technical requirements for the cultivation of *Penicillium* and *Aspergilli* gave a fundamental impulse for the development of sophisticated and powerful bioreactors, enzymes of *Trichoderma* are key players in modern biorefineries, and hairy roots of plant cells are expected to be the producer of interesting new secondary metabolites for pharmaceutical and other applications.

There exist many publications in which the specific biotechnological processes are described and evaluated. Nevertheless, a methodological compendium from the viewpoint of Bioprocess Engineers is missing. This gap shall be filled by this volume of Advance in Biochemical Engineering/Biotechnology. All reviews have a strong focus on new monitoring methods and to some extent their combination, with respect to characterize the morphological structure of the cells or organisms. The generation of mathematical multi-scale models and the application of appropriate control strategies for exploiting the high potential of these interesting and challenging biological systems are described in detail by the corresponding reviews in this issue.

The first review article by Walisko and Moench-Tegeder et al. (Institute of Biochemical Engineering, Technische Universität Braunschweig, Germany) highlights the progress in the characterisation and control of filamentous morphology with particular regard to fungal (eukaryotic) and bacterial (prokaryotic) filamentous morphologies since the mid of the 20th century. Recent strategies for the control of bioparticle shape and bioprocess performance for high product yields of filamentous systems with classical biochemical engineering parameters, targeted morphology engineering and genetic engineering will be discussed.

The contribution of Quintanilla et al. (Department of Chemical and Biochemical Engineering, Technical University of Denmark (DTU), Lyngby, Novozymes A/S, Kalundborg/Bagsværd, Denmark) outlines the important aspects concerning the ways that morphology can be measured in fungal systems. The review discusses critically the main problems to identify unique key parameters and to provide clear relationships between morphology, productivity, shear forces and rheology.

Serrano-Carreón and Galindo et al. (Departamento de Ingeniería Celular y Biocatálisis, Instituto de Biotecnología, Universidad Nacional Autónoma de México, Cuernavaca, México) present the current knowledge on the influence of hydrodynamic conditions on macroscopic morphology development in mycelial microorganisms such as filamentous fungi. Basic aspects of hydrodynamic mechanical stress and mass transfer on mycelial development and their consequences on growth and productivity of pelleted and dispersed mycelia are focused.

The most important scientific issues connected with the research of *Aspergilli*, their metabolites and their omics technologies are summarized in the contribution of Meyer et al. (Institute of Biotechnology, Department Applied and Molecular Microbiology, and Institute of Process Engineering, Department Measurement and Control, Berlin University of Technology, Berlin, Germany). Their overview illustrates the broad portfolio of methods and tools in molecular biotechnology and control engineering to quantify and model fungal morphology.

The contribution of Bizukojc and Ledakowicz (Lodz University of Technology, Department of Bioprocess Engineering, Lodz, Poland) concisely reviews the aspects in bioprocess engineering of the production of the antihypercholesterolemia drug lovastatin by the fungus *Aspergillus terreus* in the submerged cultivation in various scales. As key factors broth viscosity in conjunction with non-Newtonian behaviour of the cultivation broths, and multi-stage oxygen transfer processes were identified. Microscopic (e.g. cell concentration and morphology) and macroscopic (e.g. rheological behaviour, transfer limitation) process parameters as well as biological kinetics during *A. terreus* cultivation considering industrial realistic operating conditions are discussed.

In the review of Sugai-Guério et al. (Departamento de Engenharia Química e Engenharia de Alimentos, Universidade Federal de Santa Catarina, Florianópolis, Brazil, Departamento de Bioquímica e Biologia Molecular and Departamento de Química, Universidade Federal do Paraná, Paraná, Brazil) from the Mitchell-group, the problem of modeling the growth of filamentous fungi is addressed. Such models are necessary for a rational design and processing of solid state fermentations. The authors point out that, until now, there was not a comprehensive modeling approach in this field. They claim that a combination of biological and mass transfer phenomena should be considered in such a model, and they conclude that a discrete lattice-based model which uses differential equations to describe mass balances would be most appropriate for this application.

Solid-state fermentation is recommended for the cultivation of basidiomycetes since it mimics the natural habitat of these fungi. There are some major advantages of solid-state fermentation; however, monitoring key variables like biomass is difficult and makes process design, scale up, and control a real challenge. Steudler and Bley (Institute of Food Technology and Bioprocess Engineering, Technische Universität Dresden, Germany) give, in this review, a current overview of various direct and indirect biomass determination methods, discussing their advantages and disadvantages.

Questions and solutions in the field of monitoring and modeling the growth of hairy roots in bioprocesses are documented and discussed in the review of Lenk and

Bley (Institute of Food Technology and Bioprocess Engineering, Technische Universität Dresden, Germany). It is described that there are no online direct measurement methods for biomonitoring of those processes, but reliable indirect methods for these processes were developed. Four independent modeling architectures in this field are described and compared: continuous models, metabolic flux models, agent-based models, and neural networks. A new agent-based macroscopic with 3D visualization of simulation results is presented in detail.

Tian (Department of Plant Sciences, University of California, Davis, USA) argues that hairy root culture presents an excellent platform for producing valuable plant secondary metabolites. This chapter focuses on several major groups of secondary metabolites that are manufactured by hairy roots established from different plant species, and gives a very broad and thorough survey of this type of new bioprocesses. Additionally, the methods for preservation of hairy roots are also reviewed.

In their article Nguyen et al. (Université de Toulouse; INSA; INRA, Toulouse, France) share their broad expertise concerning the ways that morphological structure of pre-treated particulate lignocellulose suspensions can be characterized by elaborated rheological methods. These media show similar characteristics as particulate cultivation media and broths with filamentous microorganisms. Thus, the described rheological techniques are suitable tools that could be transferred to cultivation media with filamentous microorganisms and/or will be subject of further research with filamentous cultivation broths.

We would like to thank all the authors for their valuable contributions and discussions on this fascinating and inspiring topic. We also want to thank Springer for implementation of this project as well as Thomas Scheper and the Springer team for suggestions, ideas, and patience during the preparation of this volume.

Braunschweig — Rainer Krull
Dresden — Thomas Bley
March 2015

Contents

Adv Biochem Eng Biotechnol (2015) 149: 1–28
DOI 10.1007/10_2015_322

Published Online: 22 March 2015

The Taming of the Shrew - Controlling the Morphology of Filamentous Eukaryotic and Prokaryotic Microorganisms

Robert Walisko, Judith Moench-Tegeder, Jana Blotenberg, Thomas Wucherpfennig and Rainer Krull

Abstract One of the most sensitive process characteristics in the cultivation of filamentous biological systems is their complex morphology. In submerged cultures, the observed macroscopic morphology of filamentous microorganisms varies from freely dispersed mycelium to dense spherical pellets consisting of a more or less dense, branched and partially intertwined network of hyphae. Recently, the freely dispersed mycelium form has been in high demand for submerged cultivation because this morphology enhances the growth and production of several valuable products. A distinct filamentous morphology and productivity are influenced by the environment and can be controlled by inoculum concentration, spore viability, *pH* value, cultivation temperature, dissolved oxygen concentration, medium composition, mechanical stress or process mode as well as through the addition of inorganic salts or microparticles, which provides the opportunity to tailor a filamentous morphology. The suitable morphology for a given bioprocess varies depending on

---

Robert Walisko, Judith Moench-Tegeder: Equal contribution.

R. Walisko · J. Moench-Tegeder · J. Blotenberg · T. Wucherpfennig · R. Krull (✉)
Institute of Biochemical Engineering, Technische Universität Braunschweig, Gaußstraße 17, 38106 Braunschweig, Germany
e-mail: r.krull@tu-braunschweig.de

R. Walisko
e-mail: robert.walisko@tu-braunschweig.de

J. Moench-Tegeder
e-mail: j.moench-tegeder@tu-braunschweig.de

J. Blotenberg
e-mail: j.blotenberg@tu-braunschweig.de

T. Wucherpfennig
e-mail: t.wucherpfennig@tu-braunschweig.de

the desired product. Therefore, the advantages and disadvantages of each morphological type should be carefully evaluated for every biological system. Because of the high industrial relevance of filamentous microorganisms, research in previous years has aimed at the development of tools and techniques to characterise their growth and obtain quantitative estimates of their morphological properties. The focus of this review is on current advances in the characterisation and control of filamentous morphology with a separation of eukaryotic and prokaryotic systems. Furthermore, recent strategies to tailor the morphology through classical biochemical process parameters, morphology and genetic engineering to optimise the productivity of these filamentous systems are discussed.

Keywords Filamentous microorganisms · Image analysis · Morphology · Pellets · Rheology

List of Abbreviations

ALR(s)	Airlift loop reactor(s)
CLSM	Confocal laser scanning microscopy
FBRM	Focused beam reflectance
GFP	Green fluorescent protein
HGU	Hyphal growth unit
K_{BDW}	Rheological parameter, biomass independent consistency coefficient
MN	Morphology number
MPEC	Microparticle-enhanced cultivation
STR(s)	Stirred tank reactor(s)

Contents

1 Introduction

Numerous industrial bioprocesses that use filamentous eukaryotic and prokaryotic microorganisms for the production of organic acids and antibiotics have been established within the last century. The oldest large-scale industrial process exploiting filamentous microorganisms is the production of citric acid by the filamentous fungus *Aspergillus niger,* which was established in 1919 in Belgium, even without an understanding of basic molecular and biochemical engineering principles [1, 2]. Since then, fungal expression and production systems have been developed for the production of pharmaceutical compounds, proteins and especially enzymes [3–5]. For example, *A. niger* has been shown to produce 0.9 g/L of immunglobulin G [6], and 1.6 million tons of citric acid is currently produced per year [7]. Filamentous fungi are superior to other industrially applied organisms because of their extraordinary production and secretion capacities. The ability of certain filamentous fungal organisms, such as *A. niger* and *Trichoderma reesei,* to secrete large quantities of biomass-degrading enzymes provides these organisms with potential applications in biorefineries [5]. Very high titres of hydrolytic enzymes, such as amylases and proteases, can be produced by these species; for example, titres of up to 30 g/L of glucoamylase can be achieved by *A. niger* [8], and industrial strains of *T. reesei* have been shown to secrete 40 g/L of cellulases into the cultivation medium [9]. Driven by advances in equipment and analytics, laboratory-scale fungal cultivation techniques have improved substantially over the last 10–15 years, enabling the completion of reproducible and repeatable bioreactor experiments [10].

Of the bacterial filamentous organisms, the *Actinomycetes*, mostly *Streptomyces* species, are of prime industrial interest because of their production of various secondary metabolites [11]. Despite some obvious differences, the morphological characteristics of *Streptomyces* species and their impact on culture processing are similar to those of fungi, such as *A. niger* [12]. Both grow filamentously, and in submerged cultures, they may both form pellets; therefore, their growth kinetics can be expressed using the same mathematical models [11]. Although there is no general consensus as to how the morphology and other biophysical parameters of *Streptomyces* in submerged cultivation correlate with secondary metabolite production, morphology is recognised as a parameter of prime importance [13] and is, therefore, worth investigating.

An obstacle for filamentous cultivation, however, is the complex morphology during submerged growth in a liquid medium, which can range from freely growing mycelium to clumps and dense circular pellets. The morphology of the mycelium has an enormous impact on the production of enzymes and primary or secondary metabolites [14].

The description of microscopic hyphal growth and macroscopic pellet growth of filamentous microorganisms on solid media and in submerged culture originates from the systematic investigations conducted by Emerson [15], resulting in the development of the cube root law, which describes pellet growth as it occurs in the outer nutrient-supplied shell of constant width. Additional important milestones include studies by Takahashi and Yamada [16] on coagulative and non-coagulative pellet formation, Pirt and Callow [17] and Pirt [18] on growth kinetics and the control of fungal morphology and Trinci [19, 20] and Trinci and Caldwell [21] on the introduction of the hyphal growth unit (*HGU*). Even at the early research stage, Robinson and Davidson [22] addressed large-scale cultivation of eukaryotic systems and Phillips [23] examined oxygen transfer in mycelia. These efforts in the field of filamentous systems were first reviewed by Metz and Kossen [24].

Beginning in the mid-1980s, advances in new analytical methods in molecular biology and biochemical engineering and increased computer capacity reinforced filamentous systems as the focus of various research groups. The Schügerl group characterised the oxygen supply of pellets using microelectrode and microtome techniques [25]. Furthermore, automated image analysis continued growing in importance in the research of filamentous systems. Adams and Thomas [26] innovated the research of filamentous systems by implementing automated image analysis for the characterisation of pellet- and mycelium-like structures [27–30]. Additionally, morphological growth kinetics via computer simulations were studied extensively during this time period [11, 31, 32]. During the early 1990s, the Wösten group pioneered the localisation of growth and protein secretion by the filamentous fungi *A. niger* [33] and addressed the issue of morphology affected by hydrophobin complexes [34, 35]. Engineering issues such as the influence of increased viscosity on the cultivation of filamentous organisms [36–38]; the relationship between morphology and metabolite production [39, 40]; and the impact of reactor geometry, stirrer shape and size, aeration rate and subsequent energy dissipation and mechanical particle stress [41–47] have been investigated extensively. In the mid-2000s, Papagianni [48] and Grimm et al. [49] summarised the influence, kinetics and control of fungal morphology on product formation with classical biochemical engineering parameters in their often cited review articles. Because the characterisation of mycelium-like systems through Euclidean, integral geometry parameters is limited, Papagianni [50] addressed the fractal nature of these aggregates.

Microparticle-enhanced cultivation (*MPEC*) was first applied by Kaup et al. [51] as a novel method to control fungal morphology development to improve productivity in the presence of aluminium oxide and hydrous magnesium silicate particles. The Braunschweig Collaborative Research Centre, *From gene to product,* expanded the range of supplements to talc, silicate and other types of microparticles to control the morphological development of *A. niger* spp. and increase product formation [52–54]. Another sophisticated method to tailor-make fungal morphology was found to be the manipulation of the osmolality of the culture medium; increasing NaCl concentrations led to mycelial growth and enhanced productivity in *A. niger* strains [55]. However, the mycelian growth form leads to a significant increase in culture broth viscosity with a reduced mass transfer and, therefore, requires more power input into the cultivation broth. As morphology, biomass and medium viscosity affect each other, dependencies and correlations are not easy to discover [56]. A comprehensive review on morphology, rheology and productivity has been published recently [53]. It is still not clear how morphology exactly affects productivity, but both parameters are clearly correlated [55].

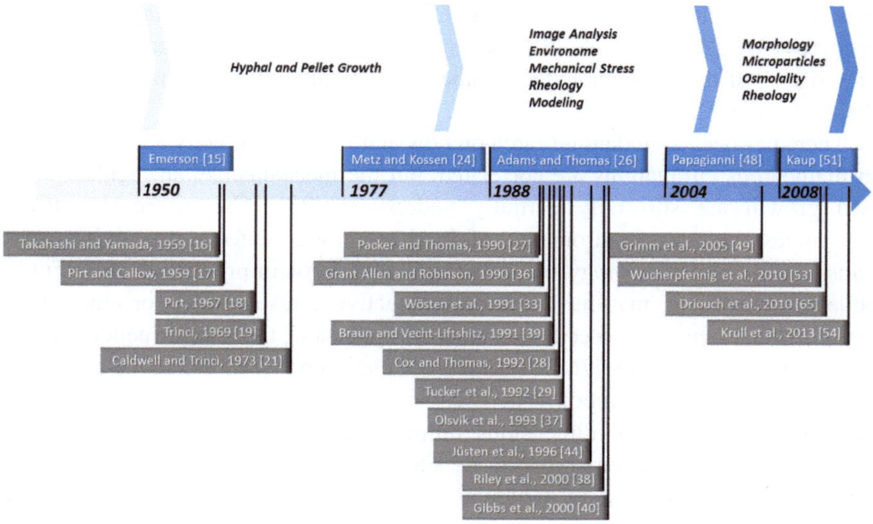

Fig. 1 Timeline of key publications on bioprocesses involving filamentous microorganisms

Several recent reviews have focused on filamentous fungi, covering morphology and productivity [53, 54, 57–60], process design [61] and molecular characteristics [3, 10, 62]. Figure 1 highlights the most frequently cited key publications on bioprocesses involving filamentous microorganisms over the last 65 years.

The aim of this review is to focus on advances in the characterisation and control of filamentous morphology with particular regard to fungal (eukaryotic) and bacterial (prokaryotic) filamentous morphologies. Furthermore, recent strategies for the control of bioparticle shape and bioprocess performance for the optimal productivity of a filamentous system with classical biochemical process parameters, targeted morphology engineering and genetic engineering will be discussed.

2 Eukaryotic Filamentous Morphology

2.1 Interdependence Between Morphology and Productivity

One of the most important industrial aspects of fungal morphology is its relation to productivity because it would be convenient to select productivity levels of cultivation batches solely based on morphologic appearance or to determine early in a batch process how much product will be formed. A relationship between morphology and productivity was therefore the focus of many experiments. Often, optimal product yield corresponded with particular morphologic phenotypes [52, 63–74]. However, not all studies revealed such a link [75, 76]. Other sources confirm a strong relationship between morphological growth form and productivity, with the mycelial morphology being favourable for the production of various enzymes [51, 52, 55, 65, 68, 70]. Generally, the morphologic appearance depends on the sum of the environmental parameters of a cultivation, often summarised as the *environome* [53, 77].

There has been abundant research on how and why morphology and productivity in filamentous cultivations are correlated, but the mechanisms that regulate polarised growth are still only partially understood [78]. The macro-morphology determines the microenvironment of hyphae through effects on mixing, mass transfer and culture rheology, which in turn affect protein production [53]. Fungal pellets, for instance, may have dense and inactive cores due to poor nutrient diffusion, which may lead to cell lysis and, thus, a loss of the interior pellet structure [79]. Wongwicharn et al. [76] revealed a correlation between the active area of the biomass and protein secretion. Microscopic morphology has other indirect effects on productivity. Hyphal dimensions influence the secretion pathway [80], and protein secretion has been shown to be situated at the tips of fungal hyphae [33]. The branching frequency, in particular, is of importance, as several studies have shown that metabolite excretion is situated mostly at the hyphal tips [33, 81, 82]. A great quantity of hyphal tips creating a dense mycelium is, therefore, often linked to increased enzyme production [83].

Recent studies seem to suggest that in addition to morphology, culture heterogeneity has a notable impact on productivity. In two studies, two populations of pellets of different sizes and gene expressions were identified [84, 85]. The population of highly expressed genes of the produced enzymes was smaller than the population with a large pellet diameter, indicating that heterogeneity and pellet size are not related [86], which was confirmed by the finding that heterogeneity in gene expression is observed between pellets with one particular diameter [86]. Unfortunately, with the exception of pellet sizes, no further morphological characterisation was conducted; thus, it remains unclear whether the subpopulations of larger and smaller pellets might have been different in their morphological form as well. Additionally, the condition of conidia used as starter cultures contributes to the heterogeneity in a submerged cultivation, with some evidence that melanin content influences the size of later formed pellets [86, 87]. Further studies showed heterogeneity within the mycelium.

Studies with an abundance of novel and ever cheaper molecular tools available, including gene expression analysis, have shown that the *RNA* composition of the central and peripheral zones of plated colonies of *A. niger* [88] and *A. oryzae* [89] is different. Furthermore, in seven-day-old colonies of *A. niger,* a large percentage of active genes showed a significant difference in *RNA* accumulation between the inner and outer zone of the colony [88]. Further studies have shown the total amount of *RNA* per hyphae to be increased by approximately 50-fold at the periphery of a 1-mm pellet compared to the pellet centre [84]. Different *RNA* profiles, however, might be explained by the absence of *RNA* shown in a recent study covering intra- and intercompartmental streaming of green fluorescent protein (*GFP*) [90].

2.2 Advances in Morphology Characterisation and Whole-Broth Modelling

Fungal morphology is often a bottleneck of productivity in many industrial processes [91], and process optimisation requires the accurate characterisation of interdependencies between process parameters and performance [92]; therefore, from the very beginning of research with filamentous organisms, it has always been a goal to characterise and model morphology [93–95]. For the quantification of filamentous morphology, early studies relied on manual measurements of hyphal growth [96, 97]. Newer studies implement more or less automated steps of computer-aided image analysis [53, 98]. However, no image analysis parameter on its own is sufficient to discriminate between different morphological forms. With the introduction of a multifunctional shape parameter termed the morphology number (*MN*), Wucherpfennig et al. [55] were able to characterise a broad distribution of morphological forms of *A. niger* using just one dimensionless process parameter. A similar approach was proposed in a review about particle shape characterisation by Blott and Pye [99], who reported that no single quantitative measure could provide a full description of a three-dimensional shape, and the use of several shape indices in combination provides the best method of grouping and discriminating between particles. Describing mycelia quantitatively essentially involves the estimation of their space-filling capacity [50], and like many other natural structures, mycelia are approximately fractal [100]. The concept was introduced by Mandelbrot as an extension of the Euclidean geometry [101] and has since been used to measure complex geometric forms [102]. Fractal dimension and branching complexity were previously found to be positively correlated with phenol-oxidase expression in *Pycnoporus cinnabarinus;* both parameters were influenced by media composition [103]. Barry et al. [104] were also able to demonstrate a relationship between the fractal dimension and the *HGU* of a *Penicillium* and an *Aspergillus* strain. Furthermore, it was demonstrated that the branching behaviour of filamentous organisms "cultivated" in silico as "virtual mycelia" can be inferred from the fractal geometry of mycelial architecture, showing the potential application of fractal analysis in morphological quantification [105]. The benefit of the in silico approach is the large amount of *mycelia* that can be generated and analysed compared to the regular growth process in the laboratory. The compliance of fractals with conventional image analysis parameters was also proven in several studies [106, 107]. Another fractal value used for the characterisation of fungal morphology is *lacunarity*, which describes the heterogeneity of a structure or the degree of structural variance within a fractal object [102]. Lacunarity was found to be especially suited to analyse and discriminate between filaments or loose clumps of *A. niger* [56].

The exact manipulation and subsequent precise characterisation of filamentous morphology enables the establishment of holistic mathematical models. Schügerl et al. [108] noted the complex interrelationships among all process characteristics that are related to growth, morphology and physiology and, through this productivity, argued that no general conclusion could be drawn without considering all relevant factors [108, 109]. However, modelling approaches with that many variables tend to become confusing and, more importantly, often lack general applicability. Kossen suggested an approach for the characterisation of productivity considering only whole-broth properties, where productivity is assessed from morphological data, rheological data and mass transfer properties [109]. Because of general problems in measuring the viscosity of filamentous organisms, it was also deemed appropriate to omit the rheological properties of the cultivation broth. Several studies showed that the complex morphology of filamentous organisms is responsible for highly viscous cultivation broths, characterised by shear-rate-dependent viscosities and by yield stress [2, 38, 110, 111]. Furthermore, the consideration of morphology and the corresponding rheology is crucial for the explanation of mass transfer in multiphase systems, as these factors influence the bubble and drop sizes of suspended air and liquids [112]. The flow behaviour of cultivation broths of filamentous organisms differs distinctly from broths of other microorganisms; this matter has been thoroughly reviewed [53]. Tucker as well as Tucker and Thomas [113, 114] were the first authors studying the influences of biomass concentration and mycelial morphology on broth rheology using an industrial strain of *Penicillium chrysogenum.* The authors found the viscosity of the cultivation broth to be dependent on clump properties and studied the effect of biomass separately from that of morphology. The rheological parameter, mostly derived from the power law, could be related to biomass concentration as dry cell weight [114]. Riley et al. [38] later enhanced that model to include morphological parameters such as clump roughness and compactness, which were found to have a definite influence on rheology. The resulting correlations were highly successful at predicting culture rheology for *P. chrysogenum* and *A. oryzae* broths [115]. Wucherpfennig et al. [56] later refined the model for application to *A. niger* culture broth using the earlier introduced *MN* and a rheological parameter termed the biomass independent consistency coefficient (K_{BDW}).

Figure 2 shows the interrelationship among the parameters morphology, rheology and productivity [107]. Here, *A. niger* morphology is shown to directly influence productivity, and with the help of the image analytical generated dimensionless parameter *MN*, it was possible to directly correlate these parameters. Using $K_{BDW,}$ a distinct influence of morphology on culture broth rheology could be shown as well. Furthermore, in an unconventional approach, a correlation between the specific activity of the produced enzyme fructofuranosidase and rheological properties of the cultivation broth was displayed [107].

2.3 Manipulation of Morphology

Distinct fungal morphologies must be generated to achieve optimum performance in biotechnological processes with filamentous organisms, and many different methods have been applied to date (Fig. 3). A classic way to direct filamentous morphology is to use defined and controlled environmental process conditions. The utilisation of different

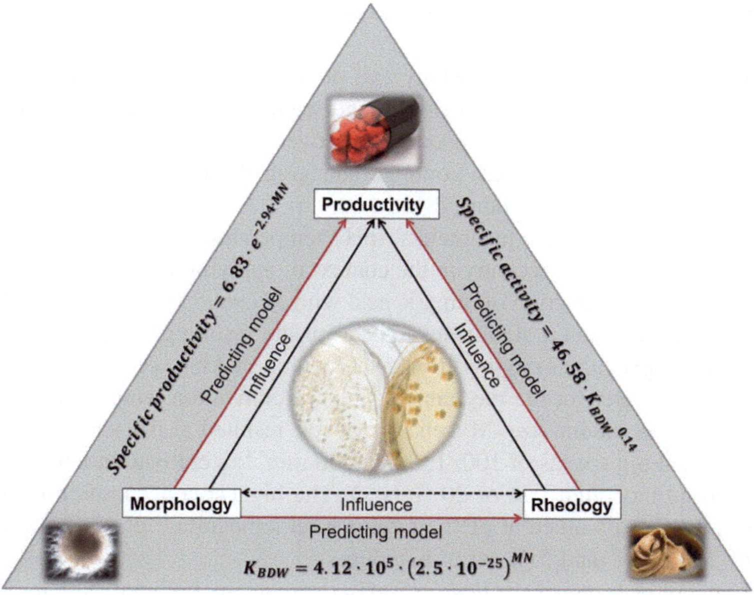

Fig. 2 Established mathematical models relating fungal morphology, rheology and productivity [56]

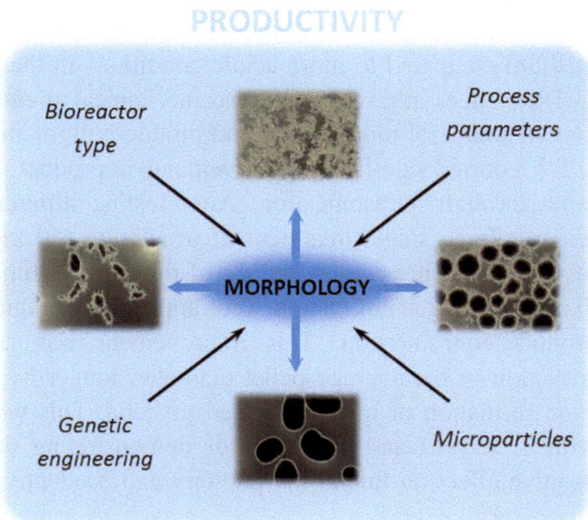

Fig. 3 Influencing parameters on filamentous morphology and the resulting productivity

reactor types, such as stirred tank reactors (*STRs*), airlift loop reactors (*ALRs*) and disposable plastic reactors, extends this concept. However, genetic methods influence microscopic growth events, macroscopic morphology development and resulting process performance. New studies intended to tailor the fungal morphology for enhanced productivity investigated the addition of inorganic salts and microparticles.

2.3.1 Classical Biochemical Process Parameters

The direct interrelation of morphology and productivity of filamentous organisms has resulted in several attempts to control and regulate the morphology of different economically relevant filamentous organisms over the years. These investigations focused on variations in the environome, changing the most important operating parameters, and are still relevant [48].

Some of these process parameters (pH, temperature, shaking intensity and inoculum) were analysed recently in the context of applying brewery wastewater as a substrate for the production of fumaric acid with *Rhizopus oryzae* [116]. All of the investigated factors had a significant impact on the resulting morphology. Pellets were formed at pH values ranging from 5 to 7 with a shaking speed of 200 min^{-1} as an indirect parameter for the mechanical stress, a cultivation temperature of 25 °C and 5 % (v/v) inoculum size. Mycelia, however, resulted at pH values between 4 and 8 with shaking speeds of 100, 150 and 250 min^{-1}; a cultivation temperature of 37 °C; and an inoculum size of 20 % (v/v). The highest concentration of fumaric acid (31 g/L) was obtained at pH 6, a cultivation temperature of 25 °C, a shaking speed of 200 min^{-1} and 5 % (v/v) inoculum.

Current cultivations of *Penicillium griseoroseum* in buffered and unbuffered mineral medium revealed improved production of polygalacturonase (EC 3.4.1.15) and significantly enhanced secretion of pectin lyase (EC 4.2.2.10) at pH values between 6.0 and 6.8, compared to more acidic conditions at pH 3 in the case of unbuffered medium [117]. Furthermore, the biomass increased, the biopellets reached a greater diameter and showed smaller hyphae on the surface in buffered medium compared to more acidic conditions in the unbuffered case.

Differences in power input and mechanical stress by varying stirring rates are a known source of morphology and productivity manipulation [118–121]. Yao et al. [122] reported significant improvements in product formation in response to power input through ultrasonication. After testing different cultivation times of *A. fumigatus* for a single treatment of irradiation (40 kHz, 500 W, 5 min), which all resulted in higher concentrations of the product fumigaclavin C, consecutive irradiations during a cultivation were applied and induced a twofold improvement in product concentration. This improvement was accompanied by a significant reduction in the average pellet diameter, longer hyphae on the pellet surface and even formation of more dispersed mycelia. This work supports previous findings with *A. terreus* and *T. versicolor* demonstrating that ultrasonication can have a positive effect on fungal morphology and productivity [123, 124].

2.3.2 Addition of Inorganic Salts or Microparticles

Effects of Inorganic Salts and Osmolality

The morphology of fungi is known to be affected by the addition of ions; polycations enhance pelleted growth, whereas polyanions suppress it, leading to the conclusion that ions influence the agglomeration behaviour of the naturally charged cell walls [39, 125, 126]. A sophisticated way to tailor fungal morphology was investigated by manipulation of osmolality within the culture medium. First experience was done by Bobowicz-Lassociska and Grajek [127] to increase protein secretion of *A. niger* mycelia by addition of KCl. Fiedurek [128] showed increased activity of *A. niger* expressed glucose oxidase twofold by adding NaCl, thus administering an osmotic shock to the fungus. Increased NaCl concentrations led also to stronger mycelial growth and enhanced productivity of two *A. niger* strains [55]. The morphology and productivity of both strains were shown to be influenced by osmotic pressure. The specific productivity of fructofuranosidase producing strain *A. niger* SKAn 1015 could be increased around 18-fold from 0.5 to 9 U/(mg h). The specific productivity of glucoamylase producing strain *A. niger* AB 1.13 could be elevated using the same procedure. Overall, it has to be noted that observed changes in productivity might not be due to the change in morphology alone. The osmolality might have affected fungal physiology and through this morphology independent of each other. The increase in the observed productivity at higher concentrations of NaCl, however, was shown to correlate with the active surface area of the fungal particles [54].

Microparticle-Enhanced Cultivations (MPEC)

A new and promising approach to control and engineer filamentous morphology without changing the original process parameters is the addition of rather insoluble microparticles to the cultivation medium [56]. Using this *MPEC*, Kaup et al. [51] were able to drastically improve the production of chloroperoxidase in the filamentous fungus *Caldariomyces fumago*. At the same time, the growth morphology of *C. fumago* was changed from pellet to mycelial growth by the addition of microparticles. This work suggested a universal mechanism occurring between microparticles and filamentous organisms because the same effect was observed for talc ($3\ MgO \cdot SiO_2 \cdot H_2O$) and aluminium oxide ($Al_2O_3$) particles.

Furthermore, a test of microparticle addition on the growth morphology of several filamentous organisms including *Penicillium* spp., *Aspergillus* spp., *Streptomyces* spp., (compare Sect. 3.2.3), *Emericella* spp., *Acremonium* spp., *Pleurotus* spp., *Rhizopus* spp. and *Chaetomium* spp. revealed the same effect as on *C. fumago* [51]. The authors suggested an improved supply of oxygen and other nutrients in the mycelium compared to the diffusion-limited pellet as a possible reason for the improved observed performance.

These results and conclusions were confirmed by the data from Driouch et al. [52, 65]. With talc microparticles, a fourfold and twofold enhancement of glucoamylase and β-fructofuranosidase was achieved, respectively. Furthermore, a *GFP*-producing strain of *Aspergillus* spp. supported the assumption that oxygen limitation might be one reason for the worse performance of pellets compared to mycelium. Thin pellet slices were analysed with a confocal laser scanning microscope (*CLSM*) and revealed that efficient *GFP* production only occurred in the oxygen-supplied outer boundary layer of the pellets. In the case of the mycelial growth morphology, which was induced by microparticle addition, biomass was more efficient in *GFP* production.

It is interesting to note that the effect induced by the microparticles on the one hand seems to be of a universal character - different particles lead to a change in morphology of different filamentous organisms (compare Table 1 in [60]), but the molecular interaction between microparticles and an organism, on the other hand, can be different. Whereas talc microparticles induce mycelial growth, titanate microparticles form small-sized pellets with an inner particle core surrounded by biomass, the so-called *core–shell pellet* [70].

Depending on the product to be produced, different morphologies of filamentous organisms can be favourable [40]. Whereas the production of enzymes, such as amylases, works best with a mycelial growth morphology of the producing strain [125, 129], the synthesis of secondary metabolites, such as citric acid or penicillin, is favourable with pellet morphology [130, 131]. Therefore, it is desirable to engineer the morphology according to the product type. Driouch et al. [52] managed to specifically fine tune the size of *A. niger* pellets by using different concentrations of talc ($3\ MgO{\cdot}SiO_2{\cdot}H_2O$) and aluminium oxide ($Al_2O_3$) microparticles. The more particles were present during the cultivation, the smaller the pellet size became. With the highest applied concentrations, no pellets were observed, but mycelial growth was present. The authors also showed that the primary conidia agglomeration shortly after inoculation was probably obstructed by the microparticles, providing a potential explanation for the effect that is induced by the microparticles.

Recently, this technique was employed in the production of lovastatin by *A. terreus* [132]. Here, however, talc microparticles were added to the pre-culture instead of the main culture and were diluted below 5 % of the initial concentration. Nevertheless, an improvement of 60 % (with significant error bars) in the product amount was observed in the main culture at talc concentrations of 12 g/L in pre-culture. These results support the findings by Driouch et al. [52] concerning particle interference with conidia in the beginning of the cultivation.

A comprehensive screening of seventeen different types of microparticles was performed by Etschmann et al. [133] in the context of secondary metabolite production with filamentous fungi. In the case of 2-phenylethanol produced by *A. niger* DSM 821, nine types of the tested microparticles improved production, whereas in the case of 6-pentyl-α-pyrone produced by *T. atroviride,* thirteen different microparticle types were capable of increasing the product yield. Interestingly, the succession of the positively acting microparticles was different for both cases. Thus, despite the general character of the particle effect, there must be case-specific conditions that make one type of particle superior to the other.

Another current publication concerning secondary metabolites addresses an improved lipid accumulation of the fungus *Mortierella isabellina* upon the addition of magnesium silicate microparticles with an average diameter of 10 μm [134]. In addition, with increasing concentrations of the microparticles up to 10 g/L in the main culture, the average pellet size decreased and mycelium morphology was induced. Similar to the findings reported by Driouch et al. [70], the product distribution within the pellets was analysed by fluorescence and *CLSM*. The results are comparable to the *GFP* expression patterns noted above.

The increasing number of publications addressing the *MPEC* of filamentous organisms and productivity enhancement indicates the relevance of this relatively new but very simple and flexible technique.

2.3.3 Genetic Engineering

Genetic analyses include microscopic events such as hyphal tip extension [135, 136], branching [137–139] and hyphal development [140–146]. Rather, rarely is the genetic influence on macroscopic morphology development and the resulting process performance investigated. Recently, a study aimed at engineering a shear-resistant morphology of *A. glaucus* for the production of aspergiolide A, an anti-tumor polyketide, via the deletion of the growth genes *AgkipA* and *AgteaR* [147]. On the microscopic scale, these deletions resulted in meandering hyphae compared to the wild type and had a different impact on the macroscopic pellet morphology. Whereas the maximum product concentration remained unchanged for all three organisms in shaking flasks, production with the *AgkipA* mutant was 24 h faster. Cultivations with the same mutant in a 5 L *STR* resulted in an improved production of aspergiolide A by 82.2 %.

In another comprehensive study, the gene *PcvelA* from *P. chrysogenum* was found to be involved in the expression of biosynthesis and developmental genes and, as a result, was responsible for mycelial growth [148]. The deletion of *PcvelA* resulted in pellet formation after 72 h and was correlated with a down-regulation of genes involved in cell wall metabolism. The authors suggested a down-regulation of genes involved in chitin degradation by the deletion of *PcvelA* as a possible reason for the observed hyper-branching phenotype, pellet formation, and delayed fragmentation and autolysis. This possibility was confirmed by microarray analyses with a *PcvelA* knockout strain in which the gene *PcchiB1*, encoding for a class V chitinase, was down-regulated [149].

2.3.4 Different Types of Bioreactors

Instead of genetically creating a shear-resistant organism, attempts have also been made to reduce shear stress by using alternatives to *STRs* to manipulate growth morphology towards a favourable form. The identification and quantification of mechanical stress, which is related to a distinct morphology of filamentous elements within the process, is one of the main application fields of research in bioprocess engineering and industrial applications. The bioreactor geometry, stirrer shape and size, gassing rate and subsequent dissipated energy are among the fluid dynamic-related criteria that affect the morphology and productivity. The stirrer in *STRs* influences cultivation in several ways: lower stirring speeds lead to inadequate nutrient distribution and gas dispersion, whereas vigorous stirring can cause cell wall damage [26, 49] or sporulation [46]. Particle stress in an *STR*, *ALR* and bubble column reactor have been investigated by varying power input and impeller geometries [44]. For *STRs*, the stress produced by different agitators can be predicted with the aid of a geometrical correlation derived from experimental particle disintegration kinetics [54].

Disposable bag reactors, primarily developed for the cultivation of mammalian cells, were applied for the cultivation of mechanical stress-sensitive basidiomycetes *Flammulina velutipes* and *Pleurotus sapidus* and to produce esterases and peptidases [150]. In both cases, the authors report significant enhancements in product concentration compared to cultivations in regular *STRs*. At the same time, the morphology of the organisms in the disposable bag reactors was directed towards small and distinct pellets with average diameters of between 0.3 and 0.5 mm, in contrast to the large and inhomogeneous lumps of biomass in *STRs*. Another approach to reducing mechanical stress on the organism is to use an *ALR*. In a direct comparison of *STR* and *ALR*, the morphology of *Pycnoporus sanguineus* was directed from compact round pellets in the case of the *STR* towards a more hairy and loose morphology in the *ALR* [151]. The efficiency of exopolysaccharide production, however, was better in the *STR*.

3 Prokaryotic Filamentous Morphology

3.1 Development of Bacterial Morphology

Filamentous bacteria such as *Streptomyces* spp., *Nocardia* spp. and *Lechevalieria* spp. produce a wide range of products, including antibiotics, immunosuppressants and other classes of biologically active substances [152]. The group *Actinomycetes* produces at least two-thirds of the known microbial antibiotics, and 80 % of these are produced by members of the genus *Streptomyces* [153].

Although these microorganisms belong to the prokaryote family, they do not behave as such and they are able to form mycelium and pellets or produce spores.

The life cycle of *Streptomyces* is marked by spore germination, the growth of vegetative mycelium and differentiation to reproductive (aerial) mycelium. The production of antibiotics and other biologically active substances mainly takes place in the stationary growth phase as secondary metabolites [152] (Fig. 4).

Germination can be divided into three different morphological steps from the dormant to vegetative state: darkening, swelling and germ tube emergence. The darkening process depends on exogenous divalent ions (Ca^{2+}, Mg^{2+} or Fe^{2+}) and endogenous energy-providing processes. The swelling requires exogenous carbon sources and is mainly based on the rehydration of the cytoplasm [154]. Germ tube emergence is based on carbon and nitrogen sources. By adding different pairs of amino acids, different nitrogen sources were tested on *S. viridochromogenes* [155] and other carbon compounds on *S. antibioticus* [154]. Shortly after entering the germination phase, the synthesis of macromolecules was observed, starting with

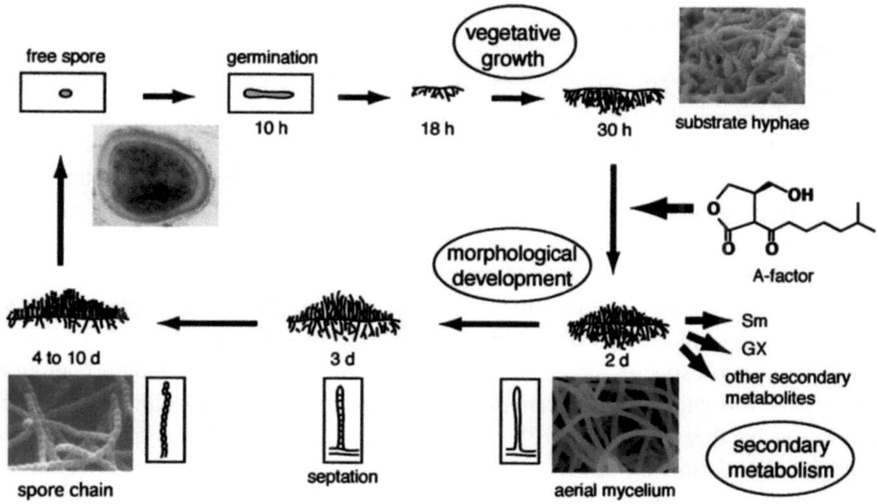

Fig. 4 Life cycle of *Streptomyces griseus* [152]

RNA followed by protein and *DNA* synthesis. The outgrowth is marked by the beginning of *DNA* synthesis [154].

On solid growth media, *Streptomyces* spp. shows two types of mycelia: vegetative and substrate mycelium and a reproductive aerial mycelium [156]. After germination of the spores, growth starts by tip extension and branching of the hyphae. New cell wall material is only synthesised at the tips of the hyphae [157]. A refined investigation showed that the development of vegetative mycelium can be further distinguished into different steps; a young, compartmentalised mycelium undergoes programmed cell death. Viable parts differentiate subsequently into multinucleated second mycelia divided by thin, infrequent and single-layer septa [158–160]. Some of these mycelia grow into the air to form aerial mycelia covered by a hydrophobic shell [161, 162]. Differentiation and septation of the apical cells lead to a chain of spores [163]. The growth of aerial mycelia is partly based on a second round of programmed cell death, where released nutrients are recycled and found in the spores [164]. Chater and Merrick [165] suggested the production of antibiotics as a defensive strategy that therefore takes place in such a late step to protect the lysed substrate against bacteria.

In submerged cultures, most *Streptomyces* spp. do not differentiate or produce spores [166]. Four morphological types have been distinguished depending on perimeter and convex perimeter, pellets, clumps, branched hyphae and unbranched hyphae [167]. The main difference between growth in submerged cultures and on solid medium is the absence of a second round of programmed death and sporulation [159]. A transition phase with transient growth arrest lies between the first vegetative compartmentalised mycelium and the second multinucleated differentiated mycelium accompanied by the first death round [159]. After the transition, the production of antibiotics occurs [168, 169].

Nevertheless, there are reports of sporulation in submerged cultures of several *Streptomyces* spp., including *S. griseus* [170] and others [171–175]. Daza et al. [166] showed the sporulation of several species after a nutritional downshift. The published record of spore survival is over 70 years [176].

According to Takahashi and Yamada [16], pellet formation for all filamentous organisms is generally described by two distinct mechanisms. The coagulative mechanisms involve early spore agglomeration; thus, several spores form one pellet. The non-coagulative mechanism is characterised by the lack of spore agglomeration; thus, one spore forms one pellet. However, regardless of the occurrence of spore agglomeration, grown out and branched hyphae tend to entangle and form clumps during the early stages of growth [27, 177]. Few filamentous bacteria species have been studied in terms of germination, agglomeration and pellet formation. For *S. tendae* and *S. aureofaciens,* non-coagulative pellet formation has been reported [178, 179], whereas *S. hygroscopicus* exhibits coagulative pellet formation [180].

Over the course of a submerged culture, hydrodynamic forces in the cultivation system lead to pellet fragmentation and erosion, thus affecting the pellet size and shape [81–184]. Furthermore, it was reported recently that submerged cultures of *Streptomyces* spp. consist of two pellet populations regardless of the examined strain and culture age: a population of pellets with small diameters of approximately 250 μm, which remained constant over time and was similar for all *Streptomyces* spp. under investigation; and a population of larger pellets with size fluctuations over time and that was dependent on the strain and media used. It was then suggested that classical parameters utilised for morphology engineering, such as pH, temperature, the composition of the medium and power input in the submerged culture, would only affect the population of large pellets, whereas the population of small pellets would remain unaffected [185].

The overall pellet stability depends on extracellular substances, which are still mostly unidentified. However, recently, it was suggested that *DNA* fragments from lysed cells, hyaluronic acid and β-glucan fibres play a major role in pellet cohesion and the architecture of *S. coelicolor* [185–187].

3.2 Similarities to and Differences from Eukaryotic Systems

3.2.1 Control of Morphology and Productivity

Similar to fungi, filamentous bacterial morphology is influenced by several process parameters, and there have been various attempts to correlate morphological characteristics with the productivity of enzymes and secondary metabolites in *Actinomycetes*. Earlier research activities in this field were summarised by Whitaker [188] and Junker et al. [189].

Interestingly, whereas the dependence between morphology and productivity was reported for several organisms and processes, different morphologies seem to be favourable for different products. Clumps and dispersed mycelia were found to be the favourable morphology for retamycin production in *S. olindensis* [190], nystatin production in *S. noursei* [191], tylosin production in *S. fradiae* [183], geldamycin production in *S. hygroscopicus* [192] and the expression of a recombinant Ala-Pro-rich O-glycoprotein in *S. lividans* [193]. However, productivities were higher in pellets than in dispersive growth forms when producing nikkomycin in *S. tendae* and avermectin in *S. avermitilis*, respectively [194, 195]. Furthermore, *S. clavuligerus* and *S. virginiae* exhibited similar productivity levels of clavulanic acid and virginiamycin, respectively, for different morphologies [177, 181]. When freely dispersed mycelia and clumps are favourable, the higher productivities can usually be attributed to an adequate supply of oxygen and other substrates, as the inner core of dense pellets is usually substrate limited. Regarding production systems which favour pelleted growth, Braun and Vecht-Lifshitz [39] postulated that the overproduction of antibiotics is affected by secreted metabolites, such as A-factor in *S. griseus* and factor 1 in *S. aureofaciens*, which can accumulate in pellets and, thus, increase productivity. Manteca et al. [159] developed this hypothesis further, suggesting that the production of secondary metabolites only occurs after mycelium differentiation from a young compartmentalised (first) mycelium to a syncytial (second) mycelium after a round of programmed cell death. This process is also supported by reports of changes in morphology during the course of cultivation, mainly during the transition from the exponential to stationary growth phase [182]. Hence, the correlation between morphology and productivity is strain and product dependent. Nevertheless, the knowledge of tools to morphologically engineer filamentous bacteria helps to further optimise product yields and process control.

3.2.2 Biochemical Process Parameters

Once a desirable morphology and, therefore, high productivity is established for a bioprocess involving filamentous organisms, the optimisation of classical environomal process parameters is a commonly utilised approach. Although several *Streptomyces* strains are used in industrial applications, mostly to produce secondary metabolites such as antibiotics, the effects of process parameters on *Actinomycetes* are not as well studied as their effects on filamentous fungi.

The impact of agitation and aeration on the morphology and productivity was demonstrated for several *Actinomycetes*. In general, high energy dissipations result in small, smooth and compact pellets, whereas lower energy dissipations lead to larger, fluffier pellets [196, 197]. Additionally, low dissolved oxygen levels promote mycelial morphologies, whereas high dissolved oxygen levels result mainly in pelleted growth [179, 195]. Therefore, the effects of agitation and aeration on *Actinomycetes* are comparable to those observed on fungi [198]. A comprehensive review on this subject was recently published by Olmos et al. [12].

In fungal morphology engineering, *pH* is a potent morphology-altering process parameter, as pellets are usually formed in more alkaline cultivation broths; thus, a decrease in *pH* results in mycelial growth [199]. Surprisingly, little is known about how changes in *pH* affect *Streptomyces* spp. morphology in submerged cultures. However, a study on *S. akiyoshiensis* suggests that filamentous bacteria react similarly to pH changes compared with fungi [200].

Another well-studied process parameter capable of altering filamentous morphology is the type and size of the inoculum. In the majority of cases, cultures are inoculated with a spore suspension. The number of viable spores in the inoculum can be directly correlated with the size of the filamentous agglomerate regardless of the strain used. Low inoculum sizes result in large, compact pellets, and pellet sizes decrease with increasing inoculum sizes [179, 201, 202]. Furthermore, vegetative mycelia and matured pellets are commonly used for inoculation in industrial cultivation processes. Such cultures tend to form clumps and pellets rather than freely dispersed hyphae [167]. These observations are consistent with those for filamentous fungi [124, 203].

The composition of growth media is known to be crucial for the development of the desired morphology. In general, complex media more often lead to dispersed mycelial morphology, whereas minimal media often promote pelleted growth [188].

The chosen carbon source has a great influence on the productivity and morphology of several *Streptomyces* spp., though no general conclusions can be drawn, as it mainly depends on the strain and the sought-after metabolite. Reviews on this rather broad field of research have been published in recent years [204, 205].

Yin et al. [195] reported that low nitrogen levels lead to decreased growth rates and therefore promote pelleted growth. Other investigations indicate that the morphology of several strains is also dependent on the type of complex nitrogen source [206–208], although admittedly, it is difficult to distinguish between effects based on the type and the concentration of nitrogen source, respectively, as the total amount of nitrogen in complex media components is usually unknown. Still, it was reported that fungal morphology is altered by different nitrogen sources [48].

Similar to the situation with fungi, a study proved that addition of ions has a distinct impact on the morphology of *S. azureus,* in which it was shown that the addition of excess Mg^{2+}, Ca^{2+} and Mn^{2+} all induced pelleted growth, although supplementation with Ca^{2+} led to smaller pellets than supplementation with Mg^{2+}. Additionally, Mn^{2+} induced pellet formation at much lower levels than Mg^{2+} and Ca^{2+} [209]. In contrast, Dobson and O'Shea [126] observed a pellet formation-enhancing effect for Ca^{2+} in cultivations of *S. hygroscopicus* but a dispersing effect for Mg^{2+}.

Finally, the supplementation of various additives to the growth media is reportedly an easily applicable tool to tailor bacterial morphology. O'Cleirigh et al. [180] investigated the influence of xanthan gum on *S. hygroscopicus* var. *geldanus*. The pellet diameters decreased significantly, presumably because of reduced spore agglomeration. The growth of the same organism was also manipulated by surfactants, namely silicone antifoam, Tween 80 and Triton X-100, leading to decreasing mean pellet diameters with increasing concentrations of surfactant [126]. These observations are contradictory to earlier studies on *S. tendae* reporting increasing pellet sizes with increasing concentrations of Triton X-100, Pluronic F68 and Brij 58 [210]. These contradictory observations can presumably be attributed to the fact that *S. tendae* forms non-coagulative pellets, whereas *S. hygroscopicus* exhibits coagulative pellet formation. Furthermore, Hobbs et al. [211] investigated the effects of the uncharged polymer polyethylene glycol (PEG) 6000 and the negatively charged polymers agar, Junlon PW110 and carboyx-methylene (Carbopol) on *S. lividans*. All of these polymers were capable of directing growth to smaller and looser pellets, although Carbopol and Junlon PW110 were significantly more effective than agar and PEG 6000. Additionally, it was proven that Junlon PW110 would promote dispersed growth in several strains of *S. lividans* as well as *S. coelicolor*. In another approach, the chelating agents bacitracin and EDTA inhibited pellet formation in *S. azureus* [209].

In the case of industrial-scale applications, *Actinomycetes* are mainly cultivated in *STRs* and *ALRs*. Only a few attempts have been made to establish alternative cultivation systems. *S. griseus* was only recently deployed in a tubular biofilm reactor. The formed biofilm was characterised and engineered for use in continuous bioprocess applications [212].

3.2.3 Addition of Microparticles

Research in this field is not as prevalent as for eukaryotic systems, but some data indicate that microparticles are capable of changing the morphology of *Streptomyces* spp., as already shown for fungi [60]. Kaup et al. [51] were able to reduce the pellet size of *S. aureofaciens* ranging between 0.9 and 2.1 mm to a range of 0.06–0.5 mm by the addition of microparticles. In microtitre plates, the optimal influence of 3-mm-diameter glass beads was tested. The average pellet size was reduced from 0.33 to 0.058 mm with significantly improved productivity [213].

3.2.4 Genetic Engineering

There are several important regulators known for *Streptomyces* spp. development and secondary metabolite production. Two early discovered and widely studied regulators are A-factor and bldA [214–217].

The A-factor, discovered by Khokhlov in 1967 as a starting point of the A-factor regulatory cascade, leads to morphological development, sporulation and secondary metabolism. The A-factor binds the A-factor-dependent receptor protein ArpA. The gene *adpA*, encoding the central transcriptional regulator AdpA, is the target of ArpA. For *S. griseus,* it was proven that without the *adpA* gene, no aerial mycelium occurred and no streptomycin was produced. A clone strain with an overexpression of *adpA* showed large amounts of streptomycin at an earlier growth phase. The streptomycin productivity is therefore dependent on the A-factor concentration [152].

Strains with mutations in *bldA* also lack aerial mycelium and antibiotic production in *S. coelicolor*. The formation of aerial mycelium and spores can be restored by adding mannitol or maltose but not the production of antibiotics [217].

It was shown that *bldA* encodes a tRNA with the leucine anticodon UUA [218]. This codon is underrepresented in *Streptomyces* but occurs in *adpA* [219]. Most morphological effects of *bldA* mutations are based on the restricted expression of *adpA* but can be corrected by changing the TTA codon in AdpA [220].

Regulation in *Streptomyces* is still not fully understood, but further research can help understand the development of the morphology and, hence, develop tools derived to improve the productivity of secondary metabolites [220].

3.3 Morphology Characterisation

The morphology of filamentous bacteria not only influences the overall productivity but also affects the rheological properties of the cultivation broth [38]. Thus, the development of suitable methods to characterise the morphology of submerged cultures has been the subject of several studies. Whereas earlier efforts mainly involved digital image analysis [177, 182, 202, 221], more recent studies focused on different techniques, including laser diffraction and focused beam reflectance [222]. Microscopic image acquisition and the subsequent analysis of the pictures provide an extensive quantitative description of the fungal morphology, both micro- and macro-morphological, and, furthermore, the application of advanced imaging techniques such as *CLSM* [223].

Laser diffraction was introduced for the quantification of fungal morphology because of its rapid, unbiased and robust results for particle size distributions as manual sample preparation is eliminated [224]. However, this method fails to report any other descriptive factors such as the roughness or aspect ratio of the particles. The measurement principal relies on the light-scattering pattern produced by suspended particles that are illuminated by a laser. The light-scattering properties are converted to a size distribution by applying mathematical models, namely the Fraunhofer approximation for particles larger than 50 μm and the Mie theory for all particles. However, these mathematical models assume that the measured particles are spherical, which might lead to inaccurate results depending on the aspect ratio and flow orientation of the clumps and pellets [225].

Rønnest et al. [225] evaluated the feasibility of laser diffraction measurements for the characterisation of *S. coelicolor* pellets and clumps by comparing them to conventional image analysis results. It was found that laser diffraction resulted in similar particle size distributions to image analysis, whereas the standard deviations were generally higher, presumably because of discrepancies between the assumptions for the mathematical model and the biological sample and statistical artefacts based on the small number of particles analysed by image analysis. Studies on the application of laser diffraction for other filamentous organisms [55, 198, 224] as well as on other biological samples [226] indicate its applicability for the investigation of size distributions in process environments.

Additionally, there have been efforts to investigate the application of the focused beam reflectance method (*FBRM*) for monitoring the growth [222] and morphological characteristics [227]. A focused laser beam is rapidly rotated to scan the particle suspension, in this case, the culture broth. Particles backscatter the laser light, which is detected. Based on the scan speed, the chord length of the particle can be calculated from the duration of backscattering light pulses. To monitor changes in particle size, shape and concentration, a chord length distribution is calculated by plotting the number of detected particles per second as a function of chord length dimension. Different from most image analysis techniques, FBRM is suitable for online monitoring. Pearson et al. [227] compared *FBRM* with traditional image analysis techniques for cultivations of *S. natalensis*. It was found that whereas *FBRM* is not applicable for characterising filamentous morphology through shape descriptors such as roughness and aspect ratio, the obtained chord length distributions enable a characterisation of morphology to some extent. However, it is not possible to directly correlate *FBRM* data to image analysis results. Additionally, it was proven that *FBRM* is a valid measurement method for biomass concentrations in cultivations involving filamentous microorganisms [222]. Furthermore, Grimm et al. deployed FBRM for studying the agglomeration kinetics of spores [228, 229].

4 Conclusions and Future Perspectives

In recent decades, suitable and powerful high-performance control strategies have been developed for a tailor-made generation of product-optimised filamentous systems. Great achievements in the different fields of biotechnology provide tools to direct the morphology of filamentous systems towards high productivity.

The major efforts in the morphological characterisation of filamentous systems show a close connection among cultivation conditions, morphology and metabolic characteristics of the individual cells. Closing the gap between the metabolic processes and the engineering level is the most interesting aspect for future research. Still, the underlying metabolic and regulatory mechanisms that lead to the formation of a highly productive agglomerate shape are far from understood. Experimental and in silico techniques provide powerful tools to gain a better understanding of the complex biological and technical aspects of filamentous systems. Recent interesting approaches include morphology engineering by osmolality variation and microparticle addition. These promising and simple techniques are currently being evaluated for many different systems and have the potential to enhance the productivity of industrial filamentous biotechnological processes.

To direct a production strain towards a suitable production candidate in industrial use, a heterogeneous filamentous system must be analysed in its detailed subpopulations. Hence, the development of advanced, powerful and automated online image analysis techniques and modelling tools are desirable for the high-quality prediction of culture productivity with minor deviations.

The key to understanding and rationally improving eukaryotic and prokaryotic filamentous systems as a common platform technology depends on overcoming the existing collaboration barriers in the fields of genetic engineering, molecular biotechnology and bioprocess engineering.

Acknowledgments The authors gratefully acknowledge financial support from Alliance Industrial Research within the project *Micro-particle based cultivation of filamentous fungi* (No. 16926/N, R. Walisko); the Lower Saxony Ministry for Science and Culture for the PhD scholarships in the Graduate Schools *Novel synthesis and formulation methods for poorly soluble drugs and sensitive biopharmaceuticals (SynFoBia)* (J. Moench-Tegeder) within the Center for Pharmaceutical Process Engineering (PVZ) and *Microbial Natural Products (MINAS)* (J. Blotenberg) at the Technische Universität Braunschweig, Germany.

References

1. Sauer M, Porro D, Mattanovich D, Branduardi P (2008) Trends Biotechnol 26:100
2. Papagianni M (2007) Biotechnol Adv 25:244
3. Meyer V (2008) Biotechnol Adv 26:177
4. Lubertozzi D, Keasling JD (2009) Biotechnol Adv 27:53
5. Baker SE (2013) Ind Biotechnol 9:105
6. Ward M, Lin C, Victoria DC, Fox BP, Fox JA, Wong DL, Meerman HJ, Pucci JP, Fong RB, Heng MH, Tsurushita N, Gieswein C, Park M, Wang H (2004) Appl Env Microbiol 70:2567
7. Berovic M, Legisa M (2007) In: El-Gewely MR (ed) Biotechnology annual review. Elsevier, pp 303–343
8. Withers JM, Swift RJ, Wiebe MG, Robson GD, Punt PJ, van den Hondel CA, Trinci APJ (1998) Biotech Bioeng 4:407
9. Gusakov A (2011) Trends Biotechnol 29:419
10. Workman M, Andersen MR, Thykaer J (2013) Ind Biotechnol 9:337
11. Nielsen J (1996) Trends Biotechnol 14:438
12. Olmos E, Mehmood N, Haj Husein L, Goergen JL, Fick M, Delaunay S (2013) Bioprocess Biosyst Eng 36:259
13. Rioseras B, López-García MT, Yagüe P, Sánchez J, Manteca Á (2014) Bioresour Technol 151:191
14. Krijgsheld P, Bleichrodt R, van Veluw GJ, Wang F, Müller WH, Dijksterhuis J, Wösten HAB (2013) Stud Mycol 74:1
15. Emerson S (1950) J Bacteriol 60:221
16. Takahashi J, Yamada K (1959) Nippon Nōgeikagaku Kaishi 33:707
17. Pirt SJ, Callow DS (1959) Nature 184:307
18. Pirt SJ (1967) J Gen Microbiol 47:181
19. Trinci AP (1969) J Gen Microbiol 57:11
20. Trinci AP (1974) J Gen Microbiol 81:225
21. Caldwell IY, Trinci APJ (1973) Arch Microbiol 88:1
22. Robinson RF, Davidson RS (1959) The large-scale growth of higher fungi. In: Umbreit WW (ed) Advances in applied microbiology, vol 1. Academic Press, US, pp 261–278
23. Phillips DH (1966) Biotechnol Bioeng 8:456
24. Metz B, Kossen NWF (1977) Biotechnol Bioeng 14:781
25. Wittler R, Baumgartl H, Lübbers DW, Schügerl K (1986) Biotechnol Bioeng 28:1024
26. Adams HL, Thomas CR (1988) Biotechnol Bioeng 32:707
27. Packer HL, Thomas CR (1990) Biotechnol Bioeng 35:870
28. Cox PW, Thomas CR (1992) Biotechnol Bioeng 39:945

29. Tucker KG, Kelly T, Delgrazia P, Thomas CR (1992) Biotechnol Progr 8:353
30. Paul G, Thomas C (1998) In: Schügerl K (ed) Advances in Biochemical Engineering/ Biotechnology, vol 60. Springer, Berlin, pp 1–59
31. Bartnicki-Garcia S, Hergert F, Gierz G (1989) Protoplasma 153:46
32. Prosser JI, Tough AJ (1991) Crit Rev Biotechnol 10:253
33. Wösten HAB, Moukha SM, Sietsma JH, Wessels JGH (1991) J Gen Microbiol 137:2017
34. de Vries OMH, Fekkes MP, Wösten HAB, Wessels JGH (1993) Arch Microbiol 159:330
35. Wösten HAB, Schuren FHJ, Wessels JGH (1994) EMBO J 13:5848
36. Grant Allen D, Robinson CW (1990) Chem Eng Sci 45:37
37. Olsvik ES, Tucker KG, Thomas CR, Kristiansen B (1993) Biotechnol Bioeng 42:1046
38. Riley GL, Tucker KG, Paul GC, Thomas CR (2000) Biotechnol Bioeng 68:160
39. Braun S, Vecht-Lifshitz SE (1991) Trends Biotechnol 9:63
40. Gibbs PA, Seviour RJ, Schmid F (2000) Crit Rev Biotechnol 20:17
41. Nienow AW (1990) Trends Biotechnol 8:224
42. Makagiansar HY, Ayazi Shamlou P, Thomas CR, Lilliy MD (1993) Bioprocess Eng 9:83
43. Jüsten P, Paul GC, Nienow AW, Thomas CR (1996) Biotech Bioeng 52:672
44. Henzler HJ (2000) In: Schügerl K, Kretzmer G (eds) Advances in Biochemical Engineering/ Biotechnology, vol 67. Springer, Berlin, pp 35–82
45. Mahnke EU, Büscher K, Hempel DC (2000) Chem Eng Technol 23:509
46. Rocha-Valadez JA, Hassan M, Corkidi G, Flores C, Galindo E, Serrano-Carreón L (2005) Biotechnol Bioeng 91:54
47. Rocha-Valadez JA, Galindo E, Serrano-Carreón L (2007) J Biotechnol 130:394
48. Papagianni M (2004) Biotechnol Adv 22:189
49. Grimm LH, Kelly S, Krull R, Hempel DC (2005) Appl Microbiol Biotechnol 69:375
50. Papagianni M (2006) Microb Cell Fact 5:5
51. Kaup B-A, Ehrich K, Pescheck M, Schrader J (2008) Biotechnol Bioeng 99:491
52. Driouch H, Sommer B, Wittmann C (2009) Biotechnol Bioeng 105:1058
53. Wucherpfennig T, Kiep KA, Driouch H, Wittmann C, Krull R (2010) In: Allen SS, Laskin I, Geoffrey MG (eds) Advances in applied microbiology. Academic Press, US, pp 89–136
54. Krull R, Wucherpfennig T, Esfandabadi ME, Walisko R, Melzer G, Hempel DC, Kampen I, Kwade A, Wittmann C (2013) J Biotechnol 163:112
55. Wucherpfennig T, Hestler T, Krull R (2011) Microb Cell Fact 10:58
56. Wucherpfennig T, Lakowitz A, Krull R (2013) J Biotechnol 163:124
57. Krull R, Cordes C, Horn H, Kampen I, Kwade A, Neu TR, Nörtemann B (2010) Adv Biochem Eng Biotechnol 121:1
58. Barry DJ, Williams GA (2011) J Microsc 244:1
59. Ward OP (2012) Biotechnol Adv 30:1119
60. Walisko R, Krull R, Schrader J, Wittmann C (2012) Biotechnol Lett 34:1975
61. Posch AE, Herwig C, Spadiut O (2013) Trends Biotechnol 31:37
62. Meyer V, Wu B, Ram A (2011) Biotechnol Lett 33:469
63. Papagianni M, Mattey M (2006) Microb Cell Fact 5:3
64. Ahamed A, Vermette P (2009) Bioresour Technol 100:5979
65. Driouch H, Roth A, Dersch P, Wittmann C (2010) Appl Microbiol Biotechnol 87:2011
66. Sitanggang AB, Wu H-S, Wang SS, Ho Y-C (2010) Bioresour Technol 101:3595
67. Choy V, Patel N, Thibault J (2011) Biotechnol Progr 27:1544
68. Driouch H, Roth A, Dersch P, Wittmann C (2011) Bioeng Bugs 2:100
69. Babič J, Pavko A (2012) J Ind Microbiol Biotechnol 39:449
70. Driouch H, Hänsch R, Wucherpfennig T, Krull R, Wittmann C (2012) Biotechnol Bioeng 109:462
71. Liu Y-S, Wu J-Y (2012) J Ind Microbiol Biotechnol 39:623
72. Singh K, Wangikar P, Jadhav S (2012) J Ind Microbiol Biotechnol 39:27
73. Tepwong P, Giri A, Ohshima T (2012) Mycoscience 53:102
74. Wucherpfennig T, Lakowitz A, Driouch H, Krull R, Wittmann C (2012) J Vis Exp 61:e4023
75. Amanullah A, Blair R, Nienow AW, Thomas CR (1999) Biotechnol Bioeng 62:434

76. Wongwicharn A, McNeil B, Harvey LM (1999) Biotechnol Bioeng 65:416
77. Deckwer WD, Hempel DC, Zeng AP, Jahn D (2006) Eng Life Sci 6:455
78. Amicucci A, Balestrini R, Kohler A, Barbieri E, Saltarelli R, Faccio A, Roberson RW, Bonfante P, Stocchi V (2011) Fungal Genet Biol 48:561
79. Hille A, Neu TR, Hempel DC, Horn H (2009) Biotechnol Bioeng 103:1202
80. McIntyre M, Müller C, Dynesen J, Nielsen J (2001) In: Nielsen J (ed) Advances in Biochemical Engineering/Biotechnology, vol 73. Springer, Berlin, pp 103–128
81. Gordon CL, Archer DB, Jeenes DJ, Doonan JH, Wells B, Trinci APJ, Robson GD (2000) J Microbiol Meth 42:39
82. Müller C, McIntyre M, Hansen K, Nielsen J (2002) Appl Env Microbiol 68:1827
83. Spohr A, Carlsen M, Nielsen J, Villadsen J (1997) Biotechnol Lett 19:257
84. de Bekker C, van Veluw GJ, Vinck A, Wiebenga LA, Wösten HAB (2011) Appl Env Microbiol 77:1263
85. van Veluw GJ, Teerstra WR, de Bekker C, Vinck A, Müller WH, Arentshorst M, van der Mei HC, Ram AF, Dijksterhuis J, Wösten HAB (2013) Stud Mycol 74:47
86. Wösten HAB, Veluw GJ, Bekker C, Krijgsheld P (2013) Biotechnol Lett 35:1155
87. Priegnitz B-E, Wargenau A, Brandt U, Rohde M, Dietrich S, Kwade A, Krull R, Fleißner A (2012) Fungal Genet Biol 49:30
88. Levin AM, de Vries RP, Conesa A, de Bekker C, Talon M, Menke HH, van Peij NN, Wösten HAB (2007) Eukaryot Cell 6:2311
89. Masai K, Maruyama J, Sakamoto K, Nakajima H, Akita O, Kitamoto K (2006) Appl Microbiol Biotechnol 71:881
90. Bleichrodt R, Vinck A, Krijgsheld P, van Leeuwen MR, Dijksterhuis J, Wösten HAB (2013) Stud Mycol 74:31
91. Colin VL, Baigorí MD, Pera LM (2013) AMB Express 3:1
92. Posch AE, Herwig C (2014) Biotechnol Prog 30:689
93. Edelstein L, Hadar Y (1983) J Theor Biol 105:427
94. Aynsley M, Ward AC, Wright AR (1990) Biotechnol Bioeng 35:820
95. Tough AJ, Pulham J, Prosser JI (1995) Biotechnol Bioeng 46:561
96. van Suijdam JC, Metz B (1981) Biotechnol Bioeng 23:111
97. Metz B, de Bruijn EW, van Suijdam JC (1981) Biotechnol Bioeng 23:149
98. Papagianni M (2014) J Microb Biochem Technol 6:189
99. Blott SJ, Pye K (2008) Sedimentology 55:31
100. Obert M, Pfeifer P, Sernetz M (1990) J Bacteriol 172:1180
101. Mandelbrot BB (1982) The fractal geometry of nature. W. H. Freeman and Co., New York
102. Smith TG, Lange GD, Marks WB (1996) J Neurosci Meth 69:123
103. Jones CL, Lonergan GT (1997) Biotechnol Lett 19:65
104. Barry DJ (2009) ISAST Trans Electron Signal Process 4:71
105. Barry DJ (2013) Biotechnol Bioeng 110:437
106. Finkler L, Ginoris Y, Luna C, Alves T, Pinto J, Coelho M (2007) World J Microb Biot 23:801
107. Wucherpfennig T (2013) Cellular morphology: a novel process parameter for the cultivation of eukaryotic cells. Cuvillier Verlag, Göttingen
108. Schügerl K, Gerlach SR, Siedenberg D (1998) Adv Biochem Eng Biotechnol 60:195
109. Kossen NW (2000) Adv Biochem Eng Biotechnol 70:1
110. Charles M (1978) In: Ghose TK, Fiechter A, Blakebrough N (eds) Advances in Biochemical Engineering/Biotechnology, vol 8. Springer, Berlin, pp 1–62
111. Olsvik ES, Kristiansen B (1992) Biotechnol Bioeng 40:1293
112. Lucatero S, Larralde-Corona CP, Corkidi G, Galindo E (2003) Biotechnol Prog 19:285
113. Tucker KG (1994) Relationship between mycelial morphology biomass concentration and broth rheology in submerged fermentations. University of Birmingham, Birmingham
114. Tucker KG, Thomas CR (1993) Trans Inst Chem Eng 71:111
115. Riley GL, Thomas CR (2010) Biotechnol Lett 32:1623
116. Das RK, Brar SK (2014) Appl Biochem Biotechnol 172:2974

117. Teixeira JA, Correa TLR, de Queiroz MV, de Araujo EF (2014) J Basic Microbiol 54:133
118. König B, Seewald C, Schügerl K (1981) Eur J Appl Microbiol Biotechnol 12:205
119. Smith JJ, Lilly MD, Fox RI (1990) Biotechnol Bioeng 35:1011
120. Ujcová E, Fencl Z, Musílková M, Seichert L (1980) Biotechnol Bioeng 22:237
121. Gomez R, Schnabel I, Garrido J (1988) Enzyme Microb Technol 10:188
122. Yao L-Y, Zhu Y-X, Jiao R-H, Lu Y-H, Tan R-X (2014) Bioresour Technol 159:112
123. Sainz-Herran N, Casas-Lopez JL, Sanchez-Perez JA, Chisti Y (2008) J Chem Technol Biotechnol 83:593
124. Wang F, Ma A-Z, Guo C, Zhuang G-Q, Liu C-Z (2013) Ultrason Sonochem 20:118
125. Elmayergi H, Scharer JM, Moo-Young M (1973) Biotechnol Bioeng 15:845
126. Dobson LF, O'Shea DG (2008) Appl Microbiol Biotechnol 81:119
127. Bobowicz-Lassociska T, Grajek W (1995) Acta Biotechnol 15:277
128. Fiedurek J (1998) J Basic Microbiol 38:107
129. Ruohang W, Webb C (1995) Biotechnol Tech 9:55
130. Clark DS, Ito K, Horitsu H (1966) Biotech Bioeng 8:465
131. Schügerl K, Wittler R, Lorenz T (1983) Trends Biotechnol 1:120
132. Gonciarz J, Bizukojc M (2014) Eng Life Sci 14:190
133. Etschmann MMW, Huth I, Walisko R, Schuster J, Krull R, Holtmann D, Wittmann C, Schrader J (2014) Yeast. doi:10.1002/yea.3022
134. Gao D, Zeng J, Yu X, Dong T, Chen S (2014) Biotechnol Bioeng 111:1758
135. Smith DJ, Payton MA (1994) Mol Cell Biol 14:6030
136. Tinsley JH, Lee IH, Minke PF, Plamann M (1998) Mol Gen Genet 259:601
137. Brunt SA, Borkar M, Silver JC (1998) Fungal Genet Biol 24:310
138. Sone T, Griffiths AJF (1999) Fungal Genet Biol 28:227
139. Veses V, Casanova M, Murgui A, Domínguez Á, Gow NAR, Martínez JP (2005) Eukaryot Cell 4:1088
140. Borgia PT, Iartchouk N, Riggle PJ, Winter KR, Koltin Y, Bulawa CE (1996) Fungal Genet Biol 20:193
141. Goh J, Kim KS, Park J, Jeon J, Park S-Y, Lee Y-H (2011) Fungal Genet Biol 48:784
142. Ichinomiya M, Motoyama T, Fujiwara M, Takagi M, Horiuchi H, Ohta A (2002) Microbiol 148:1335
143. Lettner T, Zeidler U, Gimona M, Hauser M, Breitenbach M, Bito A (2010) PLoS ONE 5: e11993
144. Wang G, Wang C, Hou R, Zhou X, Li G, Zhang S, Xu JR (2012) PLoS ONE 7:e38324
145. Zhao P-B, Ren A-Z, Xu H-J, Li D-C (2010) J Microbiol Biotechnol 20:208
146. Zhong YH, Wang TH, Wang XL, Zhang GT, Yu HN (2009) Fungal Genet Biol 46:255
147. Cai M, Zhang Y, Hu W, Shen W, Yu Z, Zhou W, Jiang T, Zhou X, Zhang Y (2014) Microb Cell Factories 13:73
148. Hoff B, Kamerewerd J, Sigl C, Mitterbauer R, Zadra I, Kurnsteiner H, Kuck U (2010) Eukaryot Cell 9:1236
149. Kamerewerd J, Zadra I, Kürnsteiner H, Kück U (2011) Microbiol 157:3036
150. Jonczyk P, Takenberg M, Hartwig S, Beutel S, Berger RG, Scheper T (2013) J Biotechnol 167:370
151. Cao J, Zhang HJ, Xu CPJ (2014) J Taiwan Inst Chem Eng 45:2075
152. Ohnishi Y, Yamazaki H, Kato J-Y, Tomono A, Horinouchi S (2005) Biosci Biotechnol Biochem 69:431
153. Bibb MJ, Buttner KF, Chater KF, Hopwood DA (eds) (2000) Practical streptomyces genetics. John Innes Foundation, Norwich
154. Hardisson C, Manzanal MB, Salas JA, Suárez JE (1978) J Gen Microbiol 105:203
155. Hirsch CF, Ensign JC (1976) J Bacteriol 126:13
156. Hodgson DA (1992) Prokaryotic structure and function. Cambridge University Press, Cambridge, pp 407–440
157. Flärdh K (2003) Curr Opin Microbiol 6:564
158. Manteca A (2005) Microbiol 151:3689

159. Manteca A, Alvarez R, Salazar N, Yague P, Sanchez J (2008) Appl Environ Microbiol 74:3877
160. Yagüe P, López-García MT, Rioseras B, Sánchez J, Manteca A (2013) FEMS Microbiol Lett 342:79
161. Hopwood DA, Glauert AM (1961) J Gen Microbiol 26:325
162. Wildermuth H (1972) Arch Mikrobiol Arch 81:309
163. Flärdh K, Buttner MJ (2009) Nat Rev Microbiol 7:36
164. Miguélez EM, Hardisson C, Manzanal MB (1999) J Cell Biol 145:515
165. Chater KF, Merrick MJ (1979) Devolpment biology of prokaryotes. In: Parish JH (ed) Studies in Microbiology, vol 1. University of California Press, Berkeley, pp 93–114
166. Daza A, Martín JF, Dominguez A, Gil JA (1989) J Gen Microbiol 135:2483
167. Denser Pamboukian CR, Guimarães LM, Facciotti MCR, Braz J (2002) J Microbiol 33:17
168. Huang J (2001) Genes Dev 15:3183
169. Neumann T, Piepersberg W, Distler J (1996) Microbiol 142:1953
170. Kendrick KE, Ensign JC (1983) J Bacteriol 155:357
171. Glazebrook MA, Doull JL, Stuttard C, Vining LC (1990) J Gen Microbiol 136:581
172. Kuimova TF, Soina VS (1981) Hindustan Antibiot Bull 23:1
173. Novella IS, Barbés C, Sánchez J (1992) Can J Microbiol 38:769
174. Rho YT, Lee KJ (1994) Microbiol Read Engl 140(Pt 8):2061
175. Rueda B, Miguélez EM, Hardisson C, Manzanal MB (2001) Can J Microbiol 47:1042
176. Hopwood DA (2007) Streptomyces in nature and medicine: the antibiotic makers. Oxford University Press, Oxford, New York
177. Yang YK, Morikawa M, Shimizu H, Shioya S, Suga K-I, Nihira T, Yamada Y (1996) J Ferment Bioeng 81:7
178. Tresner HD, Hayes JA, Backus EJ (1967) Appl Microbiol 15:1185
179. Vecht-Lifshitz SE, Magdassi S, Braun S (1990) Biotechnol Bioeng 35:890
180. O'Cleirigh C, Casey JT, Walsh PK, O'Shea DG (2005) Appl Microbiol Biotechnol 68:305
181. Belmar-Beiny MT, Thomas CR (1991) Biotechnol Bioeng 37:456
182. Treskatis SK, Orgeldinger V, Wolf H, Gilles ED (1997) Biotechnol Bioeng 53:191
183. Tamura S, Park Y, Toriyama M, Okabe M (1997) J Ferment Bioeng 83:523
184. Pinto LS, Vieira LM, Pons MN, Fonseca MMR, Menezes JC (2004) Bioprocess Biosyst Eng 26:177
185. van Veluw GJ, Petrus MLC, Gubbens J, de Graaf R, de Jong IP, van Wezel GP, Wösten HAB, Claessen D (2012) Appl Microbiol Biotechnol 96:1301
186. Kim Y-M, Kim J-H (2004) J Microbiol 42:64
187. Xu H, Chater KF, Deng Z, Tao M (2008) J Bacteriol 190:4971
188. Whitaker A (1992) Appl Biochem Biotechnol 32:23
189. Junker BH, Hesse M, Burgess B, Masurekar P, Connors N, Seeley A (2004) Appl Biochem Biotechnol 119:241
190. Pamboukian CRD, Facciotti MCR (2004) Process Biochem 39:2249
191. Jonsbu E, McIntyre M, Nielsen J (2002) J Biotechnol 95:133
192. Dobson LF, O'Cleirigh C, O'Shea DG (2008) Appl Microbiol Biotechnol 79:859
193. Gamboa-Suasnavart RA, Valdez-Cruz NA, Cordova-Dávalos LE, Martínez-Sotelo JA, Servín-González L, Espitia C, Trujillo-Roldán MA (2011) Microb Cell Factories 10:110
194. Vecht-Lifshitz SE, Sasson Y, Braun S (1992) J Appl Bacteriol 72:195
195. Yin P, Wang Y-H, Zhang S-L, Chu J, Zhuang Y-P, Chen N, Li X-F, Wu Y-B (2008) J Chin Inst Chem Eng, 39:609
196. Tough AJ, Prosser JI (1996) Microbiol 142:639
197. Bellgardt KH (1998) Adv Biochem Eng Biotechnol 60:153
198. Lin P-J, Scholz A, Krull R (2010) Biochem Eng J 49:213
199. Papagianni M, Mattey M, Berovǐ M, Kristiansen B (1999) Food Technol Biotechnol 37:165
200. Glazebrook MA, Vining LC, White RL (1992) Can J Microbiol 38:98
201. El-Enshasy HA, Farid MA, El-Sayed A (2000) J Basic Microbiol 40:333
202. O'Cleirigh C, Walsh PK, O'Shea DG (2003) Biotechnol Lett 25:1677

203. Tucker KG, Thomas CR (1992) Biotechnol Lett 14:1071
204. Ruiz B, Chávez A, Forero A, García-Huante Y, Romero A, Sánchez M, Rocha D, Sánchez B, Rodríguez-Sanoja R, Sánchez S, Langley E (2010) Crit Rev Microbiol 36:146
205. Sánchez S, Chávez A, Forero A, García-Huante Y, Romero A, Sánchez M, Rocha D, Sánchez B, Valos M, Guzmán-Trampe S, Rodríguez-Sanoja R, Langley E, Ruiz B (2010) J Antibiot (Tokyo) 63:442
206. Šťastná J, Čáslavská J, Wolf A, Vinter V, Mikulík K (1977) Folia Microbiol (Praha) 22:339
207. Choi DB, Park EY, Okabe M (1998) J Ferment Bioeng 86:413
208. Choi DB, Park EY, Okabe M (2000) Biotechnol Prog 16:525
209. Okba AK, Ogata T, Matsubara H, Matsuo S, Doi K, Ogata S (1998) J Ferment Bioeng 86:28
210. Vecht-Lifshitz SE, Magdassi S, Braun S (1989) J Dispers Sci Technol 10:265
211. Hobbs G, Brown CM, Gardner DCJ, Cullum JA, Oliver SG (1989) Appl Microbiol Biotechnol 31:272
212. Winn M, Casey E, Habimana O, Murphy CD (2014) FEMS Microbiol Lett 352:157
213. Sohoni S, Bapat P, Lantz A (2012) Microb Cell Factories 11:9
214. Chater KF (1972) J Gen Microbiol 72:9
215. Horinouchi S, Beppu T (1992) Annu Rev Microbiol 46:377
216. Khokhlov AS, Tovarova II, Borisova LN, Pliner SA, Shevchenko LN, Kornitskaia EI, Ivkina NS, Rapoport IA (1967) Dokl Akad Nauk SSSR 177:232
217. Merrick MJ (1976) J Gen Microbiol 96:299
218. Lawlor EJ, Baylis HA, Chater KF (1987) Genes Dev 1:1305
219. Chater KF, Chandra G (2008) J Microbiol Seoul Korea 46:1
220. McCormick JR, Flärdh K (2012) FEMS Microbiol Rev 36:206
221. Reichl U, King R, Gilles ED (1992) Biotechnol Bioeng 39:164
222. Pearson AP, Glennon B, Kieran PM (2004) J Chem Technol Biotechnol 79:1142
223. Park Y, Tamura S, Koike Y, Toriyama M, Okabe M (1997) J Ferment Bioeng 84:483
224. Petersen N, Stocks S, Gernaey KV (2008) Biotechnol Bioeng 100:61
225. Rønnest NP, Stocks SM, Lantz AE, Gernaey KV (2012) Biotechnol Lett 34:1465
226. Wucherpfennig T, Schilling JV, Sieblitz D, Pump M, Schütte K, Wittmann C, Krull R (2012) Eng Life Sci 12:595
227. Pearson AP, Glennon B, Kieran PM (2003) Biotechnol Progr 19:1342
228. Grimm LH, Kelly S, Hengstler J, Göbel A, Krull R, Hempel DC (2004) Biotechnol Bioeng 87:213
229. Grimm LH, Kelly S, Völkerding II, Krull R, Hempel DC (2005) Biotechnol Bioeng 92:879

Adv Biochem Eng Biotechnol (2015) 149: 29–54
DOI 10.1007/10_2015_309

Published Online: 28 February 2015

Fungal Morphology in Industrial Enzyme Production—Modelling and Monitoring

Daniela Quintanilla, Timo Hagemann, Kim Hansen and Krist V. Gernaey

Abstract Filamentous fungi are widely used in the biotechnology industry for the production of industrial enzymes. Thus, considerable work has been done with the purpose of characterizing these processes. The ultimate goal of these efforts is to be able to control and predict fermentation performance on the basis of "standardized" measurements in terms of morphology, rheology, viscosity, mass transfer and productivity. However, because the variables are connected or dependent on each other, this task is not trivial. The aim of this review article is to gather available information in order to explain the interconnectivity between the different variables in submerged fermentations. An additional factor which makes the characterization of a fermentation broth even more challenging is that the data obtained are also dependent on the way they have been collected—meaning which technologies or probes have been used, and on the way the data is interpreted—i.e. which models were applied. The main filamentous fungi used in industrial fermentation are introduced, ranging from *Trichoderma reesei* to *Aspergillus* species. Due to the fact that secondary metabolites, like antibiotics, are not to be considered bulk products, organisms like e.g. *Penicillium chrysogenum* are just briefly touched upon for the description of some characterization techniques. The potential for development of different morphological phenotypes is discussed as well, also in view of what this could mean to productivity and—equally important—the collection of the data. An overview of the state of the art techniques for morphology characterization is provided, discussing methods that finally can be employed as the computational power has grown sufficiently in the recent years. Image analysis (IA) clearly benefits most but it also means that methods like near infrared measurement (NIR), capacitance and on-line viscosity now provide potential alternatives as powerful tools for characterizing morphology.

D. Quintanilla · T. Hagemann · K.V. Gernaey (✉)
Department of Chemical and Biochemical Engineering, Technical University of Denmark (DTU), Building 229, 2800 Lyngby, Denmark
e-mail: kvg@kt.dtu.dk

T. Hagemann
Novozymes A/S, Hallas Alle 1, Building BD, 4400 Kalundborg, Denmark

K. Hansen
Novozymes A/S, Krogshoejvej 36, 2880 Bagsværd, Denmark

These measuring techniques, and to some extent their combination, allow obtaining the data necessary for supporting the creation of mathematical models describing the fermentation process. An important part of this article will indeed focus on describing the different models, and on discussing their importance to fermentations of filamentous fungi in general. The main conclusion is that it has not yet been attempted to develop an overarching model that spans across strains and scales, as most studies indeed conclude that their respective results might be strain specific and not necessarily valid across scales.

Keywords Filamentous fungi · Industrial enzymes · Morphology · Rheology · Mass transfer · Productivity · Submerged fermentations · Modelling · Monitoring · Optimizing

List of Abbreviations

AMG	Glucoamylase
CM	Capacitance Measurement
EDCF	Energy Dissipation Circulation Function
FDA	Food and Drug Administration
GMP	Good Manufacturing Practices
GRAS	Generally Recognized as Safe
IA	Image Analysis
NIR	Near Infrared Measurements
OTR	Oxygen Transfer Rate
OUR	Oxygen Uptake Rate
PAT	Process Analytical Technology
PLS	Partial Least Squares
VSC	Vesicle Supply Center

Contents

1 Introduction

Filamentous fungi are widely used in the biotechnology industry for the production of different compounds like organic acids, industrial enzymes, antibiotics, etc., for an extensive list see Papagianni [1]. The widespread use of filamentous fungi as production host is due to three main advantages which the fungi possess: (1) Filamentous fungi have an exceptional ability of secreting large amounts of proteins [2]; (2) They possess a special posttranscriptional modification machinery which allows for glycosylation, correct protein folding, etc. [3]; and, (3) A large number of species are approved by the regulatory authorities and generally recognized as safe (GRAS). Nevertheless, operating a process with filamentous fungi also has a few major disadvantages due to the unavoidable oxygen transfer and mixing limitations that occur as a consequence of the high viscosity of the medium, which is due to the combination of the high biomass concentration and the fungal morphology [4].

Filamentous microorganisms manifest two main types of morphology in submerged fermentations, usually classified as dispersed and pelletized morphology. The first category is characterized by biomass that grows in the form of freely dispersed hyphae or mycelial clumps, see Fig. 1. In the second category, pellets are highly entangled and dense spherical agglomerates of hyphae which can have diameters varying between a couple of micrometers up to several millimeters [5]. Depending on the desired product, the optimal morphology—and here optimal is to be understood as yielding the highest productivity—for a given bioprocess varies and cannot be generalized; in some cases both types of morphology are even combined in one process [6]. The pelleted morphology type is often preferred because of the resulting Newtonian fluid behavior of the medium which allows for

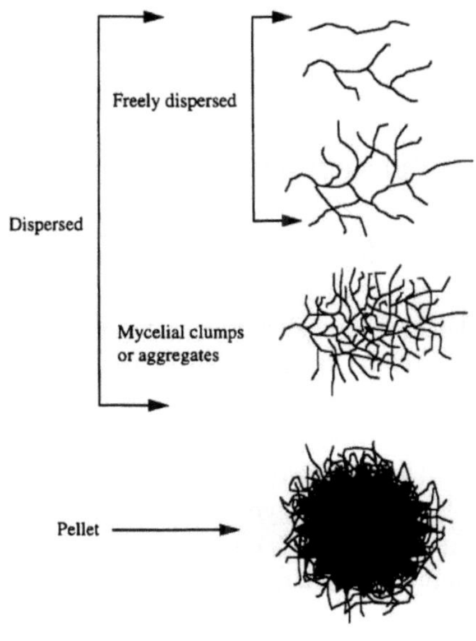

Fig. 1 Types of morphology typically found in submerged cultures of filamentous fungi [5]

better mixing and simplifies downstream processing in terms of pumping and separation of the biomass. However, the pelleted morphology results in nutrient concentration gradients within the pellet [7]. This situation is not observed in freely dispersed mycelia allowing for enhanced growth and production (provided sufficient bulk mixing capacity is available), which has been attributed to the fact that the morphology at the microscopic level has an influence on the production kinetics, e.g. on the secretion of enzymes. The latter was reported by Spohr et al. [8] who observed an increase in protein secretion from a more densely branched mutant of *Aspergillus oryzae* in comparison with the wild type. However, on the macroscopic level this type of morphology greatly affects the rheology of the fermentation broth, and therefore the transport processes in the bioreactor, and will thus increase the required power input for broth homogenization (mixing). So, the morphology of filamentous fungi is double edged, as the productivity as well as the fermentation conditions can be affected by the outer appearance of the fungus. The challenge is to separate these effects to be able to connect productivity gains to the correct phenomenon causing it. If this challenge could be overcome, the process knowledge of the fermentation scientist would be enriched tremendously, and would undoubtedly result in productivity gains.

The aim of this paper is to give a review of the research work that has been done in order to elucidate the relation between morphology and productivity and all the related variables in filamentous fungi fermentations, specifically for the production of industrial enzymes. In order to do this, an introduction to the main industrial strains is given, followed by a brief review of the morphology and physiology of filamentous fungi. A short description of the complex interaction of the different variables involved in submerged fermentations is presented. Also, this review will give an overview of the technologies that are available for morphology characterization and will discuss the latest technologies. In the final section, the paper links the capacity to characterize and model morphology to potential applications for influencing or controlling morphology as a tool for future process optimization.

2 Filamentous Fungi for Enzyme Production

2.1 Important Strains and Products

Due to their exceptionally high capacity to express and secrete proteins, filamentous fungi have become indispensable for the production of enzymes of fungal and non-fungal origin. Currently, native or recombinant industrial enzymes are mainly produced by *Aspergillus niger*, *A. oryzae* and *Trichoderma reesei* [9, 10].

The *Aspergillus* genus is one of the favorite expression systems in the production of industrial enzymes, and in particular the species *A. niger* and *A. oryzae* have been frequently used, due to their high titers of native hydrolytic enzymes, especially amylases and proteases [11]. Glucoamylase (AMG) is a homologous protein of

A. niger used for the conversion of starch to sweeteners and in the production of first generation ethanol [12]. Amylases are also added to detergents to assist in stain removal [13]. Other enzymes produced by these microorganisms include glucose oxidases, catalases, pectinases, lipases, phytases and xylanases, which are usually used in the food, detergent, textile, pulp and paper industry [14].

T. reesei is mainly known for producing cellulases, which are enzymes capable of degrading cellulose into simple sugars. They are widely used in the pulp and paper industry for the reuse of waste paper [15]. In addition, cellulases are also used within the textile industry for cotton softening and denim finishing. Another important application of these enzymes is within detergents, where they are used for color care, cleaning and anti-redeposition in washing powders [13]. Also, an enormous interest in these enzymes has arisen in the biofuel industry, as they are used in the saccharification of lignocellulosic materials which will be converted to bioethanol later on [16, 17].

2.2 Introduction to Morphology of Filamentous Fungi

Filamentous fungi are complex microorganisms constituted by complicated hyphae. A hypha is formed by one or more cells surrounded by a tubular cell wall. A hyphal element is formed by a main hypha that emerges from one spore; this main hypha is typically branched, and these branches have their own sub-branches and so on [18], as displayed in Fig. 1. *Ascomycota*, the group of organisms in which the fungi covered in this review are included, have hyphae that are divided into compartments by internal cross-walls called septa. Each septum possesses a pore large enough to allow cell organelles to flow between compartments. The collective term for the mass of hyphae is mycelium, Fig. 1. Furthermore, a hyphal element can entangle with another hyphal element and form more complex structures. The morphology of filamentous fungi is usually characterized by four variables: the length of the main hyphae (L_e), the total length of all the hyphae (L_t), the number of tips (n) and the length of a hyphal growth unit (L_{hgu}) [18]. The reader is referred to Table 1 for the definitions of additional morphological terms. The hyphal cell wall is formed by polymeric microfibrils of various biochemical composition arranged in a series of layers [19]. The microfibrils forming the hyphal wall usually consist of chitin, a polymer of *N*-acetyl-glucosamine [19].

One of the most important and interesting things to recognize in filamentous fungi is their apical extension just at the hyphae tips. This theory was established in the nineteenth century, when Reinhardt [20] proposed that fungal growth takes place by enlargement of the hyphae only at the apices. The elongation occurs by means of wall expanding according to a gradient, maximally at the extreme tip, and the materials necessary for cell wall expansion are provided by the cytoplasm [20]. Several growth models aiming to describe the exact mechanisms of how this process takes place have been proposed, and the most important ones are the steady state model [21, 22] and the hyphoid model [23]. Some other models have also

Table 1 Common morphological terms [92]

Area or projected area—area of the projection of a three-dimensional object into a two-dimensional image
Main hyphal length—length of the main hypha in a mycelium, which might be taken as the longest connected path though a mycelial tree
Branch length—length of an individual hyphal branch
Total hyphal length—sum of main hyphal length and all branch lengths in a mycelial tree
Branching frequency—number of branches (and sub-branches) in a mycelial tree
Number of tips—number of branches plus two (for the main hypha. Some tips might be extending (growing); others not
Hyphal growth unit—total hyphal length divided by the number of tips

been developed but these should rather be considered as combinations of the two former models [24, 25]. This is probably how the process is indeed carried out, since the main two theories are not self-exclusive given that they describe different features of the wall building process during tip growth [24].

In general, the theories describe hyphal growth as a consequence of a combination of wall biogenesis and turgor pressure. Wall biogenesis is ultimately an activity of the cytoplasm and that is where the building materials and necessary enzymes are synthetized; they are then later on transported in vesicles to the hyphal tip. These vesicles are accumulated in the apical dome forming a moving vesicle supply center (VSC). The VSC is an organelle from which vesicles move radially to the hyphal surface in all directions at random, and the forward migration of this pseudo-organelle is what generates the hyphoid shape [25]. Then, in the hyphal tip there are two main processes taking place, softening and hardening of the apical cell wall caused by the enzymes carried out in the vesicles. This process makes the hyphae tip more plastic and that is precisely where protein secretion takes place, carried out by a bulk flow from the cytoplasmic side to the wall. Wösten et al. showed with immunological techniques that secretion of glucoamylase in *A. niger* was carried out at the hyphal tips [26]. It is important to have this fact in mind, since it gave direction to the different research projects that were done in the area, as further described below.

2.3 Complexity of the Subject

In terms of mass transfer and rheology, filamentous fungi are very challenging hosts for the production of proteins in submerged fermentation, since their morphology is connected to these two variables, both at the microscopic and macroscopic level. Therefore, studying the relationship between fungal morphology and productivity in submerged fermentations is not an easy task due to the abundance of interrelated factors, which affect directly and indirectly the microorganism's morphology and product formation.

The particular morphology of a filamentous fungus leads to entanglements of the hyphae at high biomass concentrations (20–50 g/L); this phenomenon causes very high viscosities with non-Newtonian fluid behavior [27]. Considering a typical stirred tank reactor, it is well-known that the shear rate is at its maximum at the agitator tip, and it decreases when approaching the vessel walls. Therefore the viscosity will be low close to the impeller, and will increase towards the vessels walls. This results in a lot of problems with respect to mass transfer, moment transfer and heat transfer [27]. Metz et al. describe the viscosity as the center of the multi-directional and circular interrelation of all the factors, see Fig. 2. Thus, the broth viscosity has a major effect on the transport phenomena in the bioreactor which then will affect the process conditions [27]. Metz et al. considered the process conditions at the top of the complex interactions. These process conditions include the medium composition, mode of operation, temperature, pH, etc. The process conditions will directly affect the morphology, product formation and growth; all these variables are correlated with each other [27].

In addition to all the above-mentioned factors, there is also the fungal physiology and metabolism, which will affect morphology. For example, there is a continuous discussion with respect to shear damage of filamentous fungi; i.e. damage and fragmentation of filaments can be caused by shear forces from the impeller and by aeration. However, the aging factor of the microorganisms also plays a role making the cell walls weaker when cells grow older, and thus more susceptible to fragmentation [4]. Autolysis is another phenomenon observed in filamentous fungi which in addition contributes to hyphal fragmentation [28]. Also strain optimization plays a role, especially classical mutagenesis, since selected strains might increasingly form more single cell like structures. This results in lower viscosities, and thus better oxygen transfer, rather than higher titers [29].

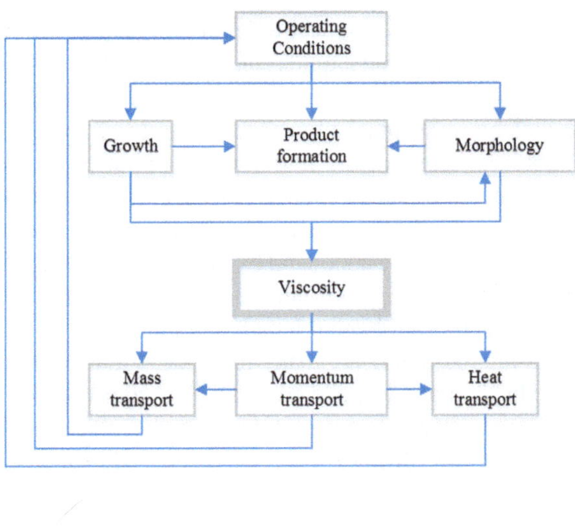

Fig. 2 Representation of all the interrelated variables in submerged fermentation. Adapted from [27]

3 Technologies for Morphology Characterization

The focus in this review paper is on on-line measurements as they have the advantage that sampling is not required. Thus, the complete content of the bioreactor is eventually measured, and there is no reduction in data quality due to sample handling or storage. Recent developments, especially in data handling, have led to the development of several approaches for rapid data generation. In the following, some methods will be discussed which benefitted greatly from the increase in calculation power to run complex algorithms which allow to interpret sensor signals with the aim of predicting biomass behavior. The latest development is Laser Scattering which in theory is able to measure crystal slurries or the shape of cells in fermentations. In practice, there is a question mark on how this technique can be employed in fermentation. Current literature describes that there might be challenges with regards to turbidity of the fermentation broth, cell density and air bubbles [30]. Despite the big potential, more mature methods will be discussed.

Image analysis
Image analysis (IA) would be the natural choice to analyze morphology as biomass literally can be seen. By taking pictures, the different morphologies (macro or micro, depending on the magnification) are digitally frozen and the obtained picture has to be processed to provide numerical characteristics of the broth sample. Therefore, IA as a method has to be divided into two phases: (1) acquiring the image; and, (2) evaluating the image. Both phases have their specific challenges.

Probes with the purpose of taking images in situ started to be described in the mid-90s. A good description of the state of the art from that time is given by Suhr et al. [31]. The challenges in *S. cereviseae* cultivation were the resolution of the resulting picture, obtaining the correct focus and the proper illumination triggered with the correct timing. Almost 20 years later, Suhr also co-authored an article in which in situ microscopy was proved adequate for measuring/monitoring animal cell cultures in perfused bioreactors [32]. The above-mentioned start-up difficulties (resolution, illumination, focus) could be overcome. Systems are described with mechanical [33] or optical definition of sample volume [34]. The latter could suffer greatly from turbid liquids and probe fouling. Some mechanical samplers/in situ microscopes feature a retractable sample chamber that could be steam cleaned and hence biomass attached to the lens could be removed [35]. Still, problems with air bubbles, turbid media and fouling of the probe, especially when dealing with filamentous organisms, have not been solved completely yet.

IA in the 90s was mainly manual meaning that tips, hyphae and/or particle diameter had to be annotated by the user (e.g. [36]). Since then, the digital processing power increased greatly and in 2011, Barry and William attested that on-line calculations of total hyphal length and number of hyphal tips amongst others now are possible [6]. There are many parameters that could be acquired and described with IA. In the case of pelleted growth, it could be the projected area of the particle, diameter, shape and circularity. Lately, also calculations via fractals have been included [37, 38]. As long as manual adjustments have to be made to acquired images, it might be faster to employ other measurement principles, i.e. laser diffraction as well/instead [39]. Else, considerable achievements were made within dedicated algorithms for IA. Please refer to Papagianni (2014) for the latest technology, which provides an overview of methods and processes, respectively [40].

Near infrared measurement (NIR)
The on-line NIR suits the regulatory requirement for process analytical technology (PAT) even for pharmaceutical purposes as it is a fast measurement, the sensors can be built in accordance to GMP regulations, and it allows running a well-defined process. The measurement principle is the absorbance of near infrared wavelength by the medium leading to specific spectra for each compound present in the media [41]. To be able to interpret the spectra correctly, deep knowledge about the process is required [42], as well as proper calibration [43] which leaves out the determination of most complex ingredients by means of NIR. It is furthermore not possible to measure reliable NIR spectra in turbid media, which includes aerated media to some extent [44].

Usually, there are disturbances in the measurement, and while the trends look similar, details of the spectra might vary and thus, pre-processing of data is applied which removes assumed artefacts of the measurement [45]. To even out some of the peaks in the measurement and to pronounce some minor effects, the second derivative of the wavelength spectrogram is the basis for further analysis [43]. The data handling can be reduced in complexity by restricting the analysis to just one major compound for just two frequencies and by use of a Fourier transformed signal [46]. In fact, the data handling is the most important part of the NIR measurement as the signal can be correlated to other parameters like biomass, product or substrate concentration. A challenge though is that the NIR spectrum has to be interpreted, and sometimes (especially when in lack of deep process knowledge), several explanations could be attributed to the same spectrum.

There currently is no literature present about the use of NIR to characterize or control fermentations with filamentous organisms. It has been successfully demonstrated that *S. cerevisiae* fermentations can be controlled and the respective responses can be predicted using a soft sensor based on the input of a PLS (Partial Least Squares) modelled NIR signal [47].

Capacitance
Starting in the early 90s, the capacitance measurement (CM) or dielectric spectroscopy [48] became increasingly popular and was tested for different biotechnological processes [49]. While originally more used for yeast fermentations [50], the attention very soon also turned to other microorganisms as well as fungi [51]. The experiments reported in the work of Krairak et al. [51] were characterized by pellet formation of the fungus—thus not really filamentous growth—and considerable disturbances by air bubbles and agitation. Even though the low pass filter for removing noise from the signal was described before [52], its technical application was first possible with the availability of increased miniaturized calculation power.

This measurement technique is extremely interesting for online bioprocess monitoring because the measurement principle differentiates between live and dead biomass: "Cells behave as capacitors due to the presence of charged molecules both inside (cytoplasm) and outside (culture broth), separated by a plasma membrane. Capacitance, measured by application of an electric field, is directly proportional to the cell concentration—Dead cells without intact membranes do not contribute to charge polarization" [53]. The same source marks this technique as "promising" for pharmaceutical GMP purposes which means that there might be good potential also for industrial fermentations aiming at production of bulk products. It is advised though to take the same approach to process control as with NIR: Verifying the CM with other, maybe even off-line, methods [54], in order to reduce the interferences to the signal from stirring and agitation.

In the special case of filamentous organisms, the publications are rather scarce. Posch et al. concluded that the currently best use of the CM would be a good building block for a soft sensor that calculates parameters related to broth rheology and organism morphology [55]. In contrast, Rønnest et al. demonstrated that good correlations between the capacitance signal and off-line dry matter measurements could be achieved for a filamentous fermentation broth [44]. With a scan through different frequencies, it was also possible to retrieve additional data about the relative size and the distribution of cells. In the case of the (non-filamentous) organism *E. coli*, a combination of a CM and a soft sensor based on first principles elemental balances successfully detected cell changes at the time of induction with subsequent predictions of the final titer [56]. Also, CM is considered to be an important tool in view of the FDA PAT initiative that was launched 2004: It is a stable technique able to provide real-time process-related information through non-destructive and non-invasive measurements [42].

The often expressed limitation of CM is the conductivity of the medium, if a medium with high salt concentrations is employed (above 50 mSiemens). The status in 2014 is though that it is possible to go as high as 100 mSiemens while still producing a signal of acceptable quality (personal communication, Bent Svanholm, Svanholm.com).

On-line Viscosity

It is often reported that different morphologies cause different fluid behavior of the broth, which is expressed in the broth viscosity. Several publications came to the conclusion that the viscosity is mainly affected by biomass concentration and morphology, stating that filamentous growth leads to an entangled network of hyphae and thus to higher viscosity [27]. Later publications confirmed this trend [36, 55, 57] and therefore, it could be concluded that an on-line viscosity measurement certainly can provide real-time information that is useful in view of controlling fermentations. The challenge is, though, that most on-line rheology measurements in industrial fermenters create a non-uniform shear field, and that parts of the broth may not undergo the same forces. This adds on to the empirical approach to calculate the shear rate based on the torque of the impeller [58]. The lack of precision makes the on-line viscosity measurement a rarely described tool for generating data that can be used in view of fermentation process control. There are works though which correlate the viscosity of the broth with the morphology, and consequently with productivity. A non-protein product example was described by Dhillon et al., namely the case of citric acid production with *A. niger* growing on apple pomace sludge, where rheology was altered by adding methanol leading to a less pseudoplastic fluid behavior [59]. In this special case, morphology is not the only factor with an influence on viscosity because also the viability of the cells is reduced by adding methanol. An interesting approach to reduce viscosity was demonstrated by Cai et al. [60]: *A. glaucus* was genetically modified so that it lacked the polarized growth of hyphae. As a result, hyphae grew curved leading to a much denser pellet and less viscous broth, and a more than 80 % increase in the production of aspergiolide A, a secondary metabolite. Most of the latter is attributed to the fact that a better oxygen transfer rate (OTR) could be achieved. This finding shows the importance of being able to monitor the broth characteristics in order to control fermentation. For the moment, however, not that many efforts are put in applying online viscosity to control (industrial) fermentation.

4 Modelling the Morphology

In an effort to achieve process improvements within the biotechnology industry, considerable focus has been put in the developing and maturing of engineering tools which facilitate process optimization. One of these tools is mathematical modelling, including both empirical and first-principles models. For a review of more engineering tools see [61]. Different research projects have been carried out with the objective to model and understand the different phenomena taking place in submerged fermentation, and to develop an improved understanding of the interactions between the variables illustrated in Fig. 2. These attempts have mainly been focused on modeling micromorphology, hyphal fragmentation and rheology, and on developing an understanding on how this affects productivity.

4.1 Micromorphology and Productivity

Since the early 90s, it has been demonstrated that protein secretion in *A. niger* was carried out mainly in the tips of fungal hyphae [26]. Therefore, several studies have been conducted aiming at correlating the number of hyphal tips with enzyme production. As an example, *A. oryzae* producing α-amylase was further investigated by comparing three different strains in batch cultivations [8]. The strains were: a wild-type (A1560), a transformant with extra copies of the coding gene and a morphological mutant (made from the transformant). By comparing the α-amylase concentration at the end of the batch and the specific branching frequency of the two recombinant strains, it was concluded that a more densely branched strain is superior in protein production (Table 2). This higher productivity might be attributed to the fact that the limiting step in high yielding protein strains is the secretion process [8].

Table 2 Final amylase concentration and specific branching frequency of the three strains of *Aspergillus oryzae*

Strain	Final concentration α-amylase (FAU/ml)	Branching frequency—k_{bran} (tip/μm/h)
Wild-type	0.63	0.0023
Transformant	2.22	0.0010
Morphological mutant	3.09	0.0021

Adapted from [8]

Bocking et al. [62] investigated the topic in the same strain (A1560). In addition, they studied a transformant strain able to produce glucoamylase along with α-amylase (AMG#13). Nine morphological mutants were generated from these two strains: 4 from A1560 and 5 from AMG#13. The mutants were screened such as to keep the same maximum specific growth rate and a lower hyphal growth unit length (more branched strains). All the strains were studied in batch, continuous and fed-batch cultivations. No clear correlation between branch frequency and ability of secreting protein was observed for the highly branched mutants in the fed-batch and continuous fermentations. They, however, found a correlation between higher branch frequency and viscosity reduction. This could lead to the conclusion that the observed productivity increase by Spohr et al. [8] was achieved due to a better OTR rather than a higher productivity. However, one should be aware that the comparisons are done in different operating modes (batch vs. fed-batch); thus different physiological state due to different grow rates might have an effect on the rate of fragmentation resulting in systems with different viscosity. Though, in the batch experiments for AMG#13 and its highly branched mutant, the latter did have an increase in glucoamylase compared with AMG#13. This might suggest that under maximum growth rate conditions there is a correlation between branch frequency and secretion, as suggested by Spohr et al. [8].

For the first time, Haack et al. reported the swelling of the hyphal tips as a consequence of high productivity in a recombinant strain of *A. oryzae* producing lipase [63]. It was suggested that tip swelling and productivity are linked, since the hyphae tips return to normal shape after the production stops due to oxygen limitations. It was indicated that these findings could help to identify the fraction of productive cells in industrial fermentation, since it is known that production heterogeneities occur in full scale due to poor mixing [61].

Most of these models are purely empirical without any structured background, and have been mainly developed using IA. Nonetheless, they have been used as practical tools for comparison. Agger et al. developed a morphologically structured model able to describe growth and product formation in batch, continuous and fed-batch cultivations for *A. oryzae* (A1560) [64]. By dividing the fungal hyphae into three different regions—extension zone representing the tips of the hyphae, active zone which is responsible for growth and product formation and an inactive hyphal region, Fig. 3—a model able to predict product formation as a function of morphology was developed. Different to the previous models, it was verified by IA

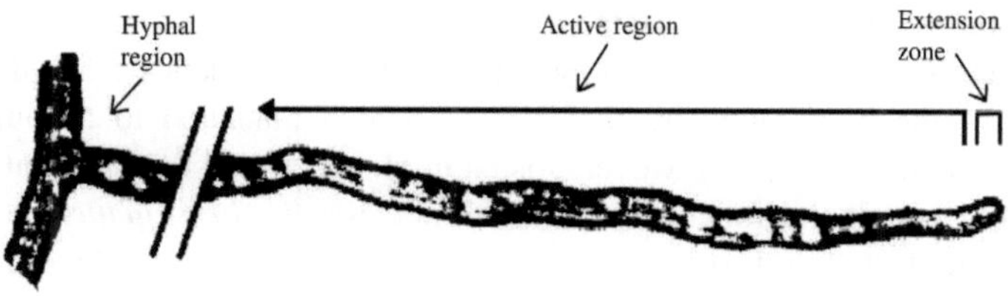

Fig. 3 Illustration of the structuring of the hyphal elements in the model of Agger et al. [64]

combined with fluorescence microscopy. The model performed well in batch and continuous cultivation. However, there seemed to be an under-prediction of product formation for a fed-batch fermentation. This difference was attributed to rheological changes or hyphae fragmentation not accounted for in the model.

As seen in Fig. 2, due to the complex interactions between the variables in a fermentation process it is also possible to affect morphology by changing the process conditions (media composition). Ahamed and Vermette [65] conducted a study where they indirectly affected morphology by varying the carbon source in the batch phase of a fed-batch process. The study was performed in *T. reesei* in the strain Rut-C30 producing cellulases. They observed a more branched morphology in the medium which also presented the highest enzyme titers. A non-linear correlation for the volumetric enzyme productivity was developed as a function of the average projected area of the total mycelia (entangled mycelia plus branched mycelia) and the number of hyphae tips. This study confirms the relation between enzyme productivity and number of tips.

4.2 Shear Stress and Morphology

One of the problems with filamentous microorganisms relies in the oxygen mass transfer limitations in the culture broth due to the high viscosities and the non-Newtonian behavior. An obvious strategy to overcome this problem is to increase agitation power. However, the question of shear damage due to fragmentation or morphological changes arises. Early works indicated that there is shear damage to the cells caused by the impeller which usually leads to reduced productivity. Nevertheless, these studies were conducted in *Penicillium chrysogenum* for penicillin production [66, 67] and are therefore out of the scope of this review. In this section, the research that has been done in order to understand the relation of shear stress caused by the impeller on morphology for filamentous fungi producing industrial enzymes will be summarized, and only for the cases where dispersed mycelium is observed (i.e. not pellets). The other focus will be the work of Jüsten et al. who developed a function capable of correlating mycelial fragmentation and power inputs at different scales and agitator types, the so called energy dissipation circulation function (EDCF) [68]. The EDCF is defined as the ratio of energy dissipation in the dispersion zone to the liquid circulation time. This model considers the specific energy dissipation and the amount of time that the cells are subject to the shear caused by the impeller (circulation frequency). The theory is an extension of the work done by Smith et al. which managed to account for the circulation frequency [67]. However, Jüsten et al. accounted for different impellers types. Even though the function was developed for *P. chrysogenum*, its applicability was later tested in *A. oryzae* managing to successfully correlate fragmentation [69]; therefore, in this review the EDCF is considered as the standard tool for comparison of the different works.

Amanullah et al. studied the effect of agitation on mycelial morphology and enzyme productivity in continuous cultures of a recombinant strain of *A. oryzae* producing α-amylase and amyloglucosidase. They found that agitation intensity has an effect on morphology. When the agitation speed was reduced by half, the mean projected area increased almost threefold. Productivity, however, was not affected [70]. A later study of the same group carried out with the same strain in fed-batch fermentations confirmed that mycelial morphology (measured as a mean projected area) is dependent on agitation intensity but that there is no correlation with enzyme production assuming a non O_2 limited process for the fed-batch phase. However, in the batch phase there was a positive correlation with agitation intensity and productivity, which could be attributed to higher biomass concentration and growth rates at higher agitation speeds [71]. An important conclusion of these two studies was that given a specific growth rate, it is possible to correlate EDCF with morphology measured as the mean projected area [71].

Fragmentation caused by different specific power inputs measured as EDCF was investigated by Li et al. for full scale fed-batch fermentations [72]. The shear forces caused by the impeller on *A. oryzae* resulted in mycelial fragmentation; however, this fragmentation was not correlated to the specific power input since a similar morphology was observed at the two levels of EDCF. Hence, fungal morphology does not seem to be dependent on mixing intensity at production scale. Effects of changes on impeller power input on productivity were not studied. In a later study this correlation was investigated. Yet again, two specific power inputs measured as EDCF were tested, and neither the morphology, productivity nor rheology differ at these two levels of EDCF under non-oxygen limited conditions [73].

Albæk et al. developed a correlation able to predict viscosity based on the mentioned EDCF and biomass measurements in fed-batch fermentation with *A. oryzae* [74]. The correlation does not take into consideration morphological data; however, the accuracy of the general model prediction demonstrates the ability of EDCF to correlate morphological data as a function of viscosity. They also observed a positive correlation with increasing agitation speed and product formation. This might be correlated to a better OTR, indicating that morphology actually does not necessarily play a major role in productivity [74].

4.3 Morphology and Rheology

The study of the rheology of culture broths has a crucial importance in the design of fermentation processes; viscosity will affect the transport phenomena taking place in a bioreactor. Fluid viscosity appears in several correlations involved in mixing and mass transfer, e.g. the Reynolds number and correlations for the volumetric mass transfer coefficient. Also, the yield stress appears in relationships used to estimate cavern sizes [75]. Therefore, it is important to study the rheology of fermentation broths, and if possible to predict it. The traditional approach for modelling the rheology of filamentous fermentations broths is based on attempts to correlate the parameters for different viscosity models (e.g. power law model) to the biomass concentration [76] (for a review of rheology see [27, 57]). However, these correlations are of a limited value [36]. Viscosity is a property that is not only dependent on biomass concentration, but is affected by the structure and extent of physical interlacing of hyphal networks [77]. Thus, several other studies have focus on correlating viscosity not only to the biomass concentration, but also to the different morphological parameters, i.e. projected area, hyphal growth unit, etc., typically characterized by IA. Other authors have characterized the morphology by other methods in order to correlate it to rheology. These attempts are further described in this section.

The direct relation between the consistency index parameter of the power law model and different morphological parameters was studied for continuous cultivations of *A. niger* [78]. The dissolved oxygen and growth rate were varied at two different biomass concentrations. It was found that at constant biomass concentration the consistency index could be linearly correlated to the roughness. By including a biomass term in a linear model, a good correlation was obtained. In this work, the correlation was not further tested for different levels of biomass concentration or for batch or fed-batch fermentations. It has to be kept in mind that this is one of the first attempts to quantitatively model the rheology of filamentous fungi based on morphology parameters. As a continuation of this work, Riley et al. developed a model to predict the consistency index based on biomass concentration and the mean mycelial maximum dimension (the longest feret across the convex area of a mycelial particle) for *P. chrysogenum*. The model was able to predict the parameter with an average root-mean-squared deviation below 30 %, which despite all the sources of errors is considered as a good result [79]. This microorganism is not included within the strains considered in this review paper. However, as an extension of this work, Riley and Thomas checked the relevance of the correlation to other fungi [80]. They reformulated the model at a different magnification for *P. chrysogenum* and tested its applicability to predict the values of the consistency index for fermentation broth of *A. niger* and *A. oryzae*. The biomass concentration and the mean mycelial maximum dimension are also used in this model as the morphological parameters. The correlation performed well for fermentation broths of *A. oryzae*, but failed to describe the consistency index for *A. niger*. No simple model was found for the flow behavior index. In fact, it was considered as a constant value, which is not always the case. Similarly, Malouf aimed at correlating the rheology properties of *T. reesei* fermentation broths to morphological parameters [81]. The Herschel-Bulkley consistency index was successfully correlated to the mean roundness by two separate correlations for the batch and fed-batch phases.

Wucherpfennig et al. studied the rheology of culture broths from *A. niger* in shake flask fermentations [37]. IA was used for morphological characterization, which was described by conventional and fractal morphological parameters. Two different fractal parameters where good at predicting rheological properties; nevertheless they were not superior to the parameters developed using conventional particle shape analysis (e.g. Morphology number [82]).

Other authors have used laser diffraction and multivariate data analysis as a tool to model the rheology of fermentation broths. Petersen et al. correlated the rheological properties of commercially relevant *A. oryzae* fermentations with respect to particle size distribution data [75]. The study was conducted in fed-batch fermentations where different feed strategies were applied, similar to the work of Bhargava et al. [83]. A partial least squares regression model was able to predict viscosity by using particle size distribution, biomass concentration and process information. In terms of practical applicability, this model is superior to the models that have been developed using IA due to the simplicity of the measurements; however, a limitation is that the model was not able to predict the rheological properties of fermentation broths of different strains and/or scales.

Another aspect which is important to consider when dealing with rheology of filamentous fungi, is how to evaluate the shear rate in the tank. Until now, it has not been possible to estimate a reliable shear rate in the fermentation tank itself. As mentioned before, in an STR it is well-known that the shear rate is at its maximum at the agitator tip and decreases towards the vessel walls. Calculating shear rate can therefore be expressed as the maximum or the average shear rate. Hitherto, it is not clear which shear rate is governing the mass transfer processes, and the way of calculating this shear rate is limited to the Metzner and Otto correlation [84]. According to Stocks [85], it should not be forgotten that this empirical correlation was developed for Reynolds numbers in the laminar and transitional regime and not for turbulent conditions where it is also frequently used in practice. Thus its applicability is limited to laboratory and pilot-scale fermenters, even though it has been typically employed for calculating shear rates in full-scale fermenters also [85]. A special challenge emerges when a reliable shear rate to evaluate the viscosity across different scales has to be estimated. This might be the reason why the developed models were not able to make predictions that apply across scales; therefore, the applicability of other correlations should be explored [86]. Adding on to that challenge is the use of different instruments to measure rheology [36]. Developing models that could predict performance of fermentations with filamentous fungi across scales with a (relatively) high accuracy is therefore still considered to be one of the major scientific challenges in this field.

4.4 Process Conditions and Morphology

Bhargava et al. [83] performed a study where they investigated the effect of different feeding strategies on morphology, protein expression and viscosity during a fed-batch fermentation. The work was done with *A. oryzae* (A1560). Tests with varying the feeding profile were carried out, keeping the same total amount of glucose, but fed in cycles (pulsed feeding). These experiments were compared with a continuously fed fermentation. The biomass, oxygen uptake rate (OUR) and total base added for pH control showed no significant difference indicating that pulsed feeding during fed-batch operation has no apparent effect on cellular metabolic activity. Neither was there a significant difference for the different cultivations in the measured extracellular protein content. Nevertheless, a considerable effect on fungal morphology (measured as average projected area) from the start of the fed-batch phase was observed. The pulsed feeding resulted in smaller hyphal elements in comparison to the elements resulting from continuous feeding. The smaller elements lead to a significant decrease in viscosity. As a consequence, the effect of the cycle time on morphology, rheology and productivity on the same strain was tested in a later work [87]. As before, no effect on biomass was observed, while the mean projected area and the viscosity decreased with the increase in cycle time. Shorter cycles resulted in constant productivity and OUR while longer cycles caused a decrease in productivity at a higher OUR—it appeared that the fungus was forced to form conidia due to starvation. This yet again shows the double sided effect of morphology, viscosity and product formation.

Other authors have also studied the effect of agitation intensity on other process variables, rather than just on morphology. Marten et al. [88] studied the relation between rheology, mass transfer and mixing for *T. reesei* in batch fermentations. They concluded that the Casson model and the Herschel-Buckley model are better in describing the rheological behavior of *T. reesei* broths in comparison with the power law model. However, a practical correlation was not obtained with respect to the effect of agitation intensity on rheology and biomass concentration.

In an attempt to understand the effect of shear stress on morphology and rheology also in *T. reesei,* Patel et al. studied the effect of agitation intensity in fed-batch fermentations [89]. With respect to shear stress and productivity, no clear correlation was observed. With respect to morphology, in the batch phase of the fermentation, no effect on agitation intensity was observed either. However, for carbon source limiting growth, there seemed to be an effect of agitation intensity on morphology, since a higher degree of fragmentation was observed as the fed-batch phase proceeded. It is not clear whether this degree of fragmentation is caused by the agitation or by self-fragmentation of the microorganisms. This however, resulted in a lower viscosity towards the end of the fermentations. A higher apparent viscosity was observed in the experiment with the highest agitation speed, which might be attributed to the higher biomass concentration. The apparent viscosity of all the experiments was evaluated at a constant shear rate and not at the shear rate in the tank, making the comparison difficult.

Table 3 Effects of morphology alteration on productivity, biomass and rheology

Authors	Microorganism	Operation mode	Growth rate (1/h)	Affected morphology parameter	Effect on productivity	Effect on growth rate/biomass	Effect on rheology	Notes
Spohr et al. [8]	*Aspergillus oryzae* (A1560)	Batch	0.18–0.27	Morphology-branching frequency	Yes. Higher productivity in the more branched strain	Yes. Lower growth rate for the more branched strain	N/A	Comparison at a specific biomass concentration
Bocking et al. [62]	*Aspergillus oryzae* (A1560)	Batch	0.28	Morphology-branching frequency	No	No	N/A	
	Aspergillus oryzae (AMG#13)	Batch	0.30	Morphology-branching frequency	Yes. Higher productivity for the more branched mutants	No	Yes. Lower viscosity for the highly branched mutants	
	Aspergillus oryzae (AMG#13)	Fed-batch	–	Morphology-branching frequency	Yes. Lower productivity for the more branched mutants	Yes. Lower biomass for the more branched mutant	N/A	Constant feed rate
	Aspergillus oryzae (AMG#13)	Fed-batch	–	Morphology-branching frequency	Yes. Slightly more productivity for one of the more branched mutants	Yes. Slightly less biomass for one of the more branched mutants	Yes. Lower viscosity for the highly branched mutants	DOT controlled feed rate

Table 4 Effects of shear stress on morphology, productivity, biomass and rheology

Authors	Microorganism	Operation mode	Growth rate (1/h)	Conditions	Effect on morphology	Effect on productivity	Effect on growth rate/biomass	Effect on rheology	Notes
Amanullah et al. [70]	*Aspergillus oryzae* (A1560)	Continuous	0.05	Agitation intensity reduction 1000–550 min^{-1}	Yes. Increase in mean projected area. Increase in hyphae length. Increase in number of tips	No	No	N/A	
Amanullah et al. [71]	*Aspergillus oryzae* (A1560)	Batch	Max	Agitation intensity 825, 675, 525 min^{-1}	Yes. Higher mean projected area at lower agitation speed	Yes. Higher productivity at higher agitation speed	Yes. Higher growth rates and biomass at higher agitation speed	N/A	
	Aspergillus oryzae (A1560)	Fed-batch	<0.02	Agitation intensity 825, 675, 525 min^{-1}	Yes. Higher mean projected area at lower agitation speed	No	No	N/A	These under no oxygen limited conditions
Li et al. [72]	*Aspergillus oryzae* (A1560)	Fed-batch	<0.03	Power input	No	N/A	No	N/A	Change in morphology as a function of time
Li et al. [73]	*Aspergillus oryzae* (A1560)	Fed-batch	<0.03	Power input	No	No	No	No	These under no oxygen limited conditions
Albaek et al. [74]	*Aspergillus oryzae* (property strain)	Fed-batch	Function of the feed flow rate	Agitation power	N/A	Yes. Higher productivity at higher agitation power	Yes. Higher biomass at higher agitation power	No significant	For the same impeller choice
	Aspergillus oryzae (property strain)	Fed-batch	Function of the feed flow rate	Impeller type	N/A	No	Yes. At some conditions higher biomass for the axial impeller	Yes. Lower viscosity for the axial impeller	At the same conditions

(continued)

Table 4 (continued)

Authors	Microorganism	Operation mode	Growth rate (1/h)	Conditions	Effect on morphology	Effect on productivity	Effect on growth rate/biomass	Effect on rheology	Notes
Marten et al. [88]	*Trichoderma reesei* Rut-C30	Batch	–	Agitation intensity 250, 400, 500 min^{-1}	N/A	Yes. Higher productivity at the higher agitation speed	No	Yes. Higher viscosity at higher agitation speed	
Patel et al. [89]	*Trichoderma reesei* Rut-C30	Batch	–	Agitation intensity	No	Not clear	Yes. Lowest growth rate at the lower agitation speed	Yes. Lower viscosity at the lowest agitation	
	Trichoderma reesei Rut-C30	Fed-batch	–	Agitation intensity	Yes. Lower clump fragmentation at the lower agitation speed	Not clear	Yes. Higher biomass at the higher agitation speed	Yes. But not clear correlation	

5 Conclusion

The key issue relies on how to incorporate all the work that has been done for the past years in order to optimize the filamentous fermentation process. As illustrated in Fig. 2, there are many variables which will influence the performance of fermentation processes, and in order to study all of them, an extensive (up to now impossible) design of experiments needs to be performed. The complex inter-correlation renders it impossible to affect a variable while keeping the rest constant, as seen in Tables 3 and 4. Thus, until now the best way of dealing with this appears to be data reconciliation for the work that has been done, which is not an easy task, due to the different set-ups used. For example, Li et al. [72, 73, 90] found no effect on the different variables from changes in specific power input; these results at full scale production differ significantly from the findings at bench scale [70, 71], which again makes it very difficult to draw a practical conclusion about the relation and interaction of all the variables involved in fermentations processes with filamentous fungi.

In addition to this, one of the biggest challenges with respect to the study of filamentous fungi in the production of enzymes is the lack of relevant industrial data. The difference between data generated by academia and the industry is enormous. For example processes studied in this review, and typically reported as a result of a study that has taken place at a university department, deal with titers of barely a couple of grams per liter of extracellular protein. In industry there are reports of titers up to hundreds of grams per liter [13]. The question remains on whether it is possible for industry to apply the results of the model studies developed by academia. It is to expect that the behavior of the industrial microorganisms would be completely different due to the stress on the host organism that is caused by such high expression levels. Therefore, if the aim is to produce results with both academic value and industrial relevance, then it is important to have a proper collaboration between industry and academia in order to overcome this issue. So, there is a need for the definition of one or more well-defined case studies that should be available publicly, with limited but sufficient industrial value to be of practical use. The case studies should allow academia to work in concentration ranges which are relevant for industry. The same is valid for the scale: Bench reactors are just much smaller and not all effects, especially regarding mixing and bubble interference to new probes likes NIR, can be studied properly with respect to industrial challenges. However, this is not an easy task, since very different—often competing—interests might be involved: indeed, a problem is that a case study with practical value does not automatically have sufficient academic value, and the other way around.

The reader might also notice that many of the articles analyzed in this review date back to the nineties. It seems like many groups stopped their work in relation to fungal morphology in submerged cultivation around the year 2000. There are some groups continuing research, also employing new technologies/measurement systems. It can just be assumed that the costs for the sensors are too high and/or the probes themselves are too bulky for bench scale reactors. In this case, the above stated/desired case promoting industry—academia collaboration would become even more important. The trade-off for companies would be releasing (confidential) process data in order to obtain e.g. new process strategies based on the potential use of the new sensor types described in Sect. 3.

Another consideration is that the development of "omics" technologies made it possible to study filamentous fungi from a different perspective [3, 91]. It might be that the solution leading to a more complete understanding of the microorganisms is to be found in an integration of the studies described in this review, and those related to the "omics" field. However, before such integration can be realized and applied in practice on industrially relevant case studies, it is quite clear that considerable additional research work will be needed in this area.

Acknowledgments The authors acknowledge financial support of the following organizations: The Novo Nordisk Foundation, Novozymes A/S and the Technical University of Denmark (DTU)

References

1. Papagianni M (2004) Fungal morphology and metabolite production in submerged mycelial processes. Biotechnol Adv 22(3):189
2. Peberdy JF (1994) Protein secretion in filamentous fungi—trying to understand a highly productive black box. Trends Biotechnol 12(2):50
3. Punt PJ, van Biezen N, Conesa A, Albers A, Mangnus J, van den Hondel C (2002) Filamentous fungi as cell factories for heterologous protein production. Trends Biotechnol 20 (5):200
4. van Suijdam JC, Metz B (1981) Influence of engineering variables upon the morphology of filamentous molds. Biotechnol Bioeng 23:111
5. Cox PW, Paul GC, Thomas CR (1998) Image analysis of the morphology of filamentous micro-organisms. Microbiology 144:817
6. Barry DJ, Williams GA (2011) Microscopic characterisation of filamentous microbes: towards fully automated morphological quantification through image analysis. J Microsc 244:1
7. Hille A, Neu TR, Hempel DC, Horn H (2005) Oxygen profiles and biomass distribution in biopellets of *Aspergillus niger*. Biotechnol Bioeng 92(5):614
8. Spohr A, Carlsen M, Nielsen J, Villadsen J (1997) Morphological characterization of recombinant strains of *Aspergillus oryzae* producing alpha-amylase during batch cultivations. Biotechnol Lett 19(3):257
9. Meyer V (2008) Genetic engineering of filamentous fungi–progress, obstacles and future trends. Biotechnol Adv 26(2):177
10. Nevalainen KMH, Te'o VSJ, Bergquist PL (2005) Heterologous protein expression in filamentous fungi. Trends Biotechnol 23(9):468
11. Lubertozzi D, Keasling JD (2009) Developing *Aspergillus* as a host for heterologous expression. Biotechnol Adv 27:53
12. Cherry B, Bashkirova EV, De Leon AL (2009) Analysis of an *Aspergillus niger* glucoamylase strain pedigree using comparative genome hybridization and real-time quantitative polymerase chain reaction. Ind Biotechnol 5(4):237
13. Cherry JR, Fidantsef AL (2003) Directed evolution of industrial enzymes: an update. Curr Opin Biotechnol 14(4):438
14. Fleissner A, Dersch P (2010) Expression and export: recombinant protein production systems for *Aspergillus*. Appl Microbiol Biotechnol 87(4):255

15. Lee SM, Koo YM (2001) Pilot scale production of cellulase using *Trichoderma reesei* Rut-c30 in fedbatch mode. Microbiol Biotechnol 11(2):229
16. Lynd LR, van Zyl WH, McBride JE, Laser M (2005) Consolidated bioprocessing of cellulosic biomass: an update. Curr Opin Biotechnol 16(5):577
17. Horn SJ, Vaaje-Kolstad G, Westereng B, Eijsink VG (2012) Novel enzymes for the degradation of cellulose. Biotechnol Biofuels 5:45
18. Kossen NWF (2000) The morphology of filamentous fungi. In: Scheper T (eds) Advance in biochemical engineering, vol 70, pp 1
19. Berry DR (1988) Physiology of industrial fungi. Blackwell Scientific Publications, Oxford
20. Reinhardt MO (1892) Das wachstum der pilzhyphen. Jahrbücher für wissenschafftliche 23:479
21. Wessels JGH (1993) Wall growth, protein excretion and morphogenesis in fungi. New Phytol 123(45):397
22. Wessels JGH(1990) Role of cell wall architecture in fungal tip growth generation. In: Heath IB (ed) Tip growth in plant and fungal cells. Academic Press Inc., London, p 1
23. Harold FM (1997) New ideas in cell biology. Protoplasma 197:137
24. Bartnicki-García S (1999) Glucans, walls, and morphogenesis: On the contributions of J. G. H. Wessels to the golden decades of fungal physiology and beyond. Fungal Genet Biol 27:119
25. Moore D (1998) Fungal morphogenesis. Cambridge University Press, Cambridge
26. Wösten HA, Moukha SM, Sietsma JH, Wessels JG (1991) Localization of growth and secretion of proteins in *Aspergillus niger*. J Gen Microbiol 137(8):2017
27. Metz B, Kossen NWF, van Suijdam JC (1979) The rheology of mould suspensions. Adv Biochem Eng 11:103
28. White S, McIntyre M, Berry DR, McNeil B (2002) The autolysis of industrial filamentous fungi. Crit Rev Biotech 22:1
29. Peter CP, Lotter S, Maier U, Büchs J (2004) Impact of out-of-phase conditions on screening results in shaking flask experiments. Biochem Eng J 17(3):205
30. Helmdach L, Schwartz F, Ulrich J (2014) Process control using advanced particle analyzing systems: applications from crystallization to fermentation processes. Chem Eng Technol 37 (2):213
31. Suhr H, Wehnert G, Schneider K, Bittner C, Scholz T, Geissler P, Jähne B, Scheper T (1995) In situ microscopy for on-line characterization of cell-populations in bioreactors, including cell-concentration measurements by depth from focus. Biotechnol Bioeng 47:106
32. Wiedemann P, Egner F, Wiegemann H, Quintana JC, Storhas W, Guez JS, Schwiebert C, Suhr H (2009) Advanced in situ microscopy for on-line monitoring of animal cell culture. In: Jenkins N, Barron N, Alves P (eds) Proceedings of the 21st annual meeting of the european society for animal cell technology (ESACT), Dublin
33. Joeris K, Frerichs JG, Konstantinov K, Scheper T (2002) In-situ microscopy: online process monitoring of mammalian cell cultures. Cytotechnology 38:129
34. Camisard V, Brienne JP, Baussart H, Hammann J, Suhr H (2002) Inline characterization of cell concentration and cell volume in agitated bioreactors using in situ microscopy: application to volume variation induced by osmotic stress. Biotechnol Bioeng 78:73
35. Höpfner T, Bluma A, Rudolph G, Lindner P, Scheper T (2010) A review of non-invasive optical-based image analysis systems for continuous bioprocess monitoring. Bioprocess Biosyst Eng 33(2):247
36. Olsvik E, Kristiansen B (1994) Rheology of filamentous fermentations. Biotechnol Adv 12:1
37. Wucherpfennig T, Lakowitz A, Krull R (2013) Comprehension of viscous morphology—Evaluation of fractal and conventional parameters for rheological characterization of *Aspergillus niger* culture broth. J Biotechnol 163(2):124
38. Barry DJ (2013) Quantifying the branching frequency of virtual filamentous microbes using fractal analysis. Biotechnol Bioeng 110(2):437
39. Rønnest N, Stocks S, Lantz A, Gernaey KV (2012) Comparison of laser diffraction and image analysis for measurement of Streptomyces coelicolor cell clumps and pellets. Biotechnol Lett 34(8):1465

40. Papagianni M (2014) Characterization of fungal morphology using digital image analysis techniques. J Microb Biochem Technol 6(4):189
41. Landgrebe D, Haake C, Höpfner T, Beutel S, Hitzmann B, Scheper T, Rhiel M, Reardon KF (2010) On-line infrared spectroscopy for bioprocess monitoring. Appl Microbiol Biotechnol 88:11
42. Marison I, Hennessy S, Foley R, Schuler M, Sivaprakasam S, Freeland B (2013) The choice of suitable online analytical techniques and data processing for monitoring of bioprocesses. Adv Biochem Eng Biotechnol 132:249
43. Petersen N, Odman P, Padrell AEC, Stocks S, Lantz AE, Gernaey KV (2009) In situ near infrared spectroscopy for analyte-specific monitoring of glucose and ammonium in Streptomyces coelicolor fermentations. Biotechnol Prog 26:263
44. Rønnest NP, Stocks SM, Lantz AE, Gernaey KV (2011) Introducing process analytical technology (PAT) in filamentous cultivation process development: comparison of advanced online sensors for biomass measurement. J Ind Microbiol Biotechnol 38(10):1679
45. Svendsen C, Skov T, van den Berg FWJ (2014) Monitoring fermentation processes using in-process measurements of different orders. J Chem Technol Biotechnol 90(2):244
46. Kenda A, Drabe C, Schenk H, Frank A, Lenzhofer M, Scherf W (2006) Application of a micromachined translatory actuator to an optical FTIR spectrometer. In: Ürey H, El-Fatatry A (eds) Proceedings of SPIE 6186, MEMS, MOEMS, and micromachining II
47. Sampaio PN, Sales KC, Rosa FO, Lopes MB, Calado CR (2014) In situ near infrared spectroscopy monitoring of cyprosin production by recombinant *Saccharomyces cerevisiae* strains. J Biotechnol 188:148
48. Markx G, Davey C, Kell D, Morris P (1991) The dielectric permittivity at radio frequencies and the bruggeman probe: novel techniques for the on-line determination of biomass concentrations in plant cell cultures. J Biotechnol 20(3):279
49. Li L, Wang ZJ, Chen XJ, Chu J, Zhuang JP, Zhang SL (2014) Optimization of polyhydroxyalkanoates fermentations with on-line capacitance measurement. Bioresour Technol 156:216
50. Mishima K, Mimura A, Takahara Y (1991) On-line monitoring of cell concentrations during yeast cultivation by dielectric measurements. J Ferment Bioeng 72(4):296
51. Krairak S, Yamamura K, Nakajima M, Shimizu H, Shioya S (1999) On-line monitoring of fungal cell concentration by dielectric spectroscopy. J Biotechnol 69:115
52. Davey CL, Davey HM, Kell DB, Todd RW (1993) Introduction to the dielectric estimation of cellular biomass in real time, with special emphasis on measurements at high volume fractions. Anal Chim Acta 279:155
53. Pohlscheidt M, Charaniya S, Bork C, Jenzsch M, Noetzel TL, Luebbert A (2013) Bioprocess and fermentation monitoring. Encycl Ind Biotechnol 1469–1491
54. Nielsen J (2010) Fermentation monitoring. Encycl Ind Biotechnol 1–20
55. Posch AE, Herwig C, Spadiut O (2013) Science-based bioprocess design for filamentous fungi. Trends Biotechnol 31:37
56. Ehgartner D, Sagmeister P, Herwig C, Wechselberger P (2014) A novel real-time method to estimate volumetric mass biodensity based on the combination of dielectric spectroscopy and soft-sensors. J Chem Technol Biotechnol 90(2):262
57. Wucherpfennig T, Kiep KA, Driouch H, Wittmann B, Krull R (2010) Morphology and rheology in filamentous cultivations. Adv Appl Microbiol 72(10):89
58. Papagianni M (2010) Rheology of filamentous microorganisms, submerged culture. Encycl Ind Biotechnol 1–23
59. Dhillon GS, Brar SK, Kaur S, Verma M (2012) Rheological studies during submerged citric acid fermentation by *Aspergillus niger* in stirred fermentor using apple pomace ultrafiltration sludge. Food Bioprocess Technol 6(5):1240
60. Cai M, Zhang Y, Hu W, Shen W, Yu Z, Zhou W, Jiang T, Zhou X, Zhang Y (2014) Genetically shaping morphology of the filamentous fungus *Aspergillus* glaucus for production of antitumor polyketide aspergiolide A. Microb Cell Fact 13:73

61. Formenti LR, Nørregaard A, Bolic A, Hernandez DQ, Hagemann T, Heins AL, Larsson H, Mears L, Mauricio-Iglesias M, Krühne U, Gernaey KV (2014) Challenges in industrial fermentation technology research. Biotechnol J 9(6):727
62. Booking SP, Wiebe MG, Robson GD, Hansen K, Christiansen LH, Trinci APG (1999) Effect of branch frequency in *Aspergillus oryzae* on protein secretion and culture viscosity. Biotechnol Bioeng 65(6):638
63. Haack MB, Olsson L, Hansen K, Lantz AE (2006) Change in hyphal morphology of *Aspergillus oryzae* during fed-batch cultivation. Appl Microbiol Biotechnol 70(4):482
64. Agger T, Spohr AB, Carlsen M, Nielsen J (1998) Growth and product formation of *Aspergillus oryzae* during submerged cultivations: verification of a morphologically structured model using fluorescent probes. Biotechnol Bioeng 57(3):321
65. Ahamed A, Vermette P (2009) Effect of culture medium composition on *Trichoderma reesei*'s morphology and cellulase production. Bioresour Technol 100(23):5979
66. Makagiansar HY, Shamlou PA, Thomas CR, Lilly MD (1993) The influence of mechanical forces on the morphology and penicillin production of *Penicillium chrysogenum*. Bioprocess Eng 9:83
67. Smith JJ, Lilly MD, Fox RI (1990) The effect of agitation on the morphology and penicillin production of *Penicillium chrysogenum*. Biotechnol Bioeng 35(10):1011
68. Jüsten P, Paul GC, Nienow AW, Thomas CR (1996) Dependence of mycelial morphology on impeller type and agitation intensity. Biotechnol Bioeng 52(6):672
69. Amanullah A, Jüsten P, Davies A, Paul GC, Nienow AW, Thomas CR (2000) Agitation induced mycelial fragmentation of *Aspergillus oryzae* and *Penicillium chrysogenum*. Biochem Eng J 5(2):109
70. Amanullah A, Blair R, Nienow AW, Thomas CR (1999) Effects of agitation intensity on mycelial morphology and protein production in chemostat cultures of recombinant *Aspergillus oryzae*. Biotechnol Bioeng 62(4):434
71. Amanullah A, Christensen LH, Hansen K, Nienow AW, Thomas CR (2002) Dependence of morphology on agitation intensity in fed-batch cultures of *Aspergillus oryzae* and its implications for recombinant protein production. Biotechnol Bioeng 77(7):815
72. Li ZJ, Shukla V, Fordyce AP, Pedersen AG, Wenger KS, Marten MR (2000) Fungal morphology and fragmentation behavior in a fed-batch *Aspergillus oryzae* fermentation at the production scale. Biotechnol Bioeng 70(3):300
73. Li ZJ, Shukla V, Wenger KS, Fordyce AP, Pedersen AG, Marten MR (2002) Effects of increased impeller power in a production-scale *Aspergillus oryzae* fermentation. Biotechnol Prog 18(3):437
74. Albaek MO, Gernaey KV, Hansen MS, Stocks SM (2011) Modeling enzyme production with *Aspergillus oryzae* in pilot scale vessels with different agitation, aeration, and agitator types. Biotechnol Bioeng 108(8):1828
75. Petersen N, Stocks S, Gernaey KV (2008) Multivariate models for prediction of rheological characteristics of filamentous fermentation broth from the size distribution. Biotechnol Bioeng 100:61
76. Allen DG, Robinson CW (1990) Measurement of rheological properties of filamentous fermentation broths. Chem Eng Sci 45:37
77. Deindoerfer FH, Gaden ELJ (1955) Effects of liquid physical properties on oxygen transfer in penicillin fermentation. Appl Microbiol 3(5):253
78. Olsvik E, Tucker KG, Thomas CR, Kristiansen B (1993) Correlation of *Aspergillus niger* broth rheological properties with biomass concentration and the shape of mycelial aggregates. Biotechnol Bioeng 42(9):1046
79. Riley GL, Tucker KG, Paul GC, Thomas CR (2000) Effect of biomass concentration and mycelial morphology on fermentation broth rheology. Biotechnol Bioeng 68(2):160
80. Riley GL, Thomas CR (2010) Applicability of *Penicillium chrysogenum* rheological correlations to broths of other fungal strains. Biotechnol Lett 32(11):1623–1629
81. Malouf P (2008) Study of the relationship of rheology, morphology and biomass concentration of *Trichoderma reesei* fermentation. MSc thesis, University of Ottawa

82. Wucherpfennig T, Hestler T, Krull R (2011) Morphology engineering—Osmolality and its effect on *Aspergillus niger* morphology and productivity. Microb Cell Fact 10:58
83. Bhargava S, Nandakumar MP, Roy A, Wenger KS, Marten MR (2003) Pulsed feeding during fed-batch fungal fermentation leads to reduced viscosity without detrimentally affecting protein expression. Biotechnol Bioeng 81(3):341
84. Metzner A, Otto RE (1957) Agitation of non-Newtonian fluids. AIChe J 3:3
85. Stocks SM (2013) Industrial enzyme production: process scale up/scale down. In: McNeil B, Archer D, Giavasis I, Harvey L (eds) Microbial production of food ingredients, enzymes and nutraceuticals. Woodhead Publishing, Cambridge, pp 144–172
86. Sánchez Pérez JA, Rodríguez Porcel EM, Casas López JL, Fernández Sevilla JM, Chisti Y (2006) Shear rate in stirred tank and bubble column bioreactors. Chem Eng J 124:1
87. Bhargava S, Wenger KS, Rane K, Rising V, Marten MR (2005) Effect of cycle time on fungal morphology, broth rheology, and recombinant enzyme productivity during pulsed addition of limiting carbon source. Biotechnol Bioeng 89(5):524
88. Marten MR, Velkovska S, Khan SA, Ollis DF (1996) Rheological, mass transfer, and mixing characterization of cellulase-producing *Trichoderma reesei* suspensions. Biotechnol Prog 12 (95):602
89. Patel N, Choy V, Malouf P, Thibault J (2009) Growth of *Trichoderma reesei* RUT C-30 in stirred tank and reciprocating plate bioreactors. Process Biochem 44(10):1164
90. Li ZJ, Shukla V, Wenger K, Fordyce A, Pedersen AG, Marten M (2002) Estimation of hyphal tensile strength in production-scale *Aspergillus oryzae* fungal fermentations. Biotechnol Bioeng 77(6):601
91. Kubicek CP (2013) Systems biological approaches towards understanding cellulase production by *Trichoderma reesei*. J Biotechnol 163(2):133
92. Thomas CR (1992) Image analysis: putting filamentous microorganisms in the picture. Trends Biotechnol 10:343

Adv Biochem Eng Biotechnol (2015) 149: 55–90
DOI 10.1007/10_2015_304

Published Online: 5 February 2015

Hydrodynamics, Fungal Physiology, and Morphology

L. Serrano-Carreón, E. Galindo, J.A. Rocha-Valadéz, A. Holguín-Salas and G. Corkidi

Abstract Filamentous cultures, such as fungi and actinomycetes, contribute substantially to the pharmaceutical industry and to enzyme production, with an annual market of about 6 billion dollars. In mechanically stirred reactors, most frequently used in fermentation industry, microbial growth and metabolite productivity depend on complex interactions between hydrodynamics, oxygen transfer, and mycelial morphology. The dissipation of energy through mechanically stirring devices, either flasks or tanks, impacts both microbial growth through shearing forces on the cells and the transfer of mass and energy, improving the contact between phases (i.e., air bubbles and microorganisms) but also causing damage to the cells at high energy dissipation rates. Mechanical-induced signaling in the cells triggers the molecular responses to shear stress; however, the complete mechanism is not known. Volumetric power input and, more importantly, the energy dissipation/circulation function are the main parameters determining mycelial size, a phenomenon that can be explained by the interaction of mycelial aggregates and Kolmogorov eddies. The use of microparticles in fungal cultures is also a strategy to increase process productivity and reproducibility by controlling fungal morphology. In order to rigorously study the effects of hydrodynamics on the physiology of fungal microorganisms, it is necessary to rule out the possible associated effects of dissolved oxygen, something which has been reported scarcely. At the other hand, the processes of phase dispersion (including the suspended solid that is the filamentous biomass) are crucial in order to get an integral knowledge about biological and physicochemical interactions within the bioreactor. Digital image analysis is a powerful tool for getting relevant information in order to establish the mechanisms of mass transfer as well as to evaluate the viability of the mycelia. This review

L. Serrano-Carreón (✉) · E. Galindo (✉) · J.A. Rocha-Valadéz · A. Holguín-Salas
G. Corkidi
Departamento de Ingeniería Celular y Biocatálisis, Instituto de Biotecnología,
Universidad Nacional Autónoma de México, Av. Universidad 2001,
62210 Cuernavaca, Mor, México
e-mail: leobardo@ibt.unam.mx

E. Galindo
e-mail: galindo@ibt.unam.mx

focuses on (a) the main characteristics of the two most common morphologies exhibited by filamentous microorganisms; (b) how hydrodynamic conditions affect morphology and physiology in filamentous cultures; and (c) techniques using digital image analysis to characterize the viability of filamentous microorganisms and mass transfer in multiphase dispersions. Representative case studies of fungi (*Trichoderma harzianum* and *Pleurotus ostreatus*) exhibiting different typical morphologies (disperse mycelia and pellets) are discussed.

Keywords Hydrodynamics · Mass transfer · Morphology · Physiology · Image analysis

Abbreviations and Symbols

C_{O_2}	Concentration of dissolved oxygen in the liquid (kg O_2 m^{-3})
d_{32}	Sauter mean diameter (μm)
D	Diameter of the impeller (m)
D_{crit}	Critical diameter (m)
D_{eff}	Diffusion diameter (m^2 s^{-1})
d_{eq}	Equilibrium diameter (μm)
d_h/d_r	Hyphal gradient in the pellet periphery (% μm^{-1})
d_i	Size of the drops/bubbles (μm)
D_{O_2}	Molecular diffusion coefficient (m^2 s^{-1})
EDCF	Energy dissipation/circulation function (kW m^{-3} s^{-1})
F_{lG}	Gaseous flow (–)
i	Volume unit
k	Constant that depends on the geometry of the impeller (–)
k_b	Number of volumes sampled (–)
k_La	Volumetric oxygen transfer coefficient (h^{-1})
L	Hyphal length (μm)
N	Stirring speed (s^{-1})
n_i	Number of drops/bubbles per volume i
P	Power supplied (kW)
P_p	Porosity of the pellet (–)
P/V_L	Volume power drawn (kW m^{-3})
q_{O_2}	Specific rate of oxygen consumption (kg O_2 kg^{-1} s^{-1})
r	Aggregate density (kg m^{-3})
R_{O_2}	Rate of oxygen consumption per unit volume (kg O_2 m^{-3} s^{-1})
t_c	Circulation time (s)
V_L	Volume of liquid (m^3)

Greek Letters

λ	Size of Kolmogorov microscale (μm)
ε	Local energy supplied (W kg^{-1})
ν	Viscosity (Pa s)

Abbreviations

6PP	6-pentyl-α-pyrone
ATP	Adenosine triphosphate
CFU	Colony-forming units
DNA	Deoxyribonucleic acid
FDA	Fluorescein diacetate
GFP	Green fluorescent protein
RNA	Ribonucleic acid
RPB	Reciprocating plate bioreactor
rpm	Radians per minute (stirring speed)
vvm	Volume of gas per volume of liquid per minute

Contents

1 Mycelial Cultures: Basic Aspects

Mycelial microorganisms play a major role in the bioprocessing industry for the production of various primary and secondary metabolites such as enzymes, antibiotics, organic acids, extracellular proteins, and flavoring compounds [1–4].

Filamentous fungi are lower eukaryotic heterotrophic microorganisms that depend on external sources of carbon and have both sexual and asexual reproduction. The mycelium is broadly composed of vegetative branched filaments (hyphal) with rigid and thick cell walls. These cell walls are composed of fibrous polysaccharides (chitin and β-glucan) and glycoproteins that determine the fibrillar structure of the wall. The hyphal growth takes place in the apical region (tips). When a new tip is formed, it

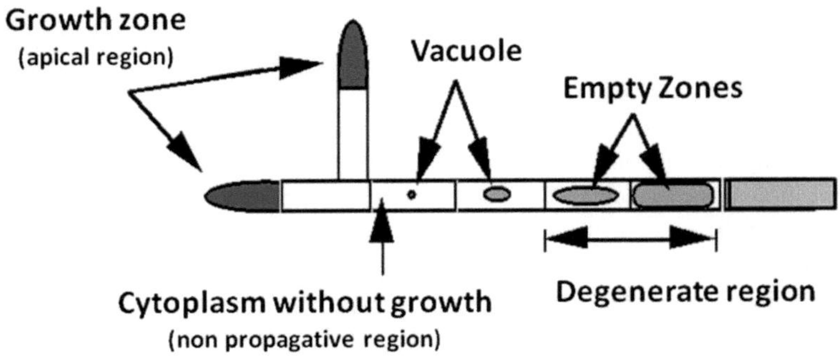

Fig. 1 Schematic diagram of the regions of cell growth and differentiation of the fungus *P. chrysogenum* (adapted from Ref. [5])

grows to a certain length, and a septum with a pore is created at the rear of the tip, creating a non-growing cell. The net result of the apical extension is the formation of non-propagative cells which, in turn, produce vacuoles. Vacuoles grow as a function of their distance from the apical region. Given that the mycelial structure is complex and multicellular, the distribution of cell ages is wide compared to unicellular microorganisms. Therefore, hyphae may exist in different physiological states and show different structural features such as growing apical cells, vacuolated regions, and lysated cells which are metabolically inactive (Fig. 1).

Mycelial cultures have different physical characteristics to unicellular cultures (bacillococci or yeast) primarily due to their large morphological and physiological diversity that develops during growth. The morphology of filamentous fungi in submerged culture depends on the environmental conditions in the bioreactor, the concentration of the inoculum, spore viability, pH, temperature, dissolved oxygen concentration, and mechanical stress [6–12]. In submerged culture, fungi develop two types of extreme macroscopic morphologies: pellets and filamentous mycelium. Pellets are compact hemispherical aggregates of hyphae in which the diffusional limitations to the center of the pellet represent the principal growth restriction. At the other hand, filamentous or sparse mycelium increases the viscosity of the broth and can drastically reduce the oxygen transfer capacity of the bioreactor [13, 14].

1.1 Cultures with Dispersed "Filamentous" Morphology (Dispersed Mycelium)

Fermentation broths are generally heterogeneous suspensions of microorganisms, culture medium components (nutrients), and various metabolites or products dispersed in a liquid phase, generally water. Unlike other broths, the rheological properties of the predominantly filamentous mycelial cultures change with time.

These changes are mainly due to the increase in cell mass, the morphological changes of the hyphae, and the interactions between hyphae [15–18].

The hyphae are usually long, thin, and branched: They are also interlocked due to their high length to diameter ratio, thereby forming a three-dimensional network. These organisms grow in a highly dispersed manner and consequently generate highly viscous and rheologically complex suspensions [19]. Most mycelial broths behave similar to pseudo-plastic fluids. Their viscosity decreases when the deformation gradient increases. Thus, in cultures carried out in mechanically stirred bioreactors, the flow characteristics change around the impeller (high-speed deformation region). This implies that the fluid surrounding the impeller has a low viscosity due to the high deformation gradient in this zone of the reactor [16, 20]. However, in areas near the walls of the bioreactor, the deformation rate applied to the fluid is lower, causing a noticeable increase in viscosity. This causes serious problems in the mixing, creating "dead" or poorly mixed areas, in which there is a reduction in the mass transfer rate and the formation of concentration gradients of nutrients and dissolved oxygen. These differences are stronger as the process progresses and the organism grows [15, 21–25]. Filamentous growth has been reported to be favorable for the production of various antibiotics or enzymes [4, 26–28].

1.2 Cultures with an Aggregate Morphology or "Pellets"

It has been reported that pellets favor the production of some metabolites as glucoamylase in *Aspergillus niger* [29]; lovastatin in *Aspergillus terreus* [30]; fumaric acid in *Rhizopus delemar* [31]; and ergothioneine in *Lentinula edodes* [32]. However, the differentiation and physiological state of the cells that carry out the synthesis of these metabolites are not yet well understood.

For practical purposes, the pellet-type morphology is more convenient in fermenters because the culture media in which pellets develop have a low viscosity and behave as Newtonian fluids, facilitating the mixing and homogenization of culture broth. For example, Cai et al. [33] obtained a mutant for a gene that affected the polarized growth of *Aspergillus glaucus*. Unlike the native strain that had a dispersed filamentous growth, the mutant grew into pellets and produced 82 % more aspergiolide A, than the native strain, and exhibiting low-viscosity broths reaching high dissolved oxygen tension during culture.

Despite the positive effects of pellet morphology on viscosity and oxygen transfer, limitations in nutrient diffusion from the broth to the center of the pellet can arise as the pellet size increases throughout fermentation. This phenomenon causes significant physiological changes in the organism. In cultures with a predominant pellet morphology (i.e., basidiomycetes), it is likely that nutritional or oxygen limitations inside the aggregates adversely affect the growth and productivity of these fungi. For this reason, even when high concentrations of dissolved oxygen are maintained in the medium, the oxygen concentration can be limited within the mycelial pellet when the aggregates grow above a critical diameter [23].

Therefore, it is essential to characterize mycelial morphology to understand the effect of process variables on the physiology of filamentous cultures. For example, Cui et al. [23] proposed an equation (Eq. 1) to determine the maximal diameter that can be reached by an aggregate (D_{crit}) without diffusional limitations at the center of the aggregate.

$$D_{crit} = \sqrt{\frac{24 \times C_{O_2} \times D_{eff}}{R_{O_2}}} \tag{1}$$

In this equation, D_{crit} is the critical diameter of the aggregates (m), C_{O_2} is the concentration of dissolved oxygen in the liquid (kg O_2 m^{-3}), D_{eff} is the effective diffusion of the oxygen in the aggregates (m^{-2} s), and R_{O_2} is the rate of oxygen consumption per unit volume (kg O_2 $m^{-3}s^{-1}$). R_{O_2} equals the product of the specific rate of oxygen consumption (q_{O_2}) and aggregate density (r). D_{eff} is the product of the molecular diffusion coefficient (D_{O_2}) and the porosity of the pellet (P_p) (Eq. 2).

$$D_{eff} = D_{o_2} \times P_p \tag{2}$$

With equations such as those proposed by Cui et al. [23], it is possible to estimate whether there is a limitation in oxygen diffusion into the pellet according to the morphological characteristics of the aggregates (porosity and diameter) and other factors such as dissolved oxygen concentration, rate of oxygen consumption, and density of the aggregates.

A negative correlation between hyphal gradient (d_h/d_r)—a morphological parameter describing the pellet periphery—and the effective diffusion coefficient has been reported [34]. The authors showed that while diffusion limitation in pellets is mainly a function of its size, the influence of the pellet periphery over the diffusion is rather high.

Furthermore, using mutant strains (GFP), the intraparticle metabolic activity can be determined and correlated it with the productivity of the process. Driouch et al. [35] demonstrated with this tool that the fructofuranosidase productivity of *A. niger* increases when titanate microparticles were used in the culture to decrease the size of the pellets. The diameter decreased from 1,700 μm (control) to 300 μm (microparticles). It was shown that, in the control culture pellets, only the surface layer (200 μm) showed metabolic activity associated with the production of fructofuranosidase. Furthermore, lavendamycin production by *Streptomyces flocculus* increased by 600 % when agitation was increased and was correlated with a decrease in the size of the pellets [36]. Pellet growth is required for efficient glucoamylase production by *A. niger*. Cultures carried out at low energy dissipation (0.063 W/kg) presented high pellet concentration and glucoamylase activity [29].

Given that morphology influences both growth and metabolite productivity in fungi, it is necessary to understand the mechanism of pellet formation and their behavior as a result of agitation at different intensities [13]. They have been generally categorized as a coagulant type or a non-coagulant type [37]. The coagulant type forms aggregates that grow to form pellets composed of spores and hyphal aggregates formed in the early stage of culture. For the non-coagulant type, a single spore grows into a pellet.

This phenomenon has been clearly described for *A. niger* in which aggregation takes place in two steps. The first one is related to surface interactions between conidia before germination, and the second one to hyphal growth [10, 38]. Aggregation of *A. niger* conidia can be manipulated by pH, which has an important influence in both first and second aggregation steps. Power input has a significant effect on pellet concentration during the second step of aggregation. Cultures carried out at pH 4 showed an increased number of pellets and specific glucoamylase productivity than those performed at pH 7. A clear correlation between product formation and the number of pellets suspended was found [10]. Furthermore, it has been shown that increasing the volumetric power input by aeration yields smaller pellets, higher pellet number, and improved glucoamylase production [39]. Production of glucoamylase seems to occur in the outer layer of the pellets, a region that is rich in single-stranded RNA, explaining the higher productivity of such cultures where small and numerous pellets are present [40].

A. niger also exhibits macro and micro colonies, which are highly heterogeneous with respect to growth, secretion, and gene expression [41]. Gene expression of 7-day colonies of *A. niger* shows that mycelia is highly differentiated, as more than 25 % of the active genes showed significant differences in expression between the inner and the outer zones of the colony [42]. More recently, the study of the heterogeneity of *A. niger* microcolonies has been reported [43]. This work shows that the gene expression of glucoamylase and ferulic acid esterase was higher in the small colonies than in the larger ones. Furthermore, RNA content per hypha was 45 times higher at the periphery than in the center of the microcolony.

Operational and culture conditions strategies as ultrasound, agitation intensity, inoculum density, substrate concentration, and manipulation of broth viscosity have been studied in order to control fungal morphology and process productivity. The manipulation of viscosity, by adding Xanthan gum, surfactants, or glass beds on the broth can be used to increase pellet count, biomass concentration and decrease pellet volume of *Streptomyces hygroscopicus*. In these cultures, specific production of geldanamycin was inversely correlated with the mean pellet diameter [44, 45].

Inoculum density also has shown to determine morphology, broth rheology, and erythromycin production by *Saccharopolyspora erythracea*. Low inoculum densities yield pellet morphology and Newtonian rheology. At the other hand, clumps yield non-Newtonian rheology and high erythromycin and were obtained at high inoculum densities [46].

The agitation intensity is also an important parameter that affects the morphology of filamentous microorganisms; and in some cases, the production of metabolites, for example, intracellular content of ergothioneine in *L. edodes* pellets is almost 3 times of that found in free filaments or clumps. In this case, the pellet size increased as agitation is reduced, favoring the highest productivity of ergothioneine [32].

Ultrasound applied during *A. terreus* cultivation influences morphology and lovastatin production; however, it does not affect fungal growth. Sonication negatively affects lovastatin production, and this was associated with the change from pellet to disperse hyphal morphology occurred at high power input values [30]. Fumiglaclavine production by *Aspergillus fumigatus* is also increased by ultrasound exposition, which produces smaller and looser pellets as compared with untreated control [47].

Initial substrate concentration influences growth morphology of *R. delemar*. At the best initial concentrations, small pellets, high biomass growth, and an increased fumaric acid production are obtained [31]. Osmolality can also affect fungal morphology and productivity. High specific productivity of fructoranosidase by *A. niger* can be achieved by the addition of sodium chloride to reach an osmolality of up to 3.2 osmol/kg. In such conditions, fructofuranosidase production is about 3 times that obtained in the control cultures [27]. In that work, a significant correlation between specific fructofuranosidase productivity and the switch between pellets to filamentous morphology were reported.

2 Hydrodynamic Stress in Mycelial Cultures

2.1 Factors Determining Fungal Morphology and Metabolite Production

The hydrodynamic conditions generated during cell culture in bioreactors can affect both shape (morphology) and cellular metabolism. When mechanical stress is applied to the wall or cell membrane, various specific receptors (such as ion channels) deform and stimulate a series of cascade reactions, which appear to affect the synthesis of proteins and RNA [48, 49]. G-protein-coupled receptors (GPCR) have been implicated in cellular responses to shear stress and morphogenesis in several fungal species. However, it is unknown how mechanical stress modulates GPCR function [50]. This phenomenon, defined as the change in the physiology of a culture depending on the shear rate gradients caused by the movement of a fluid, is commonly called hydrodynamic stress and is particularly important for mechanically stirred tanks. Specific oxygen uptake has been recently proposed as a shear stress indicator in fungal cultures [51].

There are several examples where shear stress seems to enhance fungal metabolite productivity. However, most of the reports fails to evaluate whether the effects observed are only related to shear stress and/or to an increase in the oxygen transfer rate. High lipase production by *Rhizopus chinensis* is enhanced at low inoculum density and high shear stress, which leads to a predominant pelleted morphology [52]. At the other hand, xylanase production by *Aspergillus oryzae* was not affected by agitation [53]. Impeller choice does not affect enzyme production in fed-batch cultures of *A. oryzae*; however, growth and enzyme production were enhanced at high aeration and volumetric power inputs [54]. Production of pristinamycins by *Streptomyces pristinaespiralis* is influenced by agitation. At low power input, oxygen limitation was observed and no pristinamycin production was observed. After oxygen limitation is overcome, product formation was related to power input in a bell-shaped way with a maximum at 8 W/L [55]. When this fungus was grown in the range of 0.5–6 W/L, a direct correlation between growth and maximal specific growth rate was found [56]. More recently, the same authors have shown that pristinamycins production occurred only when $k_{L}a$ was higher than 100 h^{-1}, while final product concentration and bacterial pellet diameter were correlated to *P*/*V*. A strong correlation between pristinamycins concentration and pellet interfacial area was found [57].

Negative effects of extreme shear stress on fungal productivity can also be found. High oxygen concentration and low shear stress are necessary for lovastatin production by *A. terreus* [58, 59]. Production of exopolysaccharides (EPS) by *Paecilomyces tenuipes* in airlift and stirred reactors does not differ significantly when loose aggregates are the dominant mycelial morphology. However, molecular weight and chemical composition of the product is affected by agitation in bioreactors [60]. On other hand, production of EPS by *Tremella fuciformis* is higher in airlift reactors than in stirred reactors and related to a dominant population with yeast-like morphology [61]. An increase of 40 % in the production of manganese peroxidase, between immobilized and stirred cultures of *Nematoloma frowardii*, has been reported [62]. Xylanase production by *A. terricola* seems to be affected by shear stress as enzyme production was significantly lower in stirred tanks than in airlift reactors [63]. Production of arpergiloide-A by *A. glaucus* can be increased by 322 % by using a combination of a pitched blade (upper), a disk blade turbine (lower), and n-dodecane as oxygen carrier, if compared with cultures carried out with two disk turbines [64]. Chitinases production by *Lecanicillium muscarium* is strongly affected by aeration and agitation. At low agitation and aeration, an increase on both parameters yields an increase in enzyme production, likely due to an increase in oxygen availability. Here again, enzyme production decreased once a certain threshold of shear stress is attained [65]. The use of rocking agitated disposable reactors has been reported for shear stress sensitive microorganisms. In a bag reactor, fungal cultures of *Pleurotus sapidus* and *Flammulina velutipes* grew in form of small pellets, while in stirred tanks cultures, an intense aggregation was observed [66].

Microorganism sensitivity to mechanical stress is primarily determined by the structure (presence and/or absence of a cell wall) and size of the cells [67, 68]. Cell size plays a major role in cell sensitivity to mechanical stress (Table 1). In general,

Table 1 Size and sensitivity of microorganisms to shear (adapted from Ref. [67, 77])

Microorganism	Size (μm)	Sensitivity
Bacteria	1–10	–
Filamentous fungi	>100	++
Animal cells	20	++
Animal cells in microcarriers	150	+++
Disperse plant cells	100	+
Aggregated plant cells	>1	++

the larger the cell, the greater the damage because the degree of interaction between cells and small eddies is greater. Eddies are formed from the turbulence generated in the mixing tank. Eddies are the result of turbulent flow (see further below) and are responsible for causing mechanical damage to cells [69–73]. It is hypothesized that most of the damage to biological particles is caused by eddies that are the same size as or smaller than the cell. Larger eddies carry the particles in a convective motion. Eddies of comparable or smaller sizes can act simultaneously and in opposite rotational directions on the cell surface, subjecting them to deformation gradients and therefore to shearing forces [48, 74]. As a result, the intensity in the interaction between particles and eddies determines cellular damage or mycelial aggregation during cultivation. For example, growth and pellet morphology of *A. niger* was studied under different fluid dynamic conditions. Higher shear stress negatively affected fungal growth and leads to small and dense pellets. Results showed that maximal shear stress, occurring at the vicinity of the impellers, is the main factor affecting pellet growth [75].

In mechanically stirred bioreactors, the size of terminal eddies (minimum scale with a turbulent flow) can be calculated according to the Kolmogorov microscale [76] equation (Eq. 3). In this equation, ε is the local energy supplied by the liquid mass (W kg^{-1}), λ is the size of the microscale (μm), and v is the viscosity (Pa s). According to the Kolmogorov microscale equation, if the power supplied is reduced (i.e., by decreasing the stirring speed or the diameter of the impeller) and/or the viscosity of the medium increases, the damage to the biological particles will be reduced because the eddies increase in size. λ values oscillating between 10 and 50 μm are typical in bioreactors with an energy dissipation between 0.1 and 100 W kg^{-1} and water as the fluid [77, 78]. Therefore, the most susceptible systems are those with sizes equal or superior to this range (Table 1). These systems include filamentous fungi, plant, and animal cells (>20 μm).

$$\lambda = \left(\frac{\nu^3}{\varepsilon}\right)^{1/4} \tag{3}$$

In mycelial cultures, based on the theory of isotropic turbulence [41], van Suijdam and Metz [79] proposed for the first time that the hyphal length of *P. chrysogenum* may be associated with the size of the eddies generated in a reactor and, therefore, with the operating conditions. Similarly, Ayazi-Shamlou et al. [70] and Cui et al. [77] reported that the fragmentation of the hyphae of the fungus *P. chrysogenum* was determined by the ratio of the average hyphal length (*L*) and smaller eddies whose size (i.e., Kolmogorov microscale, λ) depended on the energy supplied to the system. However, unlike observations in studies with animal cells, no association between *L* and λ was reported. This might be explained by the fact that initial physical properties were used to estimate λ. The above assumption may be valid at the beginning of the cultures or when the values of viscosity and/or density do not change significantly. However, when the rheological properties of the medium change (as in filamentous cultures), it is necessary to include any changes in viscosity in the calculation of λ. At the same time, and as a result of an increased viscosity, it is likely that the supply of energy (ε) decreases due to the higher power drop caused by the formation of gas cavities behind the impellers [80].

The importance of including the actual viscosity value and that of the power supplied (i.e., in situ measurement) in λ was shown by Li et al. [72] in mycelial cultures of *A. oryzae*. As a direct result of increased viscosity (approximately 0.1 to 1.3 Pa s), Li et al. [72] estimated significant increases in λ, which reached values of up to 900 μm. Similarly, Large et al. [81] reported significant increases in λ (100 μm at 0 h to 350 μm at 70 h) due to the increased viscosity of the medium during the filamentous cultivation of *Streptomyces clavuligerus*.

Given the important differences between typical λ values and the average hyphal length (10–50 vs. 100–300 μm, respectively [70, 77]), the shearing forces interact intensively with the surface of the hyphae and fragment them. The size of the mycelial aggregates will therefore be the final result of the balance between the fragmentation (product of hydrodynamic stress) and the apical growth of the fungus. As a result, an equilibrium diameter (d_{eq}) is established when the local shearing forces are equal to the tension forces of the hyphae. According to Li et al. [72], these forces depend on the physiological state and composition of the cell wall of the fungus. Therefore, the use and development of image analysis techniques in mycelial cultures allow us to observe in an objective and rigorous manner the morphological characteristics generated by mechanical stresses.

In mechanically stirred tanks, the morphological properties depend on the stirring speed, the number, diameter and geometry of the impellers, and the fluid characteristics. These conditions determine the hydrodynamic and mechanical forces present in the bioreactor. For instance, *T. harzianum* clump diameter showed a strong dependence on Kolmogorov microscale (λ), calculated using the specific energy dissipation rate in the impeller swept volume and the measured apparent broth viscosities. The mean clump diameter of *T. harzianum* aggregates was about 0.5 λ, which means that fungal morphology of this culture was determined by the size of eddies [73].

In general, increasing the energy supplied reduces the size of the mycelium, which favors the fragmentation of the hyphae. Amanullah et al. [82] reported a reduction in the average total length of *A. oryzae* when the mixing speed was increased (and therefore also the energy supplied) from 550 to 1,000 rpm (from 2.2 to 12.6 kW m^{-3}) in a 6-L bioreactor equipped with two Rushton turbines. Similarly, Papagianni et al. [83] noted a decrease in the average diameter and length of the hyphae by increasing the agitation speed from 200 to 600 rpm in submerged 8 L cultures of *A. niger*. Johansen et al. [84] reported the reduction of hyphal average total length by increasing the agitation speed of a Rushton turbine from 400 to 1,100 rpm (from 0.6 to 13 kW m^{-3}) in 6 L cultures of *A. awamori.* Ayazi-Shamlou et al. [71] also observed a decrease in the size (average length) of the hyphae when the power supplied increased (1.7–33 kW m^{-3}) in submerged cultures (7 and 150 L) of *P. chrysogenum*. Furthermore, the size of the aggregates and/or pellets is also significantly influenced by the energy delivered. Paul et al. [85] observed a decrease in the equivalent diameter and the area of *A. niger* aggregates when the agitation rate increased (500–800 rpm) after 17 h of culture. Meanwhile, Cui et al. [78] reported that the reduction in the average diameter of *A. awamori* pellets in 3 L cultures (grown in bioreactors with two Rushton turbines) increased with the agitation speed (from 340 to 1,050 rpm).

The morphological characteristics of the mycelial aggregates (average projected area, diameter, length, porosity, circularity, etc.) correlate with various operating parameters of a bioreactor. Tip speed has been used as a scale-up criterion for EPS production by *Ganoderma lucidum* in stirred tanks [86]. However, the most widely used parameter is the supplied volumetric power, or *P*/*V*, which has also been used as a scaling criterion in various bioprocesses [87, 90]. Volumetric power (*P*/*V*) assumes a homogeneous and constant energy dissipation throughout the tank. This factor integrates into a single parameter the speed of stirring, workload, geometry, diameter, and number of impellers, in the bioreactor. Therefore, when *P*/*V* increases (consequently causing mechanical stresses on the mycelium), the size of the aggregate decreases. This phenomenon has been reported by several authors in cultures of *P. chrysogenum* [21, 26, 70, 89], *A. oryzae* [82, 90], *A. awamori* [84, 77], and *A. niger* [83, 85].

However, Jüsten et al. [91] showed that the association between morphology and P/V is valid only for the scale studied. In their work, Jüsten et al. [91] evaluated the effect of the energy supplied to the average projected area of *P. chrysogenum* using three scales of work (1.4, 20, and 180 L) and a Rushton turbine. According to their results (Fig. 2), it was clear that there was a correlation between projected area and the delivered energy (P/V_L) for each scale of work. This suggests that there is a possible effect of tank size (volume) on energy dissipation and, therefore, on morphological characteristics. Similar results were reported by Makagiansar et al. [26] in

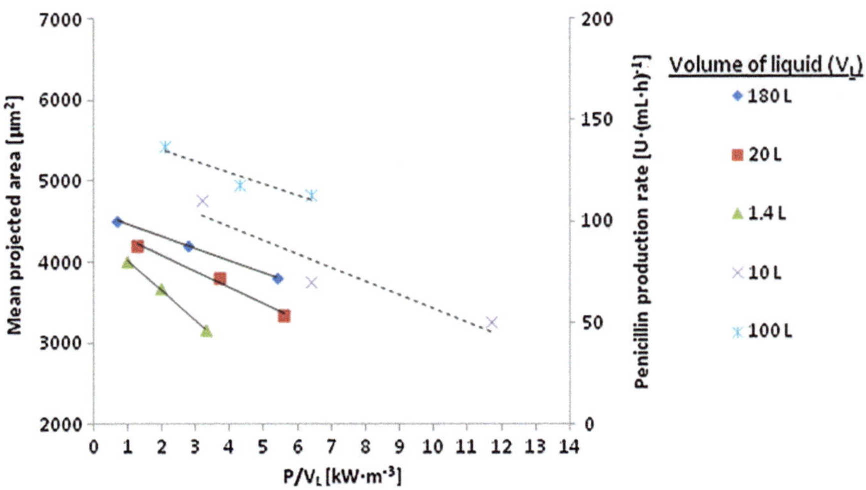

Fig. 2 Effect of the volumetric power supplied (P/V_L) on the average projected area and the rate of penicillin production in cultures of *P. chrysogenum* at different scales (adapted from Ref. [21, 91])

P. chrysogenum cultures in bioreactors of 5, 100 and 1,000 L equipped with three Rushton turbines.

Recently, Marín-Palacio et al. [92] showed that the parameter P/V does not determine the microbial growth, morphology, and recombinant protein production in cultures of *Streptomyces lividans.* These authors concluded that the transfer of oxygen in the center of the pellets could explain the changes observed between the different hydrodynamic conditions evaluated.

2.2 The Concept of "Energy Dissipation/Circulation Function" (EDCF)

Given the limited relationship between volumetric power and morphology observed in different scales of work, various reports suggest the use of the energy dissipation rate or the EDCF ("energy dissipation/circulation" function) to correlate hydrodynamic morphological characteristics with the operating variables of a bioreactor [21, 24, 26, 82, 89–91]. Smith et al. [21] developed the concept of the energy dissipation/circulation function (EDCF) based on the previous work of van Suijdam and Metz [79] and Reuss [93]. These authors proposed that fragmentation (and therefore morphological characteristics) depends on the energy dissipated in the region of the impellers (i.e., energy dispersion area) and the frequency of the passage of mycelium through the area. However, the function given by Smith et al. [21] did not allow for evaluating the effect of impellers other than the Rushton turbines. In subsequent studies, Jüsten et al. [89, 91] modified the EDCF to incorporate an additional geometric factor that was able to evaluate the hydrodynamic effect of different types of impellers (axial and radial). These authors also included the negative effect of viscosity on the power supply (power drop) in the calculation of circulation frequency. The energy dissipation rate (EDCF) is given by Eq. 4:

$$\mathrm{EDCF} = \left(\frac{P}{k \times D^3}\right) \times \left(\frac{1}{t_c}\right) \tag{4}$$

In Eq. 4, $P/(k \cdot D^3)$ is the specific energy dissipation in the impeller sweeping area (kW/m^3), P is the power supplied (kW), D is the diameter of the impeller (m), k is a constant that depends on the geometry of the impeller (dimensionless), and $1/t_c$ is the frequency of circulation of the microorganism in the area of the impeller. This variable is defined by Eq. 5:

$$\frac{1}{t_c} = \frac{F_{l_G} \times N \times D^3}{V_L} \tag{5}$$

Here, Fl_G is the gaseous flow (dimensionless), N is the stirring speed (s^{-1}), and V_L is the volume of liquid (m^3). Therefore, EDCF quantifies the energy delivered to the fluid in the area of the impellers at a given time. The correlation observed

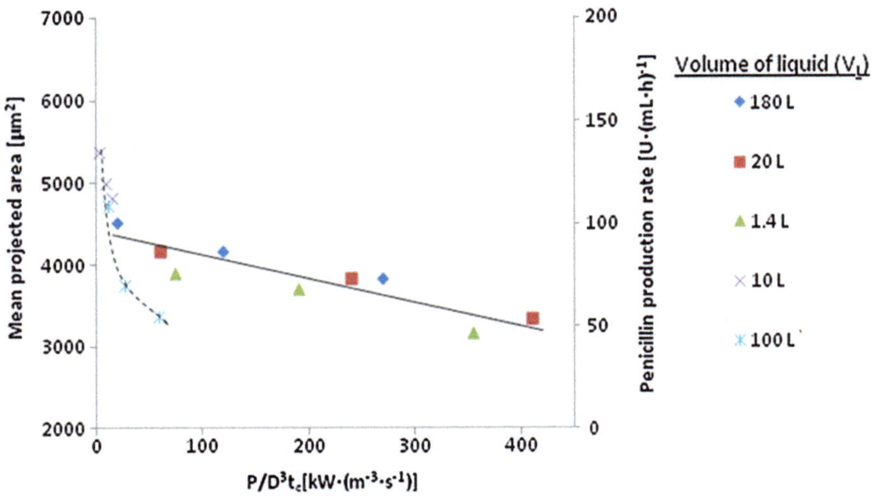

Fig. 3 Effect of the energy dissipation rate (EDCF) on the average projected area and the rate of penicillin production in cultures of *P. chrysogenum* at different scales (adapted from Ref. [21, 91])

between mycelial size and EDCF suggests that the fragmentation of the hyphae depends not only on the energy supplied but also on the frequency at which the mycelium is exposed to high levels of mechanical stress (Fig. 3).

In a very interesting study, Jüsten et al. [91] demonstrated that the mycelial morphology of *P. chrysogenum* (average projected area for the aggregates and total average length of disperse mycelia) depended on the EDCF, regardless of the scale used (1.4, 20 and 180 L). Figure 3 shows the effect of the EDCF on the average projected area. As expected, the projected area decreased when the energy supplied increased. However, unlike P/V, the EDCF correlated well with all the experimental data (Fig. 2 vs. Fig. 3). The fact that the aggregate size decreased when the EDCF increased suggests that increased mechanical damage occurs when the energy supplied increases or when the circulation time decreases (t_c). In both cases, movement of the mycelium through the impeller region (area of high mechanical stress) was favored. Thus, the mycelium was exposed to different levels of shearing stress, and the intensity depended on its position inside the tank and the fluid circulation rate. In a similar study, Makagiansar et al. [26] observed a reduction of the average hyphal length with an increase in the power supplied and/or the frequency of circulation. Therefore, fragmentation and, consequently, the size of the mycelia of *P. chrysogenum* depended on the EDCF in the three working scales used (5, 100, and 1,000 L).

Given the strong interdependence between morphology, physiology and metabolite productivity, and the notorious influence of hydrodynamic stress on morphology, different groups have used the EDCF to find associations between other biological parameters in addition to mycelial morphology and the operational conditions of a bioreactor.

In cultures of *P. chrysogenum*, Smith et al. [21] showed that the rate of penicillin production depended on the EDCF, regardless of the working scale used (10 and 100 L). However, the rate also depended on the *P*/*V* (Figs. 2 and 3). Similar results were reported by Makagiansar et al. [26] in cultures of *P. chrysogenum* in which the average hyphal length and the specific productivity of penicillin produced in bioreactors of 5, 100, and 1,000 L depended on the EDCF. Meanwhile, Jüsten et al. [89] observed that, in addition to the average projected area, the specific growth rate (μ) and the specific productivity of penicillin, depended on the EDCF function in *P. chrysogenum* cultures at different stirring rates and with three different shapes of impellers (flat blade, Rushton turbine, and pitched blade).

Rocha-Valadez et al. [73] reported that in cultures of *T. harzianum*, the energy dissipated (EDCF) has effects on the mycelial diameter. When the energy increases from 9.5 to 95.9 $kWm^{-3}\ s^{-1}$, the diameter decreases from $\sim$400 to 200 μm in cultures in which the circulation time is greater than or equal to 3 s. Furthermore, such *T. harzianum* cultures were conducted at equivalent yielding P/kD^3 conditions and developed using two different Rushton turbines diameter sets in order to study the effect of circulation time on fungal morphology. For the studied conditions, $1/t_c$ had a greater effect over mycelial clump size and growth rate than P/kD^3. Consequently, broth viscosity, and hence Kolmogorov microscale, was a function of impeller diameter, even among cultures operated at equivalent specific energy dissipation rates [73]. These effects have an impact on both, hydrodynamic and 6PP productivity, because larger cell aggregates increase viscosity by altering the rate of energy dissipation and therefore the growth rate. These effects are also reflected in the dispersion of the phases present in the fermentation (see further below), having a direct effect on the process of breaking bubbles and drops of other physical phases present in multiphase systems. Although the EDCF has been used to correlate changes in morphology and the metabolism of hydrodynamic sensitive microorganisms, this parameter allows the analysis of the shearing energy intensity only in regions close to the impeller, because this function is derived from Kolmogorov's theory of isotropic turbulence [94].

It should be pointed out that, however, the use of the EDCF to correlate metabolite productivity with hydrodynamic conditions is not universal. This was demonstrated by Amanullah et al. [82, 90] in continuously fed cultures of *A. oryzae* for the production of recombinant proteins (α-amylases and amyloglucosidases). This was probably due to fact that the fed-batch cultures were carried out under conditions of substrate limitation. As in batch cultures, without substrate limitation, biomass concentration and AMG secretion increased with increasing agitation intensity.

Although it is difficult to draw general conclusions, due to (a) the complexity of the systems and the different specific characteristics of the mycelial cultures that have been studied, (b) the somehow limited experimental strategies, which in the majority of the cases are not able to rigorously discriminate between mechanical (stress) effects and those associated with dissolved oxygen, and (c) the not clearly defined term "hydrodynamic stress," the review of the literature revealed some general points:

(a) The overall mycelial size is inversely proportional to the global energy supplied to the bioreactor.
(b) More particularly, the overall mycelial size is an inverse function of the energy dissipated in the vicinity of the impellers, as well as of the time that the biomass is present in that zone.
(c) In general, within a range of low dissipated energy, the increase in power improves the production of fungal metabolites; however, the effect could be a dissolved oxygen one.
(d) In the range of high dissipated energy, an increase in power leads to saturation or decreasing profiles in terms of metabolite production.
(e) The mechanism of mycelial damage has been shown to be associated with the length of Kolmogorov eddy size; however, very few works have reported estimated values based on experimental determination of power drawn.

2.3 The Use of Microparticles to Manipulate Fungal Morphology (and Metabolite Productivity)

The use of microparticles in stirred tanks is a useful strategy to enhance the enzyme [19, 95–100] and secondary metabolites production [101–103] by different fungal strains [98], by controlling the morphology of fungal microorganisms.

In a pioneer work [95] using talc microparticles (≤42 μm), Kaup et al. were able to control fungal morphology of *Caldariomyces fumago* from pellet to single hyphal growth, which yielded a fivefold increase in the maximum specific productivity of chloroperoxidase. In another work, glass beads of 4 mm were successfully used to produce filamentous growth and to scale up from microtiter plates to bench-scale cultures [101].

Addition of titanate microparticles (8 μm) to *A. niger* cultures yielded a sevenfold increase on the production of fructofuranosidase and glucoamylase [99]. Interestingly, microparticles were found to be part of the inner core of fungal pellets with hyphae occupying the external shell of the pellets. Using a recombinant strain expressing GFP, it was possible to demonstrate that active mycelia of control cultures is restricted to the external 200 μm surface of the pellets, while pellets obtained in microparticle containing cultures were fully active.

Lipid accumulation by *Mortierella isabellina* is enhanced by controlling fungal morphology, from pellets to dispersed mycelia, using different concentrations of magnesium silicate microparticles. Lipid cell content increased with the increase of microparticle concentration. At 10 g/L, lipid production was 2.5-fold higher than in the control experiment [102]. In another work, the production of the fungal volatiles 2-phenylethanol and 6-pentyl-α-pyrone by *A. niger* and *Trichoderma atroviride*, respectively, was also enhanced by the addition microparticles [103].

Although results obtained with the use of microparticles in fungal cultures are strain and particle specific, two are the main factors involved in its positive effects on process productivity. First, the presence of microparticles favors the reduction of aggregates size, in which the nutrient limitations occur in pellets with a diameter above 200 μm. With microparticles, fungal morphology can be tailored; therefore, mass transfer limitations are reduced when dispersed cells are obtained instead to large pellets. The second main factor affecting morphogenesis and metabolic responses of filamentous microorganisms is the mechano-sensitivity of fungal cells to shear stress. G-protein-coupled receptors (GPCR) have been implicated in cellular responses to shear stress and morphogenesis in several fungal species. However, it is unknown how mechanical stress modulates GPCR function [50].

2.4 *Filamentous Growth: The Case of Trichoderma Harzianum*

Trichoderma is a filamentous fungus that is commonly found in soil and belongs to the group *Deuteromycetes*. The fungi from this genus have the ability to synthesize different products of interest [4, 104–109] such as enzymes (i.e., cellulases, chitinases, glucanases, glucosidases, and xylanases), antibiotics (cyclosporine and trichodermine), aromatic compounds (6-pentyl-α-pyrone and γ-decalactone), organic acids (citric acid), or extracellular proteins. Additionally, they have been used as biological control agents. The growth of filamentous fungi in liquid cultures broth may generate serious deficiencies in mixing and affect the productivity of the respective bioprocesses (as already discussed in previous sections).

Trichoderma harzianum has been a good model of study because it produces metabolites of industrial interest, and the biomass itself has applications as biological control agent.

Hydrodynamics and mass transfer are particularly important in the submerged culture of *Trichoderma* sp. because mycelial growth (Fig. 4), morphology, and the

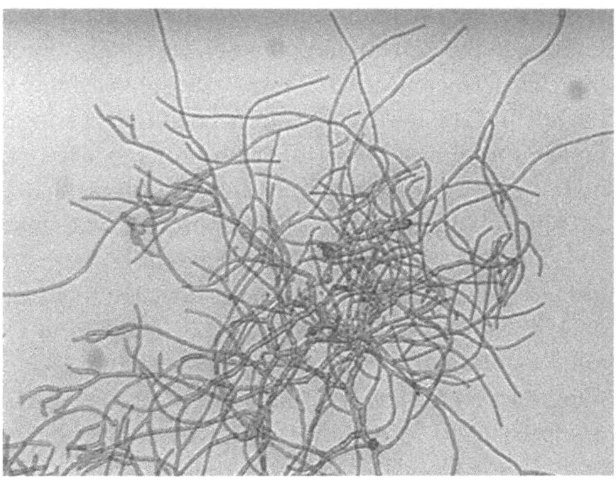

Fig. 4 Filamentous growth of *Trichoderma harzianuam* cultivated in a stirred tank (photograph obtained with a magnifier 10×)

production of various metabolites depend on the environmental conditions prevailing in the bioreactor. Several authors (see below) have observed a similar bell-shaped association between metabolite productivity (i.e., cellulases, glucanases, xylanases, or chitinases) and power or energy supplied usually manipulated through changes in stirring speed. The general explanation for this phenomenon is that the increase in stirring speed reduces oxygen limitation problems often present inside the aggregates or pellets. However, the positive effect caused by increasing the agitation rate is limited and is mainly determined by the sensitivity of fungi to hydrodynamic stress. The mechanical forces damage the mycelium and thus adversely affect the production of the metabolite of interest.

Mukataka et al. [104] observed that in 2 L cultures of *T. reesei* QM 9414 using five shaking speeds (from 100 to 500 rpm) for the production of enzymes and extracellular proteins, there was an optimal stirring condition for each metabolite. Therefore, the speed for the production of cellulases, glucosidases, and biomass was 200, 400, and 300 rpm, respectively. Meanwhile, the extracellular protein reached a similar (and maximum) concentration between 300 and 400 rpm. The authors suggest that the low production of metabolites observed in cultures grown under conditions of lower agitation was due to limited oxygen, because the pellet size was relatively large (ranging between 1 and 2 mm). However, very high agitation speeds created significant morphological changes (short and loose mycelium) and a reduction in biomass and product concentration. According to the authors, this phenomenon was the result of mycelial damage generated by an increase in shearing stress in the bioreactor. Lejeune and Baron [105], Marten et al. [110], and Felse and Panda [111] made similar observations in cultures of *T. reesei* QM 9414, *T. reesei* RUT-C30, and *T. harzianum* NCIM 1185 for the production of cellulases, xylanases, chitinases, and extracellular proteins in bioreactors of 15, 10, and 2 L, respectively. Furthermore, according to Felse and Panda [111], the energy supplied could induce cellular differentiation of *T. harzianum* because the fungi sporulated (cell differentiation in fungi that occurs when environmental conditions are not favorable for development) in cultures grown in the highest stirring conditions (>270 rpm).

The work of Godoy-Silva et al. [112] reported that, in addition to the energy supplied, the hydrodynamic conditions generated by the propellers markedly influenced the growth, morphology, and production of γ-decalactone (peach aroma) by *T. harzianum* in 10 L cultures with similar initial volumetric power drawn (0.056 kW m^{-3}). Low mechanical stress conditions (using a helical impeller) originated a higher mycelial growth in the form of pellets, compared to a system of high hydrodynamic stress (three Rushton turbines) that generated dispersed mycelium. Additionally, production of γ-decalactone was inversely proportional to the cell concentration because the highest productivities were achieved using the Rushton turbine system, where the biomass concentration was lower. Furthermore, Galindo et al. [113] observed a bell-shaped association in *T. harzianum* 500-mL flask cultures between the energy supplied and the specific productivity of 6-pentyl-α-pyrone (coconut flavor), similar to what was observed in bioreactors. Maximum productivity [0.72 mg6PP (g^{-1} h^{-1})] was obtained at a P/V of 0.9 kW m^{-3} (250 rpm, using 80 mL of medium), while the highest mycelial growth was reported at 1.13 kW m^{-3} (300 rpm). This condition presented aggregates exhibiting the smallest diameter.

Rocha-Valadez et al. [114] reported the effect of EDCF (Eq. 3) in a range between 3 and 96 kW m^{-3} s^{-1} on the morphology, mycelial growth, biomass viability, and production of 6-pentyl-α-pyrone in *T. harzianum* cultures under non-limiting dissolved oxygen conditions. It was observed that the average maximum diameter of aggregates depended on the EDCF. An increase in the EDCF [3–96 kW m^{-3} s^{-1}] reduced the diameter of the mycelial aggregates from 0.39 to 0.19 mm because fragmentation was favored. It is thought that fragmentation of the mycelium was the result of hydrodynamic stress because the average diameter of the aggregates was directly related to eddy size experimentally estimated according to the Kolmogorov microscale. The highest degree of interaction between particles and eddies was observed in the region of the impellers because the frequency of passing of the mycelium through that zone determined the aggregate size.

An increase in the EDCF caused significant changes in the metabolism of *T. harzianum* [114]. The hydrodynamic conditions resulting in higher stress reduced the specific growth rate (from 0.052 to 0.033 h^{-1}) and increased the specific C_{O_2} production, although the glucose consumption rates (0.05 ± 0.009 g g^{-1} h^{-1}) remained constant. The hydrodynamic conditions also influenced the viability and cell differentiation (i.e., sporulation) of the organism because they, respectively, decreased and increased as the EDCF increased.

The production of 6-pentyl-α-pyrone (6PP, coconut aroma) was stimulated between 7 and 10 kW m^{-3} s^{-1}) [113]. The maximum specific productivity showed a bell-shaped association with EDCF in cultures carried out at ≥20 % of dissolved oxygen. The reduction in 6PP biosynthesis [after 20 kW m^{-3} s^{-1}] was due to the effect of hydrodynamic stress that generated the largest mycelial damage (fragmentation) and cell differentiation (sporulation) of the fungus. Furthermore, it was proposed that the biosynthesis of 6PP increased due to metabolic changes seen when increasing the EDCF between 3 and 10 kW m^{-3} s^{-1}, when possible nutrient limitations (oxygen and/or glucose) were not present.

Overall, metabolite production by *Trichoderma* spp. shows a strong dependence on hydrodynamic conditions and cellular differentiation (i.e., sporulation) that occurred at high power inputs. It has been shown that metabolite productivity can be increased by the use of microparticles [103] which leads to a disperse morphology. However, there is a generalized lack of information regarding the relationships between hydrodynamics, oxygen transfer, morphology, cell differentiation, and metabolite productivity. Indeed, most of the literature does not differentiate between hydrodynamic and oxygen effects and characterization of fungal morphology as a function of these factors is scarce. Furthermore, it is crucial to understand cell differentiation within fungal aggregates in order to understand the complex relationships between physicochemical conditions prevailing in the bioreactor and metabolite productivity of the process.

2.5 Pellet Growth: The Case of Laccase Production

The production of laccases by ligninolytic basidiomycetes, also called white rot fungi, has been widely studied due to the ability of these microorganisms to grow on inexpensive substrates, excrete enzymes, and oxidize xenobiotics. *Trametes*, *Pleurotus*, *Lentinula*, *Pycnoporus*, *Botryosphaeria,* and *Phanerochaete* have been the most studied genera in recent years [115] because they can be easily cultivated in vitro and their enzymes can be easily purified.

Pleurotus is the genus of white rot fungi. In these fungi, laccases are the most important components of the extracellular lignin degradation systems [115]. The production of laccases has previously been reported for various species of the genus *Pleurotus*, including *P. pulmonarius* [116], *P. sajor-caju* [117], and *P. eryngii* [118, 119].

Pleurotus ostreatus produces laccases and manganese peroxidases but no lignin peroxidases [120]. This basidiomycete fungus is able to grow and degrade large amounts of agricultural and forestry wastes (lignocellulosic) by attacking different polymers of wood. This activity is made possible by the secretion of several laccase isoenzymes with unique physicochemical catalytic characteristics. This variety of laccase isoenzymes is related to different functions such as the synthesis and degradation of lignin, the development of fruiting bodies, pigment production, and cellular detoxification [121].

Laccase isoenzymes synthesis and secretion are strongly influenced by nutrient levels, growth conditions, stages of development, and the addition of inducers to the medium. It has been shown that most of these factors act at the transcriptional level to produce different responses between different isoforms of the same strain and in different fungal species [122].

The efficient production of laccases involves a combination of factors such as a highly productive strain, a culture medium specifically designed, the use of inductors, and the design of the bioreactor.

The production of laccases has been reported in solid state and submerged cultures, as well as in immobilized cultures [123]. However, the highest yield reported so far (300 U/mL) has been achieved in mechanically stirred fed-batch submerged cultures of *Trametes pubescens* induced with copper [124].

Several authors have reported that agitation and aeration influence the production of laccase by different fungal strains. However, the evaluation of the effect of agitation has resulted in contrasting findings, which did not clarify the role of this parameter in the production of the enzyme. On the one hand, some authors argued that high shear stresses (generated by agitation) are responsible for the low production of laccases [125–127]. On the other hand, it has been reported that the production of laccase is not affected by agitation [128, 129]. In addition, other authors have shown that laccase production is affected by agitation following a Gaussian bell-shaped distribution. This indicates that there is a stimulation effect at low agitation and a negative effect at high agitation [130].

Fenice et al. [126] evaluated the effects of agitation and aeration in submerged fermentation cultures of the fungus *Panus tigrinus* CBS 577-79. The wastewater from olive milling was used as the growth medium for the production of laccases. The study of the effect of stirring was conducted in 2 L reactors using two Rushton impellers and stirring speeds varying from 250 to 750 rpm with an airflow of 1 vvm. The effect of aeration was studied by varying the speed of aeration from 0.5 to 1.5 vvm and maintaining the impeller speed at 500 rpm. In this work, it was observed that lower enzymatic activities were achieved by increasing the stirring rate. Although these results could point out to a negative effect of the shear stress on the organism, it was also observed that the increase in flow aeration from 1 to 1.5 vvm adversely affected the specific activity of the laccases.

In 2004, Tavares et al. [128] conducted an experimental design that applied response surface methodology to optimize the production of laccases from *Trametes versicolor* in bioreactors. The rates of agitation (100–180 rpm), pH (3 and 5), and glucose concentration (0 and 9 g/L) were evaluated. In this study, the agitation speed had no effect on the maximum laccase activity, while the pH of the medium was the most important factor, followed by glucose concentration.

In a study using cultures of *Trametes versicolor* ATCC 200801, Birhanli et al. [130] described that the stirring speed had a bell-like effect on the production of laccase. Stirring speeds ranging from 0 to 250 rpm were evaluated. The highest laccase activity was obtained at 150 rpm, while lower activities were obtained at lower (0, 50 and 100 rpm) and higher (200–250 rpm) stirring speeds. The decrease in laccase activity above 150 rpm was attributed to shear stress on the pellets by high-speed stirring.

Overall, it is accepted that the shear generated by impellers negatively affects the production of laccases by superior fungi [123]. However, in the studies described above, it was not possible to separate the influence of the agitation from the effects of dissolved oxygen concentration.

A throughout study aimed at understanding the roles of hydrodynamic effects and dissolved oxygen on the mycelia growth and laccase production by *P. ostreatus* CP50, grown in a mechanically agitated bioreactor, was recently reported [131]. This study evaluated the particular effect (and possible interactions) of the initial EDCF, aeration rate, and copper induction, using a factorial experimental design. Under the experimental conditions tested (EDCF 0.9 to 5.9 kW m^{-3} s^{-1}), there was no evidence of hydrodynamic stress on the growth of *P. ostreatus*. Actually, the growth of the fungus was stimulated with increasing EDCF. However, it was found that the specific production of laccases decreased with increasing EDCF and aeration. The results show that laccase production by *P. ostreatus* CP50 is favored at a low EDCF. However, preliminary evidence indicates that the negative effect of the agitation could be related to the stimulation of protease production (decreasing enzyme activity) and may not be a direct result of the hydrodynamic stress.

2.6 Physiological Characterization of Fungal Aggregates by Image Analysis

From the previous sections, it is clear the lack of systematic information on the size of the pellets (Fig. 5) obtained during cultivation. However, the understanding of the hydrodynamic and dissolved oxygen effects on the growth and productivity of these cultures requires the accurate and quantitative characterization of fungal morphology.

The amount of viable biomass, in addition to morphology, is an important parameter that affects metabolite productivity of filamentous fungi in fermentation systems. The biomass determines the oxygen demand of the system during fermentation. The quantification of dry weight and turbidity is the most commonly used techniques to detect increases in biomass. Methods have also been used to measure the cultivability of cells by counting colony-forming units (CFUs) [132, 133]. However, these techniques do not provide information regarding the biocatalytic activity or the viability of cells. Furthermore, many of these techniques can lead to significant errors due to the effect of dilution.

Accurate viability assessment can significantly help in improving fermentation process design as well as in physiological studies and those involving monitoring and control of these bioprocesses. Fluorescence and image analysis are used jointly to analyze fungal physiology [134] and actinomycete differentiation [135, 136] and to evaluate viability through the spore germination percentage [137] or the percentage of intact membranes in the hyphal area (considered viable regions) relative to the total mycelial area [138].

Hassan et al. [139] correlated the cultivability (as the best indicator of cell viability) with the percentage of active biomass (measured by the percent of fluorescein diacetate-stained area in the hyphae) of the biological control agent *T. harzianum*. A fluorescence method based on the fact that metabolically active cells are able to hydrolyze intracellular fluorescein diacetate (FDA) producing fluorescein was used for the quantification of fungal viability. Fluorescein fluoresces at

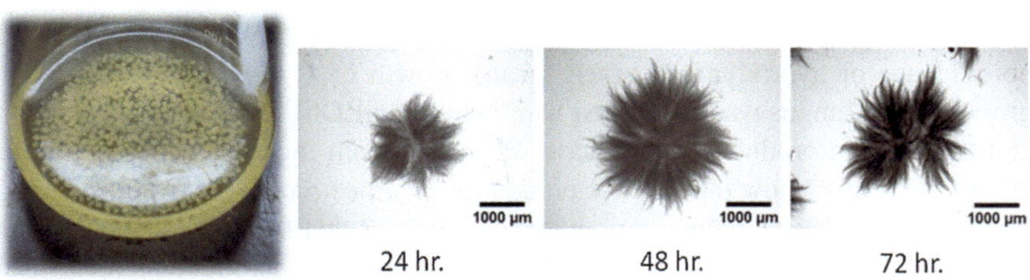

Fig. 5 Pellet growth of *Pleurotus ostreatus* in a flask and morphology at different growth time in stirred tank

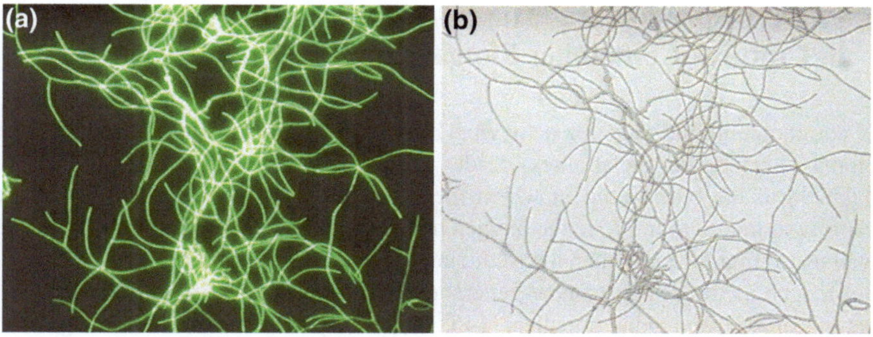

Fig. 6 Photomicrographs (10×) of *Trichoderma harzianum* with fluorescein diacetate staining. **a** Only metabolically active regions can be detected in the fluorescent image. **b** The total area of the aggregate is detected in white light microscopy [139]

490 nm and can be used to quantify the percentage of viable cells in cell clumps. The percentage of active biomass (viability) was calculated as the ratio of the stained area (in the fluorescent image) to the total area detected in the image with white light. An example of the images (before processing) is shown in Fig. 6. Pinto et al. [140] also used this last principle. Wei et al. [141] used dark-field illumination with an in situ probe inside a bioreactor to estimate the viability of the brewer's yeast *Saccharomyces cerevisiae*. They implemented two support vector machine-based classifiers for separating the cells from the background and for distinguishing live from dead cells. Their results showed very good accuracy compared with other standard methods. The method is not directly applicable to filamentous fungi, but the principles used in that work [140] may constitute a good basis to be adapted to this purpose. Lecault et al. [142] proposed a semiautomatic image analysis protocol allowing the detailed analysis of morphological parameters in addition to the assessment of the percentage viability of biomass using FDA. Their method was tested on *T. reesei* during fed-batch fermentation in a reciprocating plate bioreactor (RPB). Viability was assessed by measuring the area ratio of the fluorescent regions to the dark regions (a mask of the bright field image was subtracted from the fluorescent image to isolate the objects being analyzed). From the total area and the viable area, the percentage of viability was calculated. Choy et al. [143] used a similar method for a *Trichoderma reesei* viability estimation and found satisfactory results. Although reliable overall viability assessment of fungal cultures is now possible, it is crucial to develop techniques aimed to evaluate fungal differentiation within the mycelial aggregates. Such information has been proven to be very useful to the understanding of the effect of physicochemical environment on the growth and metabolite productivity of fungal cultures [35, 40–43, 99].

3 Phase Dispersion and Mass Transfer Characterization in Cultures Containing Fungal Biomass

A common limiting step for achieving high productivity of fungal products (the biomass itself and/or metabolites) resides in the transfer of oxygen (usually from air bubbles) to the bulk liquid in which the active biomass is suspended. There are various strategies to improve the oxygen transfer to the filamentous fermentations as well as to minimize cell damage (reducing the effect of mechanical stress by agitation), such as the design of new-generation agitators [144], using different combinations of axial and radial impellers [145, 146] and by reducing mechanical agitation by the use of ultrasound [147]. All these approaches have proven effective in a variety of models of study; however, despite the variety of studies in a wide range of operational conditions, it has not been possible—so far—to establish a general model that could describe the relationships between oxygen transfer and parameters such as scale, mycelial morphology, and operational parameters of the process [148].

The $k_{L}a$, OTR, and DOT are some of most studied parameters for analyzing the problems of oxygen transfer to the broths containing filamentous microorganisms [17, 51, 149–152]. However, these parameters are global ones and do not provide information regarding the mechanisms of mass transfer occurring at the level of single bubbles or populations of them. Furthermore, it is very limited the available information describing what happens (in terms of the size distribution of air bubbles and mass transfer area) when biomass—a suspended solid of complex morphology —is present. In addition, oils are commonly used in fungal fermentations as carbon sources, oxygen vectors, and extractive agents [64, 153–156]. However, the literature has only scarcely documented the effects of these insoluble oils in terms of the dispersion of the phases and oxygen transfer. Here, we review the relative few papers available on the field.

The presence of an organic phase in fungal cultures results in a fermentation system that requires the mixing of four phases: the culture medium (liquid phase), solvent (organic phase), mycelium (solid phase), and air (gas phase) (Fig. 7). The dispersion of these four phases occurs by continuous breaking and coalescence of air bubbles or oil droplets, which depend on the conditions of turbulence and the physical properties of the medium. Solvents have been successfully used to increase the production of aroma compounds such as 6PP, which is a coconut-like aroma compound and also a potent fungicide [155, 156].

Galindo et al. [153] published the first paper in which the distribution of air bubbles and oil droplets were characterized in a four-phase system. The Sauter mean diameter of the bubbles was smaller for increasing concentrations of castor oil and *T. harzianum* biomass. This work also reported by the first time the phenomenon of air bubbles embedded within oil drops, which has important implications in terms of the mechanism of mass transfer in such systems. Morphology of the biomass also affected the dispersion. Lucatero et al. [157] reported the effect of mycelial morphology on the dispersion of oil and air in simulated fermentation systems of *T. harzianum*. They observed that the number of bubbles trapped in

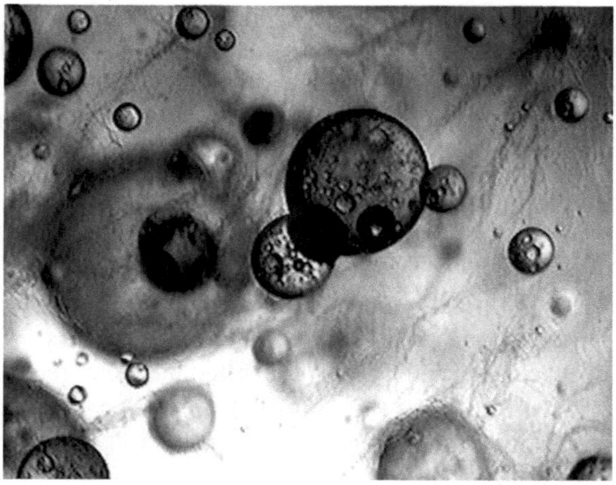

Fig. 7 Dispersion of four phases (filamentous biomass, liquid mineral media, castor oil, air) in the *Trichoderma harzianum* extractive fermentation

castor oil drops and the number of free bubbles depended mainly on the presence and concentration of dispersed mycelium. Moreover, Corkidi et al. [158] were able to document the bubble inclusion phenomenon in castor oil drops. Pulido-Mayoral and Galindo [159] conducted a study on the effects of protein concentration on the dispersion of droplets and bubbles and on the volumetric oxygen transfer coefficient ($k_L a$) in a simulated culture medium of *T. harzianum* with three phases: aqueous medium, castor oil, and air. It was observed that as the protein concentration increased, the Sauter mean diameter of the castor oil droplets decreased compared to the experiments without proteins.

In order to calculate the area of mass transfer to characterize the phases dispersions, the distribution diameter of air bubbles and oil drops must be known, which requires the measurement of a large quantity of objects. The characterization of these dispersions (partly, in terms of the size distribution of drops and bubbles in the dispersion) is crucial and necessary for understanding and optimizing these processes [160]. The bubble's size directly influences gas holdup and the interfacial area; therefore, the size distribution of bubbles is a good criterion for evaluating the efficiency of a gas–liquid contactor [161]. However, this characterization is time-consuming and tedious when manual methods are used. Furthermore, the manual analysis is influenced by subjective criteria that can change between analysts; therefore, the bubble- and drop-size distributions obtained can be biased. This situation emphasized the need for automated methods for characterizing multiphase systems. Bubble sizes in stirred dispersions have been measured by different techniques [162–164], and the results obtained have improved the understanding of important aspects of the stirred tank applications, such as mass transfer in chemical reactors. However, most of the techniques have been implemented in systems of two or three phases with low global holdup (≤ 1 %). There is little information for systems with four phases that include the biomass of filamentous fungi.

In a stirred tank, the formation of drops and bubbles depends on the balance between the process of breakage and coalescence of the phases mixed. The diameter of droplets and bubbles is usually expressed as an average diameter, such as the statistical mean diameter, volume mean diameter, or the Sauter diameter. The Sauter diameter is the most widely used measure in mass transfer studies because it links the volume of the dispersed phase with the area for transfer [165], and it is defined as:

$$d_{32} = \frac{\sum_{i=1}^{k_b} n_i d_i^3}{\sum_{i=1}^{k_b} n_i d_i^2} \tag{6}$$

where k_b is the number of volumes sampled, n_i is the number of drops/bubbles per volume i, and d_i is the size of the drops/bubbles i (Eq. 6).

Various techniques have reported the measurement of the diameter of drops and bubbles. Raman et al. [166] have reviewed the field. For the case of multiphase dispersions containing solids (i.e., biomass), in situ techniques measuring drop and bubble sizes with digital images and object recognition techniques are the only ones that are suitable [167].

In systems that include solid phases, the prediction of the mass transfer rate between gas–liquid phases is particularly complex due to the interactions between the solid particles and the other dispersed phases. Not surprisingly, the phenomena of phase dispersion and mass transfer in these systems are poorly understood at the micro level.

Some of the techniques used and reported in the literature in order to determine the diameter of droplets and bubbles are based on the acquisition and analysis of digital images. For techniques using video and digital photography, the use of cameras with high resolution and short exposure times compared to the timescale of the dynamics of the process is essential to avoid blurred images [168]. Pacek et al. [165] reported the use of advanced noninvasive video techniques in which the videos were taken from the outside of the stirred tank and were analyzed a posteriori. However, this configuration only allows for taking pictures of the processes occurring near the tank wall, giving limited information on the dynamics of the other regions within the tank.

The techniques of digital image analysis are very useful because they allow to distinguish between different phases present in the mixture. The main limitation of the techniques using digital imaging is to count with proper illumination to allow the acquisition of clear images that facilitate subsequent manual or automatic analysis. Lighting that gives depth, gives a uniform image, and allows an analysis of different regions within the stirred tank (such as near the discharge areas of the impellers) is required. Junker et al. [169] used a light source with a wavelength of 700 nm (red light), which allowed the acquisition of clear images even in high-turbidity systems.

On the other hand, several authors have described image analysis techniques that automatically recognize the circles that correspond to oil droplets or gas bubbles in the different multiphase systems (two-phase systems) [167, 170–176]. Those reports are mainly based on three techniques: (a) pattern detection using the Hough transform, (b) the use of templates, and (c) the detection of changes in a grayscale gradient. However, only Taboada et al. [177] have reported a semi-automatic image analysis method for the segmentation of bubbles and drops that was tested in two, three, and four-phase simulated fermentation model systems. The method was based on a windowed Hough transform. The method was able to reduce the total processing time up to half compared with a totally manual procedure, and the manual intervention time for the segmentation procedure was reduced to a minimum.

4 Conclusions and Perspectives

Mycelial cultures are characterized by complex interactions between the physicochemical parameters of the process (hydrodynamics, energy supply, and oxygen transfer) and growth of the biomass, morphology, and metabolite productivity. The knowledge of the effects of operational and culture conditions over the fungal growth and morphology are crucial. Deeper knowledge is required on the fundamentals of mechanosensitivity of fungal strains and its transcriptional effects on fungal morphology. On the physicochemical side, the concept of the energy dissipation circulation function (EDCF) is one of the crucial developments in the field for the understanding and predicting the effects of mechanical stress in fungal cultures. The concept of EDCF, although not universal, has been successfully applied in a wide variety of microorganisms, products, and culture conditions. Because there are continuous changes in the morphology, physiology, and mass transfer conditions of the microorganisms during fermentation, it is critical to establish an *on line* and in situ monitoring of microbial behavior and phase dispersion to control the process and to discriminate between hydrodynamic effects and those due to dissolved oxygen tension. Digital image analysis is a technique that has proven useful for characterizing multiphase dispersions containing filamentous microorganisms and for determining culture viability. Overall, what is required is to promote and to strengthen the multidisciplinary research between chemical and biochemical engineering with aspects of science and engineering of image processing in order to establish the link between these disciplines with the biological topics covered in this chapter (morphology and physiology of fungal microorganisms).

Acknowledgments DGAPA-UNAM (IT 201014 & IN 201813) and CONACyT (240438) for financial support.

References

1. Galarza Vázquez K (2014) Transforman hongos con tecnología de punta para producir compuestos industriales. Periódico: Investigación y Desarrollo del Foro Consultivo Científico y Tecnológico, AC, Mayo 2014, no. 318, año XXII, p 8
2. Braun S, Vecht-Lifshitz SE (1991) Mycelial morphology and metabolite production. TIBTECH 9:63–68
3. Janssens L, De Pooter HL, Schamp NM, Vandamme EJ (1992) Production of flavors by microorganisms. Process Biochem 27:195–215

4. Papagianni M (2004) Fungal morphology and metabolite production in submerged mycelial processes. Biotechnol Adv 22:189–259
5. Paul GC, Thomas CR (1996) A structured model for hyphal differentiation and penicillin production using *Penicillium chrysogenum*. Biotechnol Bioeng 51:558–572
6. Deckwer W-D, Jahn D, Hempel D, Zeng A-P (2006) Systems biology approaches to bioprocess development. Eng Life Sci 6(5):455–469
7. Oncu S, Tari C, Unluturk S (2007) Effect of various process parameters on morphology, rheology, and polygalacturonase production by *Aspergillus sojae* in a batch bioreactor. Biotechnol Prog 23:836–845
8. Krull R, Cordes C, Horn H, Kampen I, Kwade A, Neu TR, Nörtemann B (2010) Morphology of filamentous fungi: linking cellular biology to process engineering using *Aspergillus niger*. Adv Biochem Eng Biotechnol 121:1–21
9. Núñez-Ramírez DM, Medina-Torres L, Valencia-López JJ, Calderas F, López-Miranda J, Medrano-Roldán H, Solís-Soto A (2012) Study of the rheological properties of a fermentation broth of the fungus *Beauveria bassiana* in a bioreactor under different hydrodynamic conditions. J Microbiol Biotechnol 22(11):1494–1500
10. Grimm LH, Kelly S, Völkerding II, Krull R, Hempel DC (2005) Influence of mechanical stress and surface interaction on the aggregation of *Aspergillus niger* conidia. Biotechnol Bioeng 92(7):879–888
11. García-Soto MJ, Botello-Álvarez E, Jiménez-Islas H, Navarrete-Bolaños J, Barajas-Conde E, Rico-Martínez R (2006) Growth morphology and hydrodynamics of filamentous fungi in submerged cultures. In: Guevara-González G, Torres-Pacheco I (eds) Advances in agricultural and food biotechnology, Chap. 2. Research Signpost, Kerala, pp 17–34
12. Krull R, Cordes C, Horn H, Kampen I, Kwade A, Neu TR, Nörtemann B (2010) Morphology of filamentous fungi: linking cellular biology to process engineering using *Aspergillus niger*. Adv Biochem Eng/Biotechnol 121:1–21
13. Nielsen J, Johansen CL, Jacobsen M, Krabben P, Villadsen J (1995) Pellet formation and fragmentation in submerged cultures of *Penicillium chrysogenum* and its relation to penicillin production. Biotechnol Prog 11(1):93–98
14. Gibbs PA, Seviour RJ, Schmid F (2000) Growth of filamentous fungi in submerged culture: problems and possible solutions. Crit Rev Biotechnol 20(1):17–48
15. Olsvik E, Kristiansen B (1994) Rheology of filamentous fermentations. Biotechnol Adv 12:1–39
16. Riley GL, Tucker KG, Paul GC, Thomas CR (2000) Effect of biomass concentration and mycelial morphology on fermentation broth rheology. Biotechnol Bioeng 68:160–172
17. Gabelle JC, Jourdier E, Licht RB, Chaabane B, Henaut I, Morchain J, Augier F (2012) Impact of rheology on the mass transfer coefficient during the growth phase of *Trichoderma reesei* in stirred bioreactor. Chem Eng Sci 75:408–417
18. Dhillon GS, Kaur Brar S, Kaur S, Verma M (2013) Rheological studies during submerged citric acid fermentation by *Aspergillus niger* in stirred fermentor using apple pomace ultrafiltration sludge. Food Bioprocess Technol 6:1240–1250
19. Wucherpfennig T, Kiep KA, Driouch H, Wittmann C, Krull R (2010) Morphology and rheology in filamentous cultivations, Chap. 4. In: Advances in applied microbiology, vol 72. Elsevier Inc., London, pp 89–136
20. Wittler R, Baumgartl H, Lübbers DW, Schügerl K (1986) Investigations of oxygen transfer into *Penicilium chrysogenum* pellets by microprobe measurements. Biotechnol Bioeng 28:1024–1036
21. Smith JJ, Lilly MD, Fox RI (1990) The effect of agitation on the morphology and penicillin production of *Penicillium chrysogenum*. Biotechnol Bioeng 35:1011–1023
22. Nienow AW (1990) Agitators for mycelial fermentations. TIBTECH 8:224–233
23. Cui YQ, van der Lans RGJM, Luyben KChAM (1998) Effects of dissolved oxygen tension and mechanical forces on fungal morphology in submerged fermentation. Biotechnol Bioeng 57:409–419

24. Li ZJ, Shukla V, Fordyce AP, Pedersen AG, Wenger KS, Marten MR (2000) Fungal morphology and fragmentation behavior in a fed-batch *Aspergillus oryzae* fermentation at the production scale. Biotechnol Bioeng 70:300–312
25. Olmos E, Mehmood N, Haj Husein L, Georgen JL, Fick M, Delaunay S (2013) Effect of bioreactor hydrodynamic on the physiology of *Streptomyces*. Bioprocess Biosyst Eng 36:259–272
26. Makagiansar HY, Ayazi-Shamlou P, Thomas CR, Lilly MD (1993) The influence of mechanical forces on the morphology and penicillin production of *Penicillium chrysogenum*. Bioprocess Eng 9:83–90
27. Wucherpfennig T, Hestler T, Krull R (2011) Morphology engineering-osmolality and its effect on *Aspergillus niger* morphology and productivity. Microb Cell Fact 10(58):2–15
28. van Veluw GJ, Petrus MLC, Gubbens J, de Graaf R, de Jong IP, van Wezel GP, Wösten HAB, Claessen D (2012) Analysis of two distinct mycelia populations in liquid-grown *Streptomyces* cultures using a flow cytometry-based proteomics approach. Appl Microbiol Biotechnol 96:1301–1312
29. Kelly S, Grimm LH, Hengstler J, Schultheis E, Krull R, Hempel DC (2004) Agitation effects on submerged growth and product formation of *Aspergillus niger*. Bioprocess Biosyst Eng 26:315–323
30. Sainz Herrán N, Casas López JL, Sánchez Pérez JA, Chisti Y (2008) Effects of ultrasound on culture of *Aspergillus terreus*. J Chem Technol Biotechnol 83:593–600
31. Zhou Z, Du G, Hua Z, Zhou J, Chen J (2011) Optimization of fumaric acid production by *Rhizopus delemar* based on the morphology formation. Bioresource Technol 102:9345–9349
32. Tepwong P, Giri A, Ohshima T (2012) Effect of mycelia morphology on ergothioneine production during liquid fermentation of *Lentinula edodes*. Mycosci 53:102–112
33. Cai M, Zhang Y, Hu W, Shen W, Yu Z, Zhou W, Jiang T, Zhou X, Zhang Y (2014) Genetically shaping morphology of the filamentous fungus *Aspergillus glaucus* for production of antitumor polyketide aspergiolide A. Microb Cell Fact 13:73–83
34. Hille A, Neu TR, Hempel DC, Horn H (2009) Effective diffusivities and mass fluxes in fungal biopellets. Biotechnol Bioeng 103(6):1202–1213
35. Driouch H, Hänsch R, Wucherpfennig T, Krull R, Wittmann C (2012) Improved enzyme production by bio-pellets of *Aspergillus niger*: targeted morphology engineering using titanate microparticles. Biotechnol Bioeng 109(2):462–471
36. Xia X, Lin S, Xia XX, Cong FS, Zhong JJ (2014) Significance of agitation-induced shear stress on mycelium morphology and lavendamycin production by engineered *Streptomyces flocculus*. Appl Microbiol Biotechnol 98:4399–4407
37. Carlsen M, Spohr AB, Nielsen J, Villadsen J (1996) Morphology and physiology of an α-amylase producing strain of *Aspergillus oryzae* during batch cultivations. Biotechnol Bioeng 49:266–276
38. Colin VL, Baigorí MD, Pera LM (2013) Tailoring fungal morphology of *Aspergillus niger* MYA 135 by altering the hyphal morphology and the conidia adhesion capacity: biotechnological applications. AMB Express 3(27):1–13
39. Lin P-J, Scholz A, Krull R (2010) Effect of volumetric power input by aeration and agitation on pellet morphology and product formation of *Aspergillus niger*. Biochem Eng J 49:213–220
40. El-Enshasy H, Kleine J, Rinas U (2006) Agitation effects on morphology and protein productive fractions of filamentous and pelleted growth forms of recombinant *Aspergillus niger*. Process Biochem 41:2103–2112
41. Wösten HAB, van Veluw GJ, de Bekker C, Krijgsheld P (2013) Heterogeneity in the mycelium: implications for the use of fungi as cell factories. Biotechnol Lett 35:1155–1164
42. Levin AM, de Vries RP, Conesa A, de Bekker C, Talon M, Menke HH, van Peij NNME, Wösten HAB (2007) Spatial differentation in the vegetative mycelium of *Aspergillus niger*. Eukaryot Cell 6(12):2311–2322

43. de Bekker C, van Veluw GJ, Vinck A, Wiebenga A, Wösten HAB (2011) Heterogeneity of *Aspergillus niger* microcolonies in liquid shaken cultures. Appl Environ Microbiol 77 (4):1263–1267
44. O'Cleirigh C, Casey JT, Walsh PK, O'Shea DG (2005) Morphological engineering of *Streptomyces hygroscopicus* var. *geldanus*: regulation of pellets morphology through manipulation of broth viscosity. Appl Microbiol Biotechnol 68:305–310
45. Dobson LF, O′Cleirigh CC, O′Shea DG (2008) The influence of morphology on geldanamycin production in submerged fermentations of *Streptomyces hygroscopicus var. geldanus*. Appl Microbiol Biotechnol 79:859–866
46. Ghojavand H, Bonakdarpur B, Heydarian SM, Hamedi J (2011) The inter-relationship between inoculums concentration, morphology, rheology and erythromycin productivity in submerged cultivation. Braz J Chem Eng 28(4):565–574
47. Yao L-Y, Zhu Y-H, Jiao R-H, Lu Y-H, Tan R-X (2014) Enhanced production of fumigaclavine C by ultrasound stimulation in a two-stage culture of *Aspergillus fumigatus* CY018. Bioresource Technol 159:112–117
48. Prokop A, Bajpai R (1992) The sensitivity of biocatalysts to hydrodynamic shear stress. Adv Appl Microbiol 37:165–232
49. Chisti Y (2001) Hydrodynamic damage to animal cells. Crit Rev Biotechnol 21:67–110
50. Kumamoto CA (2008) Molecular mechanisms of mechanosensing and their roles in fungal contact sensing. Nature Rev Microbiol 6:667–673
51. Garcia-Ochoa F, Escobar S, Gomez E (2015) Specific oxygen uptake rate as indicator of cell response of *Rhodococcus erythropolis* cultures to shear effects. Chem Eng Sci 122:491–499
52. Teng Y, Xu Y, Wang D (2009) Changes in morphology of *Rhizopus chinensis* in submerged fermentation and their effect on production of mycelium bound lipase. Bioprocess Biosyst Eng 32:397–405
53. Chipeta ZA, du Preez JC, Christopher L (2008) Effect of cultivation pH and agitation rate on growth and xylanase production by *Aspergillus oryzae* in spent sulphite liquor. J Ind Microbiol Biotechnol 35:587–594
54. Albaek MO, Gernaey KV, Hansen MS, Stocks SM (2011) Modeling enzyme production with *Aspergillus oryzae* in pilot scale vessels with different agitation, aeration, and agitator types. Biotechnol Bioeng 108(8):1828–1840
55. Mehmood N, Olmos E, Marchal P, Georgen J-L, Delaunay S (2010) Relation between pristinamycins production by *Streptomyces pristinaespiralis*, power dissipation and volumetric gas-liquid mass transfer coefficient, k_La. Process Biochem 45:1779–1786
56. Mehmood N, Olmos E, Georgen J-L, Blanchard F, Ullisch D, Klöckner W, Büchs J, Delaunay S (2011) Oxygen supply controls the onset of pristinamycins production by *Streptomyces pristinaespiralis* in shaking flasks. Biotechnol Bioeng 108(9):2151–2161
57. Mehmood N, Olmos E, Goergen J-L, Blanchard F, Marchal P, Klöckner W, Büchs J, Delaunay S (2012) Decoupling of oxygen transfer and power dissipation for the study of the production of pristinamycins by *Sreptomyces pristinaespiralis* in shaking flasks. Biochem Eng J 68:25–33
58. Casas López JL, Sánchez Pérez JA, Fernández Sevilla JM, Rodríguez Porcel EM, Chisti Y (2005) Pellet morphology, culture rheology and lovastatin production in cultures of *Aspergillus terreus*. J Biotechnol 116:61–77
59. Rodríguez Porcel EM, Casas López JL, Sánchez Pérez JA, Fernández Sevilla JM, García Sánchez JL, Chisti Y (2006) *Aspergillus terreus* broth rheology, oxygen transfer, and lovastatin production in gas-aireated slurry reactor. Ind Eng Chem Res 45:4837–4843
60. Xu CP, Kim SW, Hwang HJ, Yun JW (2006) Production of exopolysaccharides by submerged culture of an enthomopathogenic fungus, *Paecilomyces tenuipes* C240 in stirred-tank and airlift reactors. Bioresource Technol 97:770–777
61. Cho EJ, Oh JY, Chang HY, Yun JW (2006) Production of exopolysaccharides by submerged mycelia culture of a mushroom *Tremella fuciformis*. J Biotechnol 127:129–140

62. Rogalski J, Szczodrak J, Janusz G (2006) Manganese peroxidase production in submerged cultures by free and immobilized mycelia of *Nematoloma frowardii*. Bioresource Technol 97:469–476
63. Michelin M, Teixeira de Morales Polizeli ML, Pereira da Silva D, Santos Ruzene D, Vicente AA, Jorge JA, Terenzi HF, Teixeira JA (2011) Production of xylanolytic enzymes by *Aspergillus terricola* in stirred tank and airlift tower loop bioreactors. J Ind Microbiol Biotechnol 38:1979–1984
64. Cai M, Zhou X, Lu J, Fan J, Niu C, Zhou J, Sun X, Kang L, Zhang Y (2011) Enhancing aspergiolide A production from a shear-sensitive and easy-foaming marine derived filamentous fungus *Aspergillus glaucus* by oxygen carrier addition and impeller combination in a bioreactor. Bioresource Technol 102:3584–3586
65. Fenice M, Barghini P, Selbmann L, Federici F (2012) Combined effects of agitation and aeration on the chitinolytic enzymes production by the antartic fungus *Lecanicillium muscarium* CCFEE 5003. Microb Cell Fact 11(12):1–10
66. Jonczyk P, Takenberg M, Hartwig S, Beutel S, Berger RG, Scheper T (2013) Cultivation of shear stress sensitive microorganisms in disposable bag reactor system. J Biotechnol 167:370–376
67. Tramper J, van't Riet K. (1991). Basic bioreactor design. Marcel Dekker Inc., New York, pp 136–180
68. Joshi J, Elias C, Patole M (1996) Role of hydrodynamic shear in the cultivation of animal, plant and microbial cells. Biochem Eng J 62:121–141
69. Croughan MS, Hamel JF, Wang DIC (1987) Hydrodynamic effects on animal cells grown in microcarrier cultures. Biotechnol Bioeng 29:130–141
70. Ayazi-Shamlou P, Makagiansar HY, Ison AP, Lilly MD, Thomas CR (1994) Turbulent breakage of filamentous microorganisms in submerged culture in mechanically stirred bioreactors. Chem Eng Sci 49:2621–2631
71. Thomas CR, Zhang Z (1998) The effect of hydrodynamics on biological materials. In: Galindo E, y Ramírez OT (eds) Advances in Bioprocess Engineering II. Kluwer Academic Publishers, Berlin, pp 137–170
72. Li ZJ, Shukla V, Wenger K, Fordyce A, Pedersen AG, Marten M (2002) Estimation of hyphal tensile strength in production-scale *Aspergillus oryzae* fungal fermentations. Biotechnol Bioeng 77:601–613
73. Rocha-Valadez JA, Galindo E, Serrano-Carreón L (2007) The influence of circulation frequency on fungal morphology: a case study considering Kolmogorov microscale in constant specific energy dissipation rate cultures of *Trichoderma harzianum*. J Biotechnol 130:394–401
74. Zhang Z, Al-Rubeai M, Thomas CR (1993) Estimation of disruption of animal cells by turbulent capillary flow. Biotechnol Bioeng 42:987–993
75. Kelly S, Grimm LH, Bendig C, Hampel DC, Krull R (2006) Effects of fluid dynamic induced shear stress on fungal growth and morphology. Process Biochem 41:2113–2117
76. Kolmogorov AN (1941) The local structure of turbulence in incompressible viscous fluids for very large Reynolds numbers. Dokl Akad Nauk SSSR 30. Proc R Soc Lond A [trans: Levin V (1991)] 434:9–13
77. Cui YQ, van der Lans RGJM, Luyben KCAM (1997) Effect of agitation intensities on fungal morphology of submerged fermentation. Biotechnol Bioeng, 55:715–726
78. Li ZJ, Bhargava S, Marten MR (2002) Measurements of the fragmentation rate constant imply that the tensile strength of fungal hyphae can change significantly during growth. Biotechnol Lett 24:1–7
79. van Suijdam JC, Metz B (1981) Influence of engineering variables upon the morphology of filamentous molds. Biotechnol Bioeng 23:111–148
80. Nienow AW, Ulbrecht JJ (1985) Gas-liquid mixing and mass transfer in high viscosity liquids. In: Ulbrecht JJ, Patterson GE (eds) Mixing of liquids by mechanical agitation. Gordons and Breach, New York, pp 203–235

81. Large KP, Ison AP, Williams DJ (1998) The effect of agitation rate on lipid utilization and clavulanic acid production in *Streptomyces clavuligerus*. J Biotechnol 63:111–119
82. Amanullah A, Blair R, Nienow AW, Thomas CR (1999) Effects of agitation intensity on mycelial morphology and protein production in chemostat cultures of recombinant *Aspergillus oryzae*. Biotechnol Bioeng 62:434–446
83. Papagianni M, Mattey M, Kristiansen B (1999) Hyphal vacuolation and fragmentation in batch and fed-batch culture of *Aspergillus niger* and its relation to citric acid production. Process Biochem 35:359–366
84. Johansen CL, Coolen L, Hunik JH (1998) Influence of morphology on product formation in *Aspergillus awamori* during submerged fermentations. Biotechnol Prog 14:233–240
85. Paul GC, Priede MA, Thomas CR (1999) Relationship between morphology and citric acid production in submerged *Aspergillus niger* fermentations. Biochem Eng J 3:121–129
86. Tang Y-J, Zhang W, Liu R-S, Zhu L-W, Zhong J-J (2011) Scale-up study on the fed batch fermentation of *Ganoderma lucidum* for the hyperproduction of ganoderic acid and *Ganoderma* polysaccharides. Process Biochem 46:404–408
87. Charles M (1985) Fermentation design and scale-up. In: Moo-Young M (ed) Comprehensive Biotechnology, vol 2. Pergamon Press, Oxford, pp 120–150
88. Gamboa-Suasnavart RA, Marín-Palacio LD, Martínez-Sotelo JA, Espitia C, Servín-González L, Valdez-Cruz NA, Trujillo-Roldán MA (2013) Scale-up from shake flasks to bioreactor, based on power input and *Streptomyces lividans* morphology, for the production of recombinant APA (45/47 kDa protein) from *Mycobacterium tuberculosis*. World J Microbiol Biotechnol 29:1421–1429
89. Jüsten P, Paul G, Nienow AW, Thomas C (1998) Dependence of *Penicillium chrysogenum* growth, morphology, vacuolation, and productivity in fed-batch fermentations on impeller type and agitation intensity. Biotechnol Bioeng 59:762–775
90. Amanullah A, Christensen LH, Hansen K, Nienow AW, Thomas CR (2002) Dependence of morphology on agitation intensity in fed-batch cultures of *Aspergillus oryzae* and its implications for recombinant protein production. Biotechnol Bioeng 77:815–826
91. Jüsten P, Paul GC, Nienow AW, Thomas CR (1996) Dependence of mycelial morphology on impeller type and agitation intensity. Biotechnol Bioeng 52:672–684
92. Marín-Palacio LD, Gamboa-Suasnavart RA, Valdéz-Cruz NA, Servín-González L, Córdova-Aguilar MS, Soto E, Klöckner W, Büchs J, Trujillo Roldán MA (2014) The role of volumetric power input in the growth, morphology, and production of a recombinant glycoprotein by *Streptomyces lividans* in shake flasks. Biochemical Eng J 90: 224–233
93. Reuss M (1988) Influence of mechanical stress on the growth of *Rhizopus nigricans* in stirred bioreactors. Chem Eng Technol 11:178–187
94. Zou X, Xia J, Chu J, Zhuang Y, Zhang S (2012) Real-time fluid dynamics investigation and physiological response for erythromycin fermentation scale-up from 50 L to 132 m^3 fermenter. Bioprocess Biosyst Eng 35:789–800
95. Kaup B-A, Ehrich K, Pescheck M, Schrader J (2007) Microparticle-enhanced cultivation of filamentous microorganisms: increased chloroperoxidase formation by *Caldariomyces fumago* as an example. Biotechnol Bioeng 99(3):491–498
96. Driouch H, Roth A, Dersch P, Wittmann C (2010) Optimized bioprocess for production of fructofuranosidase by recombinant *Aspergillus niger*. Appl Microbiol Biotechnol 87:2011–2024
97. Driouch H, Sommer B, Wittmann C (2010) Morphology engineering of *Aspergillus niger* for improved enzyme production. Biotechnol Bioeng 105:68–1058
98. Walisko R, Krull R, Schrader J, Wittmann C (2012) Microparticle based morphology engineering of filamentous microorganisms for industrial bio-production. Biotechnol Lett 34:1975–1982
99. Driouch H, Roth A, Dersch P, Wittmann C (2011) Filamentous fungi in good shape. Microparticles for tailor-made fungal morphology and enzyme production. Bioeng Bugs 2:2, 100–104

100. Krull R, Wucherpfennig T, Esfandabadi ME, Walisko R, Melzer G, Hempel DC, Kampen I, Kwade A, Wittmann C (2013) Characterization and control of fungal morphology for improved production performance in biotechnology. J Biotechnol 163:112–123
101. Sohoni SV, Bapat PM, Lantz AE (2012) Rubust, small-scale cultivation platform for *Streptomyces coelicolor*. Microb Cell Fact 11(9):1–10
102. Gao D, Zeng J, Yu X, Dong T, Chen S (2014) Improved lipid accumulation by morphology engineering of oleaginous fungus *Mortierella isabellina*. Biotechnol Bioeng 111(9):1758–1766
103. Etschmann MMW, Huth I, Walisko R, Schuster J, Krull R, Holtmann D, Wittmann C, Schrader J (2014) Improving 2-phenylethanol and 6-pentyl-α-pyrone production with fungi by microparticle-enhanced cultivation (MPEC). Yeast. Published online 9 Jul 2014. doi:10.1002/yea.3022
104. Mukataka S, Kobayashi N, Sato S, Takahashi J (1988) Variation in cellulase-constituting components from *Trichoderma reesei* with agitation intensity. Biotechnol Bioeng 32:760–763
105. Lejeune R, Baron GV (1995) Effect of agitation on growth and enzyme production of *Trichoderma reesei* in batch fermentation. Appl Microbiol Biotechnol 43:249–258
106. Apsite A, Viesturs U, Steinberga A, Toma M (1998) Morphology and antifungal action of the genus *Trichoderma* cultivated in geometrically dissimilar bioreactors. World J Microbiol Biotechnol 14:23–29
107. Serrano-Carreón L, Flores C, Galindo E (1997) γ-Decalactone production by *Trichoderma harzianum* in stirred bioreactors. Biotechnol Prog 13:205–208
108. Serrano-Carreón L, Flores C, Rodríguez B, Galindo E (2004) *Rhizoctonia solani*, an elicitor of 6-pentyl-α-pyrone production by *Trichoderma harzianum* in a two liquid phases, extractive fermentation system. Biotechnol Lett 26:1403–1406
109. Hjeljord L, Tronsmo A (1998) *Trichoderma* and *Gliocladium* in biological control: an overview. In: Harman E, Kubicek CP (eds) *Trichoderma* and *Gliocladium*, vol 2. Taylor and Francis, London, pp 131–151
110. Marten MR, Velkovska S, Khan SA, Ollis DF (1996) Rheological, mass transfer, and mixing characterization of cellulase-producing *Trichoderma reseei* suspensions. Biotechnol Prog 12:602–611
111. Felse PA, Panda T (2000) Submerged culture production of chitinase by *Trichoderma harzianum* in stirred tank bioreactors-the influence of agitator speed. Biochem Eng J 4:115–120
112. Godoy-Silva RD, Serrano-Carreón L, Ascanio G, Galindo E (1997) Effect of impeller geometry on the production of aroma compounds by *Trichoderma harzianum*. In: Nienow AW (ed) Proceedings of 4th International Conference on Bioreactor and Bioprocess Fluid Dynamics. BHR Group, Bedford, pp 61–72
113. Galindo E, Flores C, Larralde-Corona P, Corkidi G, Rocha-Valadez JA, Serrano-Carreón L (2004) Production of 6-pentyl-α-pyrone by *Trichoderma harzianum* cultured in unbaffled and baffled shake flasks. Biochem Eng J 18:1–8
114. Rocha-Valadez JA, Hassan M, Corkidi G, Flores C, Galindo E, Serrano-Carreón L (2005) 6-pentyl-α-pyrone production by *Trichoderma harzianum*: the influence of energy dissipation rate and its implications on fungal physiology. Biotechnol Bioeng 91:54–61
115. Baldrian P (2006) Fungal laccases—occurrence and properties. FEMS Microbiol Rev 30:215–242
116. De Souza C, Tychanowics G, De Souza D, Peralta R (2004) Production of laccase isoforms by *Pleurotus pulmonarius* in response to presence of phenolic and aromatic compounds. J Basic Microbiol 44:129–136
117. Soden DM, Dobson ADW (2001) Differential regulation of laccase gene expression in *Pleurotus sajor-caju*. Microbiology-Sgm 147:1755–1763
118. Muñoz C, Guillén F, Martínez AT (1997) Induction and characterization of laccases in the ligninolytic fungus *Pleurotus eryngii*. Curr Microbiol 34:1–5

119. Ueda M, Shintani K, Nakanishi-Anjyuin A, Nakazawa M, Kusuda M, Nakatani F, Kawaguchi T, Tsujiyama S, Kawanishi M, Yagi T, Miyatake K (2012) A proteina from *Pleurotus eryngii* var. *tuoliensis* C.J. Mou with strong removal activity against the natural steroid hormone, estriol: purification, characterization, and identification as a laccase. Enzyme Microb Technol 51(6–7):402–407
120. Giardina P, Palmieri G, Scaloni A, Fontanella B, Faraco V, Cennamo G, Sannia G (1999) Protein and gene structure of a blue laccase from *Pleurotus ostreatus*. Biochem J 341:655–663
121. Lettera V, Del Vecchio C, Piscitelli A, Sannia G (2011) Low impact strategies to improve ligninolytic enzyme production in filamentous fungi: the case of laccase in *Pleurotus ostreatus*. C R Biol 11:781–788
122. Piscitelli A, Giardina P, Lettera V, Pezzella C, Sannia G, Faraco V (2011) Induction and transcriptional production of laccases in fungi. Current Genomics 12:104–112
123. Rodríguez Couto S, Toca Herrera JL (2007) Laccase production at reactor scale by filamentous fungi. Biotechnol Adv 25:558–569
124. Galhaup CD, Wagner H, Hinterstoisser B, Haltrich D (2002) Increased production of laccase by the wood-degrading basidiomycete *Trametes pubescens*. Enz Microb Technol 30:529–536
125. Hess J, Leitner C, Galhaup C, Kulbe KD, Hinterstoisser B, Steinwender M (2002) Enhanced formation of extracellular laccase activity by the white-rot fungus *Trametes multicolor*. Appl Biochem Biotechnol 98:229–241
126. Fenice M, Sermanni GG, Federici F, D'Annibale A (2003) Submerged and solid-state production of laccase and Mn-peroxidase by *Panus tigrinus* on olive mill wastewater-based media. J Biotechnol 100:77–85
127. Silvério SC, Moreira S, Milagres AMF, Macedo EA, Teixeira JA, Mussatto SI (2013) Laccase production by free and immobilized mycelia of *Peniophora cinérea* and *Trametes versicolor*: a comparative study. Bioprocess Biosyst Eng 36:365–373
128. Tavares APM, Coelho MAZ, Agapito MSM, Coutinho JAP, Xavier AMRB (2006) Optimization and modeling of laccase production by *Trametes versicolor* in a bioreactor using statistical experimental design. Appl Biochem Biotechnol 134:233–248
129. Babic J, Pavko A (2011) Enhanced enzyme production with the pelleted form of *D. squalens* in laboratory bioreactors using added natural lignin inducer. J Ind Microbiol Biotechnol 39:449–457
130. Birhanli E, Yesilada O (2010) Enhanced production of laccase in repeated-batch cultures of *Funalia trogii* and *Trametes versicolor*. Biochem Eng J 52:33–37
131. Tinoco-Valencia R, Gómez-Cruz C, Galindo E, Serrano-Carreón L (2014) Toward an understanding of the effects of agitation and aeration on growth and laccases production by *Pleurotus ostreatus*. J Biotechnol 177:67–73
132. Jolicoeur M, Williams RD, Chavarie C, Fortin JA, Archambault J (1999) Production of *Glomus intraradices* propagules, an arbuscular mycorrhizal fungus, in an airlift bioreactor. Biotechnol Bioeng 63:224–232
133. Muñoz GA, Agosin E, Cotoras M, Martin RS, Volpe D (1995) Comparison of aerial and submerged spore properties for *Trichoderma harzianum*. FEMS Microbiol Lett 125:63–70
134. Vanhoutte B, Pons MN, Thomas CR, Louvel L, Vivier H (1995) Characterization of *Penicillium chrysogenum* physiology in submerged cultures by color and monochrome image analysis. Biotechnol Bioeng 48:1–11
135. Drouin JF, Louvel L, Vanhoutte B, Vivier H, Pons MN, Germain P (1997) Quantitative characterization of cellular differentiation of *Streptomyces ambofaciens* in submerged culture by image analysis. Biotechnol Tech 11:819–824
136. Reichl U, Yang H, Gilles ED, Wolf H (1990) New improved method for measuring the interseptal spacing in hyphae of *Streptomyces tendae* by fluosescence microscopy coupled with image processing. FEMS Microbiol Lett 67:207–210
137. Paul GC, Kent CA, Thomas CR (1992) Viability testing and characterization of germination of fungal spores by automatic image analysis. Biotechnol Bioeng 42:11–23

138. Sebastine IM, Stocks SM, Cox PW, Thomas CR (1999) Characterisation of percentage viability of *Streptomyces clavuligerus* using image analysis. Biotechnol Techniques 13:419–423
139. Hassan M, Corkidi G, Galindo E, Flores C, Serrano-Carreón L (2002) Accurate and rapid viability assessment of *Trichoderma harzianum* using fluorescence-based digital image analysis. Biotechnol Bioeng 80(6):677–684
140. Pinto LS, Vieira LM, Pons MN, Fonseca MMR, Menezes JC (2004) Morphology and viability analysis of *Streptomyces clavuligerus* in industrial cultivation systems. Bioprocess Biosyst Eng 26:177–184
141. Wei N, You J, Friehs K, Flschel E, Nattkemper TW (2007) An in situ probe for *on-line* monitoring of cell density and viability on the basis of dark field microscopy in conjunction with image processing and supervised machine learning. Biotechol Bioeng 97(6):1489–1500
142. Lecault V, Patel N, Thibault J (2007) Morphological characterization and viability assessment of *Trichoderma reesei* by image analysis. Biotechnol Prog 23:734–740
143. Choy V, Patel N, Thibault J (2011) Application of image analysis in the fungal fermentation of *Trichoderma reesei* RUT-C30. Biotechnol Prog 27(6):1544–1553
144. Zhu H, Sun J, Tian B, Wang H (2014) A novel stirrer design and its application in submerged fermentation of the edible fungus *Pleurotus ostreatus*. Bioprocess Biosyst Eng. Published online: 19 Sept 2014. doi: 10.1007/s00449-014-1290-6
145. Zou X, Xia JX, chu J, Zhuang Y, Zhang S (2012) Real-time fluid dynamics investigation and physiological response for erythromycin fermentation scale-up from 50 L to 132 m^3 fermenter. Bioprocess Biosyst Eng 35:789–800
146. Núñez-Ramírez DM, Valencia-López JJ, Calderas F, Solís-Soto A, López-Miranda J, Medrano-Roldán H, Medina-Torres L (2012) Mixing analysis for a fermentation broth of the fungus *Beauveria bassiana* under different hydrodynamic conditions in a bioreactor. Chem Eng Technol 35(11):1954–1961
147. Liu R, Sun W, Liu C-Z (2011) Computational fluid dynamics modeling of mass-transfer behavior in a bioreactor for hairy root culture. II. Analysis of ultrasound-intensified process. Biotechnol Prog 27(6):1672–1679
148. Formenti LR, Nørregaard A, Bolic A, Quintanilla Hernandez D, Hagemann T, Heins A-L, Larsson H, Mears L, Mauricio-Iglesias M, Krühne U, Gernaey KV (2014) Challenges in industrial fermentation technology research. Biotechnol J 9:727–738
149. Galaction A-I, Cascaval D, Oniscu C, Turnea M (2004) Prediction of oxygen mass transfer coefficients in stirred bioreactors for bacteria, yeasts and fungus broths. Biochem Eng J 20:85–94
150. Galaction A-I, Cascaval D, Oniscu C, Turnea M (2005) Evaluation and modeling of the aerobic stirred bioreactor performances for fungus broths. Chem Biochem Eng Q 19(1):87–97
151. Mishra P, Srivastava P, Kundu S (2005) A competitive evaluation of oxygen mass transfer and broth viscosity using Cephalosporin-C production as case strategy. World J Microbiol Biotechnol 21:525–530
152. Singh D, Kaur G (2014) Swainsonine, a novel fungal metabolite: optimization of fermentative production and bioreactor operations using evolutionary programming. Bioprocess Biosyst Eng 37:1599–1607
153. Galindo E, Pacek AW, Nienow AW (2000) Study of drop and bubble sizes in a simulated mycelial fermentation broth of up to four phases. Biotechnol Bioeng 69:213–221
154. Quijano G, Revah S, Gutiérrez-Rojas M, Flores-Cotera LB, Thalasso F (2009) Oxygen transfer in three-phase airlift and stirred tank reactors using silicone oil as transfer vector. Process Biochem 44:619–624.
155. Serrano-Carreón L, Balderas-Ruíz K, Galindo E, Rito-Palomares M (2002) Production and biotransformation of 6-pentyl-α-pyrone by *Trichoderma harzianum* in two-phase culture systems. Appl Microbiol Biotechnol 58:170–174
156. Kalyani A, Prapulla SG, Karanth NG (2000) Study on the production of 6-pentyl-α-pyrone using two methods of fermentation. Appl Microbiol Biotechnol 53:610–612

157. Lucatero S, Larralde-Corona CP, Corkidi G, Galindo E (2003) Oil and air dispersion in a simulated fermentation broth as a function of mycelial morphology. Biotechnol Prog 19 (2):285–292
158. Corkidi G, Rojas A, Pimentel A, Galindo E (2012) Visualization of compound drop formation in multiphase processes for dentification of factors influencing bubble and water droplet inclusions in oil drops. Chem Eng Res Design 90:1727–1738
159. Pulido-Mayoral N, Galindo E (2004) Phases dispersion and oxygen transfer in a simulated fermentation broth containing castor oil and proteins. Biotechnol Progress 20:1608–1613
160. Córdova-Aguilar MS, Sánchez A, Serrano-Carreón L, Galindo E (2001) Oil and fungal biomass dispersion in a stirred tank containing a simulated fermentation broth. J Chem Technol Biotechnol 76(11):1101–1106
161. Bouaifi M, Hebrard G, Bastoul D, Roustan M (2001) A comparative study of gas hold-up, bubble size, interfacial area and mass transfer coefficients in stirred gas-liquid reactors and bubble columns. Chem Eng Process 40:97–111
162. Alves SS, Maia CI, Vasconcelos JMT, Serralheiro AJ (2002) Bubble size in aerated stirred tanks. Chem Eng J 89:109–117
163. Laakkonen M, Moilanen P, Alopaeus V, Aittamaa J (2007) Modelling local bubble size distributions in agitated vessels. Chem Eng Sci 62:721–740
164. Montante G, Horn D, Paglianti A (2008) Gas-liquid flow and bubble size distribution in stirred tanks. Chem Eng Sci 63:2107–2118
165. Pacek AW, Man CC, Nienow AW (1998) On the Sauter mean diameter and size distributions in turbulent liquid/liquid dispersions in a stirred vessel. Chem Eng Sci 53:2005–2011
166. Raman AAA, Abidin MIIA, Nor MIM (2013) Review on measurement techniques for drop size distribution in a stirred vessel. Ind Eng Chem Res 52:16085–16094
167. Maaß S, Rojahn J, Hänsch R, Kraume M (2012) Automated drop detection using image analysis for online particle size monitoring in multiphase systems. Comput Chem Eng 45:27–37
168. Laakkonen M, Moilanen P, Miettinen T, Saari K, Honkanen M, Saarenrinne P, Aittamaa J (2005) Local bubble size distributions in agitated vessel. Comparison of three experimental techniques. Chem Eng Res Design 83(A1):50–58
169. Junker B, Maciejak W, Darnell B, Lester M, Pollack M (2007) Feasibility of an in situ measurement device for bubble size and distribution. Bioprocess Biosyst Eng 30:313–326
170. Alban FB, Sajjadi S, Yianneskis M (2004) Dynamic tracking of fast liquid-liquid dispersions processes with a real-time *in-situ* optical technique. Chem Eng Res Design 82(A8):1054–1060
171. Zabulis X, Papara M, Chatziargyriou A, Karapantsios TD (2007) Detection of densely dispersed spherical bubbles in digital images based on a template matching technique. Application to wet foams. Colloids Surf A: Physicochem Eng Aspects 309:96–106
172. Brás LMR, Gomes EF, Ribeiro MMM, Guimaraes MML (2009) Drop distribution determination in a liquid–liquid dispersion by image processing. Int J Chem Eng 1–6. Article ID: 746439:1–6
173. Khalil A, Puel F, Chevalier Y, Galvan J-M, Rivoire A, Klein J-P (2010) Study of droplet size distribution during an emulsification process using in situ video probe coupled with an automatic image analysis. Chem Eng J 165:946–957
174. Maaß S, Wollny S, Voigt A, Kraume M (2011) Experimental comparison of measurement techniques for drop size distributions in liquid/liquid dispersions. Exp Fluids 50:259–269
175. Nataliya S, Matas J, Eerola T, Lensu L, Kälviäinen H (2012) Detection of bubbles as concentric circular arrangements. In: Proceedings of 21st International Conference on Pattern Recognition (ICPR 2012), Tsukuba, Japan, pp 2655–2659, 11–15 Nov 2012
176. Rojas-Domínguez A, Holguín Salas A, Galindo E, Corkidi G (2015) Gradient-detection-pattern transform—application to automated measurement of oil drops in images of multiphase dispersions. Chem Eng Technol 38(00):1–10
177. Taboada B, Vega-Alvarado L, Córdova-Aguilar MS, Galindo E, Corkidi G (2006) Semi-automatic image analysis methodology for the segmentation of bubbles and drops in complex dispersions occurring in bioreactors. Exp Fluids 41(3):383–392

Adv Biochem Eng Biotechnol (2015) 149: 91–132
DOI 10.1007/10_2014_297

Published Online: 24 January 2015

The Cell Factory *Aspergillus* Enters the Big Data Era: Opportunities and Challenges for Optimising Product Formation

Vera Meyer, Markus Fiedler, Benjamin Nitsche and Rudibert King

Abstract Living with limits. Getting more from less. Producing commodities and high-value products from renewable resources including waste. What is the driving force and quintessence of bioeconomy outlines the lifestyle and product portfolio of *Aspergillus*, a saprophytic genus, to which some of the top-performing microbial cell factories belong: *Aspergillus niger*, *Aspergillus oryzae* and *Aspergillus terreus*. What makes them so interesting for exploitation in biotechnology and how can they help us to address key challenges of the twenty-first century? How can these strains become trimmed for better growth on second-generation feedstocks and how can we enlarge their product portfolio by genetic and metabolic engineering to get more from less? On the other hand, what makes it so challenging to deduce biological meaning from the wealth of *Aspergillus* -omics data? And which hurdles hinder us to model and engineer industrial strains for higher productivity and better rheological performance under industrial cultivation conditions? In this review, we will address these issues by highlighting most recent findings from the *Aspergillus* research with a focus on fungal growth, physiology, morphology and product formation. Indeed, the last years brought us many surprising insights into model and industrial strains. They clearly told us that similar is not the same: there are different ways to make a hypha, there are more protein secretion routes than anticipated and there are different molecular and physical mechanisms which control polar growth and the development of hyphal networks. We will discuss new conceptual frameworks derived from these insights and the future scientific advances necessary to create value from *Aspergillus* Big Data.

V. Meyer (✉) · M. Fiedler · B. Nitsche
Department Applied and Molecular Microbiology, Institute of Biotechnology,
Berlin University of Technology, Gustav-Meyer-Allee 25, 13355 Berlin, Germany
e-mail: vera.meyer@tu-berlin.de

R. King
Department Measurement and Control, Institute of Process Engineering,
Berlin University of Technology, Hardenbergstr. 36a, 10623 Berlin, Germany

Keywords *Aspergillus* · Protein secretion · Morphology · Modelling · Physiology · -omics · Database · Genome-scale model · Metabolism · Secondary metabolite

Contents

1 Introduction

Because *Aspergilli*, like many other filamentous fungi, first digest and then ingest their food, they have astounding capabilities in secreting enzymes hydrolysing biopolymers such as starch, (ligno-)cellulose, pectin, xylan, proteins and lipids. This feature is exploited by many biotech companies to manufacture enzymes for use in different industries such as paper and pulp, food and feed, laundry and textile, and biofuel. Moreover, *Aspergilli* produce and secrete a variety of bioactive compounds including organic acids and secondary metabolites which outcompete cohabitant bacteria and fungi and which are beneficial for human health and consumption. Many recent reviews have comprehensively discussed the wealth and variety of marketed *Aspergillus* products and listed the respective manufacturers [1–4].

What more is to come? New products! Commodities which are competitive with current petrochemical-derived products, smarter enzyme mixtures which faster and fully degrade lignocellulosic biomass and pharmaceuticals which have not been produced with *Aspergilli* so far. Better strains! Which produce more from cheaper substrates and which are genetically streamlined for better rheological performance under industrial fermentation conditions? These are very challenging goals, which need to be accomplished to help to reduce the heavy dependence of the modern world on fossil resources and to move it towards a post-petroleum future. In the following chapters, we will discuss current concepts, tools and approaches to reach these ambitious goals. To come to the point, some recent publications refuted long-lasting dogmas: (i) 'the key to the fungal hypha lies in the apex', i.e. protein secretion occurs exclusively at the hyphal apex [5, 6], (ii) industrial fungi such as *Penicillium chrysogenum* and *Trichoderma reesei* do not have sex [7, 8]. Others paved the way for new product classes which can be manufactured with *A. niger*: (i) nonribosomal peptides for medical applications [9], (ii) itaconic acid for use in bioplastics and biofuel industry [10, 11].

Although the post-genomic era has offered us holistic insights into the genomic and physiological landscape of *Aspergillus* species, it left us also with a flood of data. Decisions that were previously intuitive or based on data from laborious trial-and-error experimental approaches can now be made based on bioinformatics data itself. However, the 'Big Data tsunami' [12] poses major challenges. Managing and unbiased mining of the sheer size of -omics data is one. Not leaving them under-analysed in databases is the other. But deducing meaningful knowledge and new hypothesis is the most difficult one. What are the success stories in the field of systems biology of industrial filamentous fungi? Which hurdles hinder the move from intuitive to knowledge-based strain optimisation strategies? We will highlight in this review some examples to illustrate the current state of the art of *Aspergillus* systems biology with a focus on fungal growth, morphology, physiology and product formation. For further information, the reader is directed to recent reviews [13, 14]. Last but not least, we will summarise the state of the art of micro- or macro-morphological models for filamentous growing fungi such as *Aspergillus* to describe growth and geometrical appearance of pellets and mycelia. Recent approaches and mathematical concepts will be highlighted and the challenges discussed to develop biologically better founded models.

2 More from Less: Enlarging the Product Portfolio of *Aspergillus*

The pioneering role of *Aspergillus* species for microbial biotechnology is undisputed. Since almost 100 years, citric acid is produced on an industrial scale with *A. niger* and used as preservative and flavour enhancer in food and beverages. Due to its long history of safe use and its capability to withstand harsh industrial fermentation conditions, it later became leading for fermentation-derived enzyme production [4]. *A. oryzae* is the exclusive producer of another acid, kojic acid, which is applied as bleaching agent in cosmetics and as building block for biodegradable plastics. More influential, *A. oryzae* has traditionally been used to ferment starch-rich grains, like rice and soybeans, for the production of Asian food and beverages. The importance of *A. terreus* is attributed to its product lovastatin, a cholesterol-lowering agent which became a blockbuster after Merck released it for marketing in 1987 (trade name Mevacor, [15]). There are legitimate reasons to believe that industrial *Aspergillus* strains could keep and extend their pioneering role in microbial biotechnology. The following examples substantiate this prognosis.

2.1 Aspergillus Trimmed for Second-Generation Feedstocks

To achieve economic and ecological sustainable microbial production processes, it is crucial that feedstocks do not compete with the food supply chain but rely on non-edible crop residues such as lignocellulose. Current commercial cellulase cocktails for conversion of lignocellulosic biomass to bioethanol are derived from filamentous fungi (e.g. Cellic CTec2 from Novozymes and Accellerase from DuPont) and are a blend of three classes of enzymes: cellulases, hemicellulases and β-glucosidases. Whereas the first two are produced by *T. reesei* fermentation, the latter is a product of *A. niger*. Although *T. reesei* is the 'undisputable champion' in cellulose hydrolysis (*T. reesei* secretes cellulases up to 100 g/l), its cellulose preparations have to be supplemented with *A. niger* β-glucosidases to achieve efficient cellulose saccharification ([16] and references therein). Recently, a two-stage biorefining strategy was reported which utilised wheat straw as lignocellulosic raw material and only *A. niger* enzymes for saccharification [17]. Note that wheat straw is the second most abundant lignocellulosic crop residue in the world [18]. In the first stage, a solid-state fermentation of *A. niger* was performed on wheat straw for the production of (hemi)cellulases and β-glucosidases. These enzymes were extracted and used in the second stage to hydrolyse the fermented wheat straw. Remarkably, the freshly extracted enzymes of *A. niger* performed the same or even better than the commercial CTec2 solution [17]. Another report demonstrated that enzyme cocktails derived from submerged co-cultivation of *T. reesei* with *Aspergillus* species can compete with blended extracts from corresponding single cultures. This is an interesting alternative as it lowers production cost due to the integration of two fermentations into a single bioreactor and due to the elimination of an enzyme blending step [19]. *A. oryzae* is a poor secretor of cellulases. However, Yamada and co-workers demonstrated that overexpression of the kojic acid gene cluster specific transcription factor *kojT* in combination with heterologous expression of three fungal cellulases in *A. oryzae* led the fungus to produce kojic acid directly from cellulose. Although the yield was very low, the proof of concept was given that *A. oryzae* can be trimmed to produce kojic acid when cultivated on second-generation feedstocks [20].

2.2 A. niger Engineered for New Products

Itaconic acid stems from the citric acid cycle and is one of the most promising and flexible new platform chemicals derived from cellulosic biomass (for a review see [21]). Although it is currently produced by *A. terreus* with yields up to 76 % of the theoretical maximum, there are efforts to establish *A. niger* as itaconate producer. Several considerations make *A. niger* interesting as an alternative production host: first, its enormous citric acid production capacity could potentially be redirected to itaconate synthesis. Second, the fermentation can be performed at low pH which allows autosterile production and simplifies downstream processing. However, it is currently impossible to predict whether *A. terreus* can indeed be replaced by *A. niger*. Although the transfer of itaconate synthesis genes and reduction of by-product formation has been achieved, limited intracellular transport has been identified as severe bottleneck [10, 21–23]. Further studies are necessary to fully lift the potential of *A. niger* as heterologous itaconate producer.

Most recently, it was demonstrated for the first time that *A. niger* is a potent expression host for heterologous nonribosomal peptides [9]. For the proof of concept, the nonribosomal peptide synthetase gene *esyn1* from *Fusarium oxysporum* was expressed under control of the tunable bacterial–fungal hybrid promoter Tet-On [24] in *A. niger*. In *F. oxysporum*, ESYN catalyses the synthesis of the cyclic depsipeptide enniatin, an antibiotic used to treat bacterial throat infections. The strong inducibility of the Tet-On system combined with controlled bioreactor cultivation allowed impressively high enniatin yields in *A. niger* (about 5 g/l), not reported so far for any heterologous enniatin production host. As also an autonomous *A. niger* expression host was engineered, being independent from feeding with the enniatin precursor D-2-hydroxyvaleric acid, the production cost was lowered considerably [9].

2.3 Aspergillus is Becoming Transparent

As of 2010, ten genomes of different *Aspergillus* species were publically accessible [25]. As of 2014, the number increased to 21 fully sequenced *Aspergillus* species, which are accessible through MycoCosm, a fungal genomics portal (http://jgi.doe.gov/fungi) developed by the US Department of Energy Joint Genome Institute [26]. In addition, dozens of *Aspergillus* strains are becoming currently resequenced including *A. niger* strains which have been transformed with bacterial cellulases, which are enzyme hyperproducers or spontaneous mutants (http://jgi.doe.gov/fungi). Subsequent comparative genome analyses will likely make *Aspergillus* one of the best studied and understood fungal genera. A tremendous opportunity, which will help to distinguish between common and strain-specific genomic features as exemplarily shown for two *A. niger* strains, the citric acid-producing strain ATCC 1015 and the enzyme-producing strain CBS 513.88 [27] (see Chap. 5).

3 The Life Cycle of *Aspergillus* and Its Link to Product Formation

To fully lift the potential of *Aspergillus* as a multi-purpose cell factory, a key question needs to be addressed: which metabolites, proteins and biopolymer-degrading enzymes are predominantly produced during which phase of its life cycle? As nature has evolutionary optimised *Aspergillus* already to produce certain compounds under specific habitat and nutrient conditions, one 'simply' needs to understand how the native metabolic properties of *Aspergillus* are linked to its lifestyle and life cycle in order to further optimise *Aspergillus* by rational means. Clearly, certain product classes are associated with different stages of its life cycle, such as secondary metabolite synthesis in *A. niger* with sporulation [28]. Likewise, its saprophytic lifestyle is the basis for the tight regulation of polysaccharide hydrolase expression and secretion [29].

3.1 Morphogenesis and Product Formation

The life cycle of *Aspergillus* and filamentous fungi starts with breaking dormancy of (a)sexual spores (Fig. 1). In *A. niger*, the resulting onset of spore germination, which is characterised by isotropic spore swelling, becomes only initiated when a carbon source, a nitrogen source and a germination trigger are present in the environment [30]. D-glucose, 2-deoxy-D-glucose, D-mannose, D-xylose and L-amino acids such as L-asparagine, L-glutamine and L-tryptophan were identified as germination triggers. A full list of triggers and non-triggers can be found in [30, 31]. Interestingly, dormant spores (named conidia) of *A. niger* contain a high level of transcripts from about 40 % of *A. niger* genes. These transcripts include genes involved in protein synthesis and respiration and probably prime the conidia for the onset of germination [32, 33]. Note that the *A. niger* genome has been predicted to contain about 14,000 genes [34]. Interestingly, also a considerable amount of

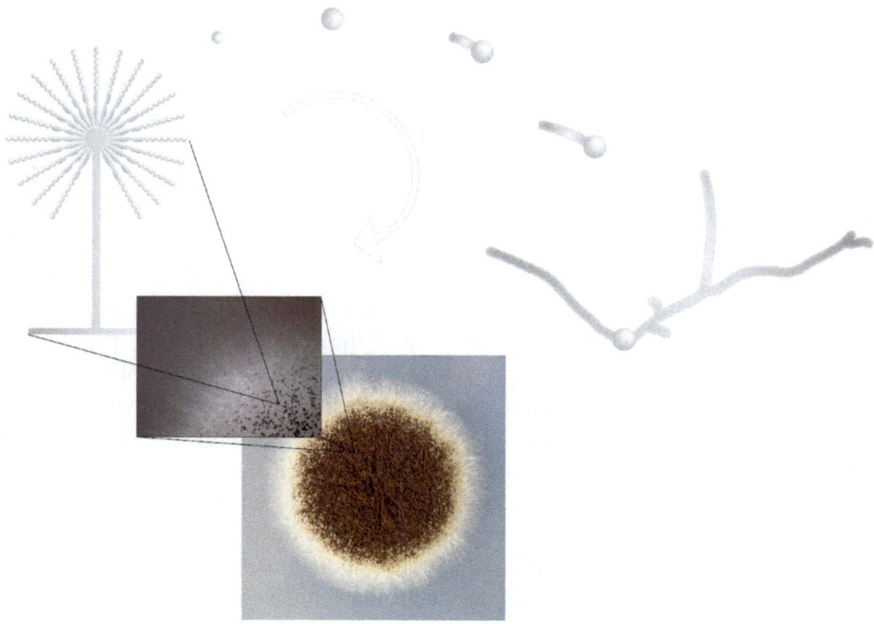

Fig. 1 The life cycle of *Aspergillus niger*. The different stages of spore swelling, germ tube outgrowth, germ tube elongation and branch formation are schematically shown. These processes lead to the formation of a true circular colony, which appears *brownish black* due to *brownish-black*-coloured conidia. The conidia become constricted from a conidiophore (schematically shown)

antisense transcripts (10 %) were detected, which are assumed to prevent expression of proteins that are not required during spore swelling [32].

The first stage of spore swelling is followed by the outgrowth of a germ tube. This break in cell symmetry results from the establishment of a polarity axis. If this polarity axis becomes maintained, sustained germ tube elongation and the formation of a hypha can be observed. In higher fungi, such as Ascomycetes and Basidiomycetes, this hypha becomes compartmentalised by septa, which contain a central pore that allows inter-compartmental streaming [35]. Lateral and/or apical branch formation in this hypha eventually leads to the formation of an interconnected hyphal network, the mycelium (Fig. 1, [36]). In consequence, the age of the mycelium increases from the periphery towards the central zone, whereas substrate concentrations decrease. This spatial heterogeneity results in profound differences on the level of the transcriptome [37] and secretome [38] in different zones of *A. niger* colonies and supports the idea that medium substrates become sequentially degraded. Notably, a large fraction of the secreted proteins remains trapped in the fungal cell wall from which they become only slowly released. In turn, a remarkable fraction of proteins secreted within a more central zone of a mycelium was synthesised earlier during growth, i.e. when the respective protein-producing hyphae were at the periphery of the colony. As an example, partial degradation of fungal cell walls in the central zones allowed the detection of ~120 secreted proteins, but of only ~60 in the untreated control colony [38]. Such retention phenomena bring additional complexity to the heterogeneous spatio-temporal substrate degradation pattern observed in *A. niger* colonies and most likely also occur in submerged pellet-forming mycelia of *A. niger* during industrial fermentation processes.

Beside this interzonal heterogeneity, intrazonal hyphal heterogeneity is also a characteristic attribute of *A. niger*. Although the periphery of *A. niger* colonies or pellets is the active growth zone and the primary zone of protein secretion [39], neighbouring hyphae differ with respect to RNA and protein levels [40–42]. Whilst it may well be assumed that such heterogeneities could be beneficial for *A. niger* under natural habitat conditions, an increase of the actively secreting hyphal population would be a constructive approach to improve the productivity of submerged industrial processes [43]. However, heterogeneity of neighbouring hyphae is less pronounced in submerged pellets compared to surface cultures, which is likely linked to a more homogeneous substrate distribution in liquid compared to solid media [42]. Last but not least, carbon starvation-induced heterogeneity can be observed in dispersed mycelia of *A. niger* when cultivated under submerged conditions. Two morphologically different hyphal populations appear when the carbon source is depleted (Fig. 2), on the one hand thick but non-growing hyphae with many empty compartments and on the other hand thin but growing hyphae. Growth of the latter is likely fuelled by carbon recycling via autophagy of the older hyphae [44, 45]. This is in fact an elegant strategy of filamentous fungi, to make use of internal resources, i.e. degradation products, before a compartment collapses for the survival of a neighbouring compartment [46].

3.2 Asexual Development and Product Formation

Surface cultures of *Aspergillus* start to form aerial hyphae and conidiophores after achieving developmental competence (Fig. 1), a process which can be triggered by the liquid–air interface, chemical signals and light. At the end of this developmental process, conidia are formed via mitotic events. For details on key morphogenetic genes and regulatory networks involved, the reader is directed to [47, 48]. Notably, conidiophore structures are morphologically informative enough to distinguish between the genera *Aspergillus* and *Penicillium* [49]. Although the process of conidiation is repressed under submerged conditions, it is possible to induce sporulation by severe carbon limitation in retentostats, where growth rates approaching zero can be adjusted [50].

Several studies analysed the transcriptomes of conidiating *A. niger* and *T. reesei* cultures during liquid [44, 50] and solid-state cultivations [51, 52] and demonstrated that nutrient starvation and conidiation are tightly linked. The fraction of differentially expressed genes ranged between 10 and 50 %, depending on the platform and statistics applied, and clearly highlighted that the transition from vegetative hyphae to developmental structures is accompanied by complex physiological adaptations.

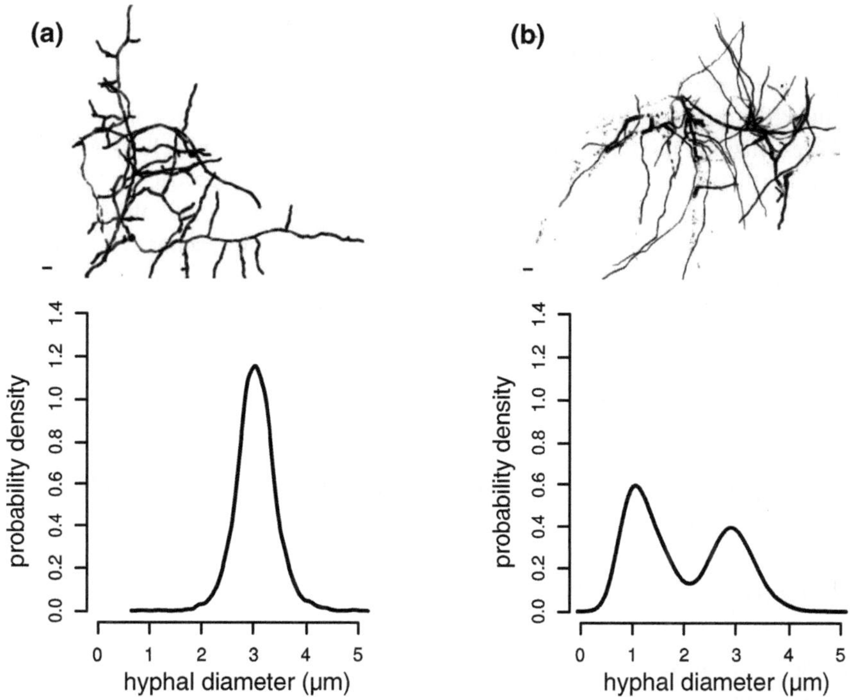

Fig. 2 Two hyphal populations can be observed during carbon starvation of *A. niger* when cultivated under submerged conditions. Hyphal diameters can be used to distinguish both populations: old hyphae (formed during exponential growth phase) and young hyphae (formed during starvation). An image analysis algorithm developed by [44] analyses microscopic pictures from samples of various cultivation time points (**a**, **b**) and plots probability density curves for the distributions of hyphal diameters. Diameters from exponentially growing hyphae (**a**) display a normal distribution with a mean of approximately 3 μm. The second population of thinner hyphae with a mean diameter of approximately 1 μm start to emerge during starvation (**b**). During prolonged starvation, the ratio of thin/thick hyphae gradually increases [44]. Bar, 10 μm

A major starvation-induced response linked with asexual development is activation of secondary metabolite genes and gene clusters, which are under control of different signalling pathways including epigenetic control mechanisms [53–55]. These processes can easily be tracked and studied in retentostat cultivations [44, 50].

Another major response during carbon starvation is induction of carbohydrate-active enzymes (CAZymes), including enzyme classes degrading the fungal cell wall and plant-derived polysaccharides. Both are of major interest for industrial applications. Whereas the first class is unwanted as it supposedly reduces the fraction of actively growing hyphae, the second class contains (un)known candidate enzymes for new biorefining strategies (see Chap. 1 and [17]). Delmas et al. [29] and Van Munster et al. [56] uncovered that carbon starvation induces de-repression of genes encoding sugar transporters and a comprehensive set of CAZymes, which become secreted as 'scouting enzymes'. Depending on the actual polysaccharide(s) present in the surroundings, they catalyse the liberation of specific sugar monomers, which in turn trigger expression and secretion of the specific enzyme set necessary to fully degrade the polysaccharide(s) present in the environment. The requirement of complex enzyme mixtures for efficient saccharification of lignocellulosic biomass underlines the importance of genome-wide and systematic approaches to comprehensively decipher the polysaccharide degradation potential of *Aspergillus*. Recently, structural data from 16 major plant-derived polysaccharides have been combined with a list of 188 predicted CAZymes from *A. niger* [57]. The resulting analysis framework can now be queried systematically with data from genome-wide expression profiling analyses such as obtained from *A. niger* cultivated on sugarcane bagasse [58] or wheat straw [29].

4 Polarised Growth of *Aspergillus* and Its Link to Protein Production

4.1 There Are More Ways to Make a Hypha

Although sustained polarised growth is the defining attribute of filamentous fungi, hyphal filaments exhibit a striking diversity in organisation. In some fungi, the nuclei are distributed dot-like along the hypha (e.g. *Aspergillus nidulans*); in others, they appear in bulk (e.g. *A. niger*). Some fungi form filaments, which are dikaryotic and largely composed of vacuoles (e.g. *Ustilago maydis*, [59]), some form hyphal networks which are remarkably heterogeneous with respect to gene expression and physiology as described in Chap. 2 (e.g. *A. niger* and *Neurospora crassa*, [60, 61]). Recently, Woronin bodies, originally thought to be emergency patches which seal septal pores of filamentous ascomycetes upon mechanical injury, were shown to maintain hyphal heterogeneity in intact hyphae by impeding cytoplasmic streaming. Cytoplasmic GFP expression was heterogeneous in neighbouring hyphae of *A. oryzae* which formed Woronin bodies but was homogeneous in a strain that could not form these organelles. Closure of septa is a dynamic and reversible process, resulting on average in 40 % closed apical septa of exploring *A. oryzae* hyphae [62].

4.2 Microscale and Macroscale Polarity

Why does a hypha grow in a polarised manner? Because of pressure! On the one hand, cytoskeletal structures (microtubuli, actin) form the polarised architecture of

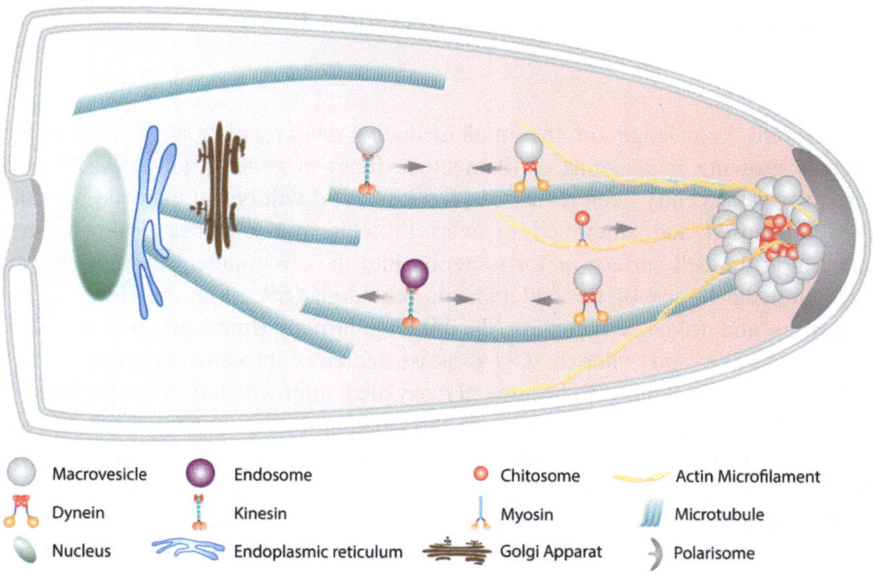

Fig. 3 Subcellular structures define and maintain the architecture of a growing hypha. Cytoskeletal-based transport of vesicles and endosomes is realised by different motor proteins (kinesins, dyneins and myosins) which move along microtubules and actin cables in different directions. For the sake of brevity, the endocytotic actin ring is not shown. For details, see text

growing hyphae. Along these, secretory vesicles, containing lipids, cell wall proteins, building blocks and proteins destined for secretion, are actively transported by motor proteins towards the hyphal tip (Fig. 3, [63]). On the other hand, internal hydrostatic pressure (turgor) is created. Water flows into the growing hyphal tip due to the uptake of solutes and the biosynthesis of osmotically active solutes within the cell. According to the model after Lee [64], water inflow is greatest at the tip and declines exponentially behind the tip, thus causing the cell wall to expand. Stretching of the cell membrane and cell wall provokes a tip-localised calcium gradient that eventually mediates vesicle fusion and thus cell expansion. Next to this microscale polarity, macroscale polarity exists as well. Mycelia can be viewed as 'hydraulic networks' because of a laminar mass flow of cytoplasm towards the colony edge. The driving force for this cytoplasmic movement is thought to result from maximum uptake of nutrients behind the foraging hyphae, leading to increased water inflow and increased pressure [64]. It has been proposed that this intrahyphal transport is regulated by the closure of septa, which become plugged by Woronin bodies [65]. It is very likely that this pressure-driven long-distance transport complements the short-distance transport mediated by cytoskeletal motor proteins [66].

4.3 Microscale Polarity and the Secretory Route of Aspergillus

Our current knowledge on the implementation and regulation of vesicle flux towards growing hyphal tips of filamentous fungi in general and *Aspergillus* in particular has recently been reviewed [4, 67, 68] and will be summarised here in a very condensed and simplified manner: Proteins and enzymes destined for a function at the cell surface or for secretion into the environment become translocated into the lumen of the endoplasmic reticulum (ER) (Fig. 3). Here, several chaperons and foldases including the binding protein BipA, protein disulphide isomerase PdiA and calnexin ClxA assist secretory proteins in their folding. Many secretory proteins also become glycosylated, after which they are packed into vesicles and transported to and through the Golgi complex via COPII-vesicles and are then delivered to the Spitzenkörper via a microtubule-mediated transport. The Spitzenkörper is a multi-component, vesicle-rich, pleomorphic structure at the hyphal tip and viewed as the main choreographer of hyphal tip growth (for a detailed review see [69]). It translocates secretory vesicles to actin cables, which transport them to the plasma membrane. Vesicles fuse with the plasma membrane, thus releasing their cargo. Whilst secretory proteins become released into the environment, GPI-anchored proteins and cell wall-synthesising proteins (e.g. chitin and glucan synthases) become retained at the membrane to catalyse cell wall polymers such as chitin, glucans and galactomannans, which eventually expand the cell wall (for see review [70]). The vectorial flux of vesicles from the Spitzenkörper to the plasma membrane is orchestrated by the polarisome, which mediates the nucleation of actin cables and together with cell end markers directs the vesicle flux towards the apex (Fig. 3, [63, 68, 71]).

It is currently a matter of debate, whether exocytosis and endocytosis are spatially coupled. Retrieval of excess membrane material resulting from vesicle fusion with the membrane and removal of cell wall synthetising enzymes from subapical membrane regions by endocytosis are thought to be essential to maintain the fungal shape and its polarity [63, 72–74]. In *A. nidulans*, empty vesicles are proposed to become recycled by endocytosis at a subapical actin collar, transported back to the Golgi, where they become reloaded with new protein cargo destined for secretion [63]. Hence, protein secretion should be viewed as a concerted and balanced activity of exo- and endocytotic events. Furthermore, restriction of cell wall enzymes to the very tip in *C. albicans* has been shown to ensure polarised cell wall expansion as deduced from experimental and 3D modelling data [74]. However, a mathematical model proposed for the vesicle flux to and from the Spitzenkörper in *Rhizoctonia solani* predicts that removal of membrane material is not significant to compensate a potential excess of membrane added by exocytosis [75]. Is this further evidence that there are more ways to make a hypha? So far, experimental data from industrial relevant *Aspergilli* are not available which do illuminate any potential link between their exocytotic and endocytotic machineries and which could refute or validate either model. Demonstrated so far is that tubulin and actin cytoskeletal elements are key for the transport of secretory vesicles to the hyphal tips of *A. niger* and *A. oryzae*, whereby secretory vesicles were also occasionally observed at septa, suggesting that secretion can also occur at sites of septa [6, 76].

In addition to protein and membrane turnover via endocytosis, pioneering work in the model systems *A. nidulans* and *U. maydis* has shown that bi-directional movement of early endosomes along actin cables and microtubuli actively disperses mRNA and ribosomes, i.e. the protein translation machinery along hyphal compartments, thus also contributing to hyphal growth (Fig. 3, for details see reviews [77, 78]). Hence, a complex interplay of cytoskeletal structures, polarity factors, cell end markers, and exocytotic and endocytotic events assure the secretion of proteins and polarised hyphal growth [68, 79].

4.4 A Holistic View on the Secretory Route of Aspergillus

Filamentous fungi are extraordinary in their capacities to secrete endogenous proteins, whereby *A. niger* (up to 20 g/l) and *T. reesei* (up to 100 g/l) are outstanding [80, 81]. We and others have most recently reviewed the current strategies to exploit these capacities for the expression of heterologous proteins in *Aspergillus* and the systems biology approaches that have been undertaken to study the physiological and regulatory aspects related to homologous and heterologous protein production. In order to not exceed the framework of this review, the reader is directed to these reviews [4, 14, 25, 82]. What have we learnt so far? First, *A. niger* and *A. oryzae* flexibly adapt the protein flux through the secretory pathway. If genetically forced protein overexpression causes an overload of the secretory route, both ER-associated degradation and the unfolded proteins response UPR aim to restore cellular homeostasis. In addition, many genes become transcriptionally down-regulated, which are obviously less important for growth and survival under the specific growth condition. This phenomenon is called repression under secretion stress (RESS) and was first observed in *T. reesei* [83]. Second, under identical specific growth rates realised in chemostat cultivations, a glucoamylase-overproducing strain of *A. niger* has to modify the expression of about 1,500 genes compared to the wild-type strain in order to secrete about seven times more glucoamylase. The up-regulated gene set contained statistically enriched genes functioning in ER translocation, protein glycosylation, vesicle transport and ion homeostasis [84]. A comparative transcriptomics analysis of six independent overexpression and UPR stress studies in *A. niger* uncovered that a key set of about 40 genes is crucial for *A. niger* to overexpress a protein and cope with this cellular burden. The complete gene list can be found in [84] and includes genes encoding ER-resident chaperones and foldases (e.g. *clxA*, *pdiA*, *bipA*), as well as genes important for ER translocation, protein glycosylation and COPII-based vesicle trafficking to name but a few. Third, the amino acid sequence of secretory proteins somehow affects the level of secretion. By comparing the extracellular protein yield of about 600 homologous and 2,000 heterologous proteins in *A. niger*, van den

Berg and colleagues identified a positive effect for tyrosine and asparagine, whereas methionine and lysine hampered high protein yields [85]. Fourth, we are far away from rationally and predictively designing *A. niger* or *A. oryzae* as cell factories for protein production. Despite more than 30 years of research on these strains, the quantitative biology behind protein secretion is completely unknown. For example, how many vesicles can a hyphal tip of *A. niger* accommodate per unit of time? How many vesicles are necessary to ensure maximum transport of the protein of interest whilst not interfering with the transport of other vesicles carrying proteins and building blocks for normal growth? How long does the protein of interest need to become properly folded? What does this mean with respect to how many proteins can be channelled through the secretory pathway in order to provide each protein molecule sufficient time to become correctly folded? What would then be the optimum transcription rate for the respective encoding gene and the half-life of its mRNA? And finally, how is the capacity for protein secretion related to primary metabolism which provides the secretory route with ATP, redox equivalents and amino acids? Although *Aspergillus* already entered the post-genomic era, we are far away from answering these questions and still work on the inventory of its secretory machinery. We have no idea yet, who guides the secretory route, who keeps all players in homeostatic balance and what might be the mechanistic explanation(s) for the phenomenon that *A. niger* is such an outstanding protein secretor.

4.5 There Are More Ways to Secrete a Protein

There is accumulating evidence that filamentous fungi secrete proteins not only at hyphal tips via the secretory pathway described above. These unconventional secretion pathways (or 'unconventional membrane trafficking pathways') have also been postulated for plant and mammalian systems (for reviews see [86–90]). The first report for filamentous fungi goes back to 2011, where it was demonstrated that exocytosis constitutively occurs in *A. oryzae* at hyphal septa [6]. Originally, it was thought that cytoskeleton-mediated transport of enzymes and materials for cell wall synthesis towards septa is required only during their formation. However, it was shown that exocytosis continues to occur after septation. Although the representative secretory enzyme of *A. oryzae* α-amylase was mainly transported to the hyphal apex, plasma membrane transporters involved in purine and amino acid uptake were mainly transported to septa [6]. Similarly, a plasma membrane proton pump of *N. crassa* is incorporated in distal hyphal regions independently of the Spitzenkörper at the hyphal tip [91]. Although these observations were unexpected and refute the dogma that 'the key to the fungal hypha lies in the apex', i.e. protein secretion occurs exclusively at the hyphal apex [5], they basically link microscale polarity with macroscale polarity. If nutrients are preferentially taken up in regions more central in the colony, they will generate the small pressure gradient necessary to establish the mass flow of cytoplasmic material. We are coming closer to unveil the secret of fungal micro-fluidics!

Other examples for unconventional secretion mechanisms in filamentous fungi have been reviewed most recently in [86] and are schematically summarised in Fig. 4. First, a brefeldin A-insensitive protein secretion pathway is likely to exist. Brefeldin A is a compound which inhibits vesicle trafficking between the ER and the Golgi. However, addition of brefeldin A does not inhibit chitin synthase secretion in *N. crassa* [92], suggesting that secretory proteins can leave the ER and reach the plasma membrane independently of the Golgi apparatus [86]. Another example is secretion of proteins which do not contain a signal peptide and thus do not enter the ER. This has for example been shown for an endochitinase from *U. maydis* [93]. A third example for unconventional secretion is based on the multi-vesicular bodies, which are late endosomal compartments which take up Golgi-derived vesicles. Upon fusion with the plasma membrane, they release these vesicles into the extracellular space [86]. Hence, secretion in filamentous fungi can be realised by more than a single pathway, it would not be surprising if additional routes will become uncovered in the future.

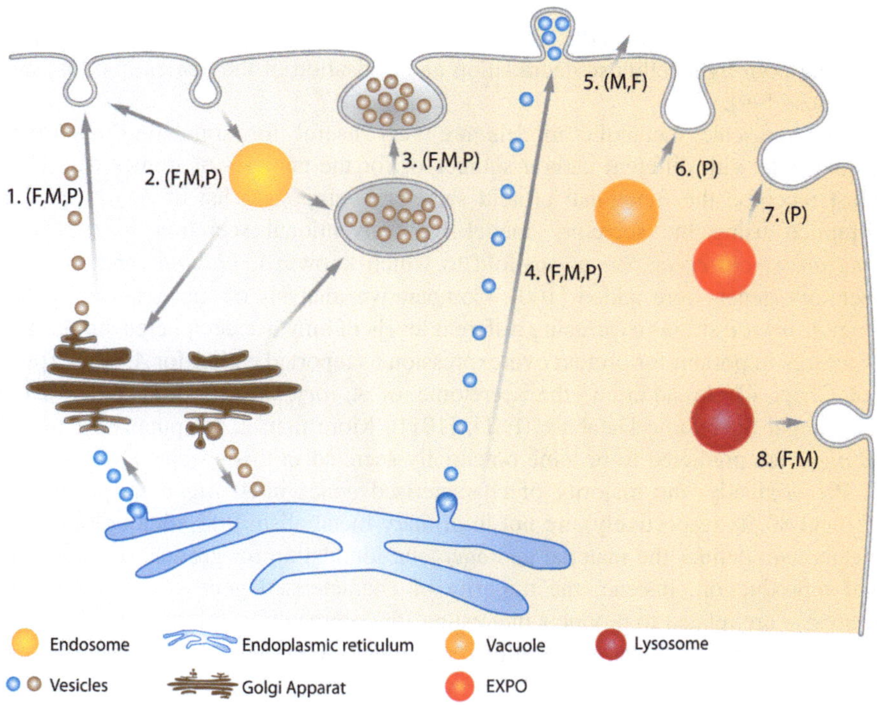

Fig. 4 Different protein secretion routes executed in eukaryotic cell systems including fungi (F), plants (P) and mammalian cells (M). For details, see text

5 The Genetics and Physiology Underlying *Aspergillus* Morphology and Productivity

5.1 From Genomes to Gene Function

The *Aspergillus* Genome Database (AspGD; http://www.aspgd.org/) is a freely accessible Web resource providing an outstanding comparative genomics toolbox for the fungal community. AspGD houses sequence data from currently 20 *Aspergillus* genomes and improves gene annotation using publically available transcriptomics data [94, 95]. As of September 2014, 14,056 genes have been predicted for *A. niger* strain CBS 513.88 (232 of which have a verified function), 11,902 for *A. oryzae* (199 of which have a verified function) and 10,678 for the model species *A. nidulans* (1,149 of which have a verified function). Hence, the industrial species have 98 % uncharacterised genes. However, comparative genomics in tandem with the reconstruction of genome-scale metabolic networks can successfully be used for an initial assignment of gene functions as shown for *A. oryzae* and *A. niger*. This resulted in the tentative assignment of gene functions to 1,314 and 871 genes, respectively [96–98]. Comparative genomics also revealed that *Aspergilli* have evolutionary well adapted for naturally performing both lignocellulose degradation and utilisation of the pentoses xylose and arabinose [99].

Genome-scale metabolic models are very useful for simulating maximum growth rate with different carbon sources or for the analysis of -omics data [13]. Most recently, the functional protein secretory component list of *A. oryzae* was compiled using the secretory model for conventional secretion in the yeast *Saccharomyces cerevisiae* as scaffold to which known *A. nidulans* and *A. niger* secretory genes were added [100]. Comparative analysis of transcriptomics data from *A. oryzae* strains expressing different levels of amylase deciphered similar key processes important for protein overexpression as reported earlier for *A. niger* ([84], see Chap. 4). In addition, the secretome of *A. oryzae* was defined by using the Fungal Secretome Database (FSD; [101]). More than 2,200 putative genes of *A. niger* are predicted to become potentially secreted in this species [100].

Paradoxically, the majority of characterised genes in *A. niger* and *A. oryzae* (50 and 80 %, respectively) are not in primary metabolism [98], although primary metabolism defines the material and energetic foundation for growth, development and reproduction. Instead, the majority of characterised genes in *A. niger* and *A. oryzae* are related to products interesting for commercial exploitation: secondary metabolites [54, 102] and extracellular enzymes (see above).

5.2 *Citric Acid or Protein Production—What Makes the Difference?*

The genome sequence of the *A. niger* strain CBS 513.88 used for industrial glucoamylase production was published in 2007 [34]. Four years later, the genome of the acidogenic wild-type strain ATCC 1015 was made publically available and compared to strain CBS 513.88 [27]. Comparative genomics and transcriptomics uncovered a remarkable diversity between both strains. First, about 400/500 genes are unique in CBS 513.88/ATCC 1015, the most notable difference is two alpha-amylases which are only present in strain CBS 513.88. Both genes originate from *A. oryzae* and were most likely acquired through horizontal gene transfer. Second, a high number of polymorphisms were found (8 ± 16 SNPs/kb) between both strains, many of which were clustered in hypervariable regions. Third, transcription of about 4,800 genes differed in both strains (with an almost equal number up-regulated in each strain) when cultivated under the same growth condition in glucose-based batch cultures. Increased transcription of genes in strain CBS 513.88 was related to glycolysis, amino acid synthesis (including the entire biosynthetic pathways of amino acids which are over-represented in the glucoamylase, i.e. threonine, serine and tryptophan) and tRNA-aminoacyl synthase formation. In contrast, the citric acid-producing strain ATCC 1015 displayed higher expression of genes with a function in the transport of electrons, carbohydrates and organic acids. Interestingly, nitrogen source utilisation is impaired in strain CBS 513.88, an observation which could also be relevant to protein production. This study by Andersen and colleagues thus elegantly showed the power of comparative -omics analysis for the identification of factors contributing to the different product portfolio of *A. niger* strains.

5.3 *Moulding the Mould—The Genetic Basis of Aspergillus Morphology*

The number of genes functionally implicated in hyphal morphogenesis and development of the model fungus *A. nidulans* has been estimated to be about 2,000 [36], a number which most likely can be assumed also for other filamentous fungi including the industrial *Aspergillus* strains. However, the function of only about 5 % of the assumed morphogenetic genes has been understood so far, mainly from studies in the model organisms *A. nidulans*, *N. crassa* or *U. maydis*. It is beyond the scope of this review to provide a comprehensive overview on the function and regulation of the morphogenetic genes and modifiers known so far. For the interested reader, we recommended the following most recent reviews for further reading: polarity factors and cell end markers [67, 69], membrane trafficking [78, 103], septum function [104], calcium regulation [104, 105], cytoskeletal structures [68], cell wall formation [70, 106], asexual development [48, 49] and fungal microfluidics [64]. Instead, we will discuss in the following some recent findings which could help to illuminate the connection between hyphal morphology, vesicle trafficking and protein secretion in the industrial strains *A. niger* and *A. oryzae*.

When cultivated under submerged conditions, filamentous fungi adopt two macroscopic morphologies—a pelleted morphology or a dispersed morphology. As both interfere with the productivity, rheology and downstream processing of industrial fermentation processes (see Chap. 7), a mechanistic framework is sought since a long time in order to describe and control the optimum macroscopic morphology under industrial conditions. A so far unsolved problem as too many process variables including type and concentration of carbon and nitrogen source, pH, temperature, fermenter geometry, agitation systems, culture mode [107] as well as biological parameters including conidial adhesion capacity [108] and branching frequency [109] affect the macroscopic morphology of filamentous fungi. With respect to the genetic basis of filamentous fungal morphology and its link to protein secretion, two hypotheses are currently under discussion: (i) the higher the branching frequency, the higher the probability that macroscopic pellets and not dispersed mycelia will be formed. Often, the tendency to form pellets under submerged conditions can be assessed during solid-state cultivations on agar plates—the colonies formed are more compact, have a reduced colony diameter and are sometimes delayed in sporulation (Fig. 5), (ii) the higher the number of hyphal tips, the more the proteins become secreted. The latter hypothesis has been addressed in

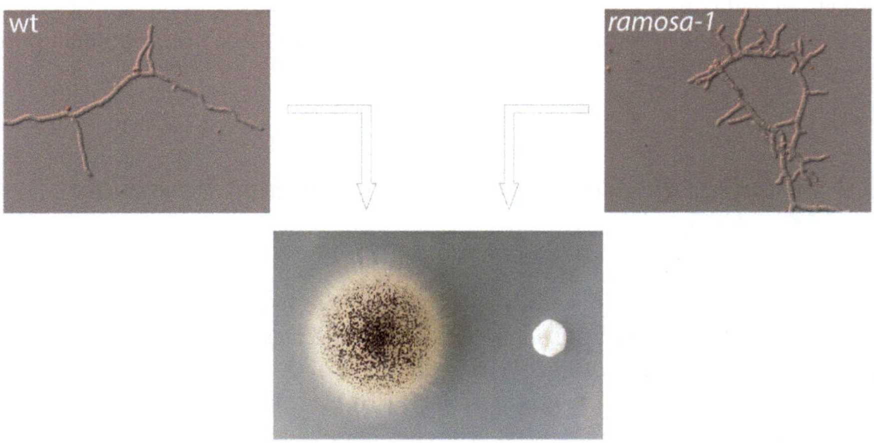

Fig. 5 Branching frequency determines macroscopic morphology of *A. niger*. Shown are young hyphae from a wild-type strain (wt) and from the conditional hyperbranching mutant *ramosa-1* of the same age. In this mutant, a point mutation within the TORC2 component RmsA provokes excessive apical branching and increased septation under restrictive conditions, leading to drastically reduced colony growth rate on agar plates and the formation of a very compact colony [113]. Note that *ramosa-1* is also impaired in sporulation—a defect, which is not a general feature of hyperbranching mutants

several studies in *A. niger* and *A. oryzae*; however, only contradictory results have been reported. Some studies proposed a positive correlation between branching frequency and protein secretion yield; others did not [110–114]. It is therefore still a matter of debate which is the best macroscopic morphology of *Aspergillus* with respect to productivity and rheology during protein production.

Hyphal branching is the basis for mycelial development. Whereas branches usually arise from basal regions (lateral branching), branch formation can also occur via tip splitting (apical branching). Tip splitting is thought to result from the accumulation of secretory vesicles at the tip because the capacity for vesicle fusion with the membrane has been exceeded. To accommodate the abnormal accumulation of the vesicles, the tip divides into two new branches [115]. What, then, are the underlying cellular events and signalling networks of (sub)apical branch formation? Several transcriptomics and functional genomics approaches undertaken in *A. niger* have shed some light on these processes. Central to these studies was controlled manipulation of the actin cytoskeleton, which resulted in defined mutants with a hyperbranching or apolar growth phenotype [109, 114, 116, 117]. The transcriptomic signatures of these actin (hyper- or de-) polarisation mutants uncovered that several regulatory and metabolic pathways likely participate in morphogenetic events of *A. niger*: phospholipid signalling, sphingolipid signalling, TORC2 signalling, calcium signalling and cell wall integrity signalling. These pathways supposedly induce different physiological changes during branch formation including changes in sterol metabolism, amino acid metabolism, ion transport (Na^+, K^+, Ca^{2+}, Fe^{2+}, Zn^{2+}, Co^{2+}) and protein trafficking [109, 114]. A key conclusion of these comparative transcriptomics analyses is the prediction that lipid molecules such as phosphatidic acid, diacylglycerol, phosphatidylinositol-4,5-bisphosphate, inositolphosphates and sphingosin-1-phosphate occupy a central position in the polar growth control of *A. niger*. Their function as secondary messengers has been proven for many model eukaryotes; however, almost nothing is known about their cellular roles in *Aspergilli*. Thus, only an integrated systems biology approach which combines transcriptomics data with lipidomics, proteomics and metabolomics data is inevitable to comprehensively inventory the morphogenetic network of *A. niger*.

These recent studies in *A. niger* also carved out further very interesting findings: First, the function of some molecular switches controlling actin polarisation (e.g. Rho GTPases) differs in *A. niger* and *A. nidulans* [117, 118], giving further evidence that members of the same genus employ alternative strategies to make a hypha and that caution is thus required when extrapolating findings of one (model) organism to another [119]. Second, simply increasing the number of hyphal tips does not per se result in increased protein secretion. If the amount of secretory vesicles remains unchanged, the vesicles become merely distributed to more tips, and the amount of vesicles per individual hyphal tip lowers accordingly [114]. Third, it is possible to generate hyperbranching mutants (e.g. by deleting the Rho GTPase gene *racA*) which form a more compact macroscopic morphology but pellets and display the same growth rate as the wild-type strain—thus representing ideal systems for future studies to investigate the relationship between intracellular protein secretion capacities and rheological properties under controlled conditions [114].

What more is to come? Although comparative genomics and transcriptomics are so far the dominant approaches studying the genetic basis of *Aspergillus* morphology, we can soon expect more insights into the translatome and phosphoproteome of *Aspergilli* during polarised growth. First reports have already been published which pave the way to a more holistic understanding of translational control mechanisms during protein secretion stress [120] and which will help to decode phosphorylation-mediated regulatory networks controlling fungal morphogenesis and secretion [121].

5.4 Moulding the Mould—Linking Aspergillus Morphology with Primary Metabolism

Although we are far away from comprehensively deciphering transcriptomic signatures of *A. niger* morphological mutants, it is apparent that they suggest a connection between polar growth control and primary metabolism. For example, the *A. niger* TORC2 complex is not only central for the polarisation of actin at the hyphal tip but also of vital importance for energy metabolism, viability and salt balance of *A. niger* [109, 116], the function of the Rho GTPase RacA lies not only in the control of actin polarisation but is also somehow linked to carbon and nitrogen source utilisation [114], the transcriptional repressor TupA links morphogenesis, development and nitrogen metabolism [122]. What might be the connection between these different processes?

Of course, cells do only grow if sufficient amount of nutrients are present in the medium; hence, the most obvious interdependence lies between the nutritional/ energetic status of a cell which is the prerequisite for any kind of growth. Our current understanding of the main fungal signalling pathways balancing growth and nutritional status is based upon the paradigms established in the models *S. cerevisiae* and *A. nidulans*. From these systems, we know that the cAMP-protein kinase A and Snf1 signalling pathways are key for carbon sensing [123], whereas TORC1 signalling is important for nitrogen sensing [124]. However, a very recent publication on the regulation of dimorphism (i.e. the switch between a yeast-like apolar growth mode to a filamentous polar growth mode) in *U. maydis* has provided for the first time the evidence of a physical interaction of a high-affinity ammonium transporter (Ump2) with a Rho GTPase (Rho1) at the plasma membrane of a filamentous fungus. This interaction occurs under low-ammonium conditions and inhibits Rho1 from negatively regulating Rac1 (the homologue of the *A. niger* Rho GTPase RacA), thereby inducing the switch to filamentous growth [125]. This is a very compelling study which without any doubt clearly demonstrates that the nutritional status can also directly control polarised growth.

6 Integration and Analysis of *Aspergillus* Genomic Data

The fungal scientific community witnessed a revolution during the last two decades. Since the publication of the first fungal genome (*S. cerevisiae*) in 1996 [126], the number of publically available fungal genome sequences increased exponentially reaching more than 250 by 2014 [26, 127]. For the genus *Aspergillus*, the first sequencing projects started in the late 1990s leading to the publication of the first *Aspergillus* genome, namely *A. nidulans* in 2005 [128] followed by other milestone publications for *A. oryzae* [129], *Aspergillus fumigatus* [130] and *A. niger* [34]. As of 2014, the number of fully sequenced *Aspergillus* species reached 21, all being publically accessible via the database MycoCosm (http://jgi.doe.gov/fungi; [26]). Thanks to increased throughput, reduced cost and improved accuracy of next-generation sequencing technologies, there is no end of this trend within sight. A project with the ambitious title '1,000 Fungal Genomes' is currently running at the US Department of Energy Joint Genome Institute to sequence all major branches of the entire fungal tree of life [26], an initiative which can only be successful if the fungal scientific community continues to jointly develop powerful Web-based and wet-lab tools and pipelines for gene annotation and gene function analyses [131].

6.1 Opportunities and Challenges of Computational Genomics for Aspergillus

The availability of different genome sequences constitutes a rich resource for comparative genomics that generates new knowledge linking genotypes with phenotypic traits. Such comparative genomic studies have provided insights into the genetic basis for citric acid or protein production of *A. niger* strains as discussed in Chap. 5 [27]. Other comparative genomic studies disclosed CAZymes relevant for lignocellulose degradation in *A. nidulans*, *A. niger* and *A. oryzae* [132] and the closely related white-rot and brown-rot basidiomycetes *Ceriporiopsis subvermispora* and *Phanerochaete chrysosporium* [133] or predicted secondary metabolite gene clusters in *Aspergilli* and plant pathogenic filamentous fungi [134, 135].

The wealth of sequence data will without doubt further significantly enrich the portfolio of proteins, enzymes and metabolites to become produced by industrial *Aspergilli* or other industrial filamentous fungal production hosts. However, the challenge for the next decade is definitely no longer in sequence generation but data interpretation, data integration and methods for the high-throughput functional characterisation of predicted genes [136]. In the near future, next-generation sequencing approaches will be important for the correction of structural gene annotations 'just' by sequencing fungal transcriptomes [95], in addition, microarray-assisted transcriptomics will continue to be a powerful technology for the study or prediction of gene function as recently shown for the accurate assignment of secondary metabolite gene clusters in *A. nidulans* [137].

A list of currently available databases dedicated to the analysis of fungal genomes with a focus on *Aspergillus* projects is summarised in Table 1. These resources structure sequence information, provide annotations at different levels and connect gene and protein predictions with functional genomics data. What is important to make these databases sustainable? An important issue here is data conformity, in particular for functional annotations, which should have been generated with the same annotation pipeline in order to provide meaningful results in comparative genomics studies [26]. User-friendly intuitive interfaces are required to extend the acceptance and frequent use of already existing resources. Clearly, conversion of data types, identifiers and formats is a formidable task for scientists with a limited computational background, which will of course limit the use of existing powerful resources. Besides, a major risk for genomic resources can be seen in funding reductions. Expiration of funded projects periods or the lack of long-term community support will result in diversified and redundant resources, which in turn makes data conformity difficult.

In order to move towards conceptual models describing and predicting the link between *Aspergillus* growth, morphology and product formation, the integration of multiple -omics data types is key. Examples of association studies linking transcript profiles with proteomics, metabolomics data and genome-scale models have been given in the chapters above, some of which also applied guilt-by-association and enrichment/overrepresentation analyses to identify co-expression and gene regulatory networks in *A. niger* and *A. oryzae*. Further examples can be found in [14, 109, 138–141] to name but a few.

Computational methods guiding protein redesign have also been successfully used to improve protein production levels in *A. niger*. The proof-of-concept study was recently provided for two enzymes by van den Berg and colleagues [142, 143], who applied a machine-learning approach to identify DNA and protein features which discriminate between low- and high-level secretion rates based on the data for over 2,600 homologous or heterologous proteins expressed in *A. niger*. Using this strategy, both proteins were redesigned according to the calculated features of high-level secreted proteins. Up to 45 mutations were introduced which eventually led to a 10-fold increase in extracellular enzyme concentrations. A very impressive work, which can certainly become transferred to other proteins or features as the underlying methodology is generic [143].

6.2 Opportunities and Challenges of Functional Genomics for Aspergillus

98 % of the *A. niger* and *A. oryzae* genes are uncharacterised [98], urgently calling for high-throughput experimental techniques to elucidate their function. The systematic inactivation of predicted open reading frames by gene deletion or gene disruption and/or their controlled overexpression are two complementing approaches to study the function of genes. A respective genome-wide deletion

Table 1 Selected Web-based resources for *Aspergillus* genomics (as of 23 September 2014)

Genomics databases
• *Aspergillus* Genome Database (AspGD) [95] – http://www.aspgd.org/ – Genome, orthology and transcriptome data for comparative genomics of 20 *Aspergillus* species – Comprehensive manually curation of Gene Ontology annotation for selected *Aspergillus* species (*A. niger, A. nidulans, A. oryzae* and *A. fumigatus*)
• Central *Aspergillus* Data REpository (CADRE) [196] – http://www.cadre-genomes.org.uk/index.htm – Genome, orthology and transcriptome data for comparative genomics of nine *Aspergillus* genomes (eight species)
• Fungal genome initiative – http://www.broadinstitute.org/scientific-community/science/projects/fungal-genome-initiative/fungal-genome-initiative – Including a comparative database for eight *Aspergillus* species
• Ensemble fungi [197] – http://fungi.ensembl.org/index.html – Genome data for 52 fungal genomes including nine *Aspergillus* genomes from eight species
• MycoCosm [26] – http://genome.jgi-psf.org/programs/fungi/index.jsf – Genomics resource for the US Department of Energy Joint Genome Institute's '1,000 fungal genome project' – Comparative data for 394 fungal genomes including 21 *Aspergillus* genomes (19 species)
• FungiDB [198] – http://fungidb.org/fungidb/ – Comparative data for 64 fungal organisms including ten *Aspergillus* genomes from nine species
• Comparative fungal genomics platform [199] – http://cfgp.snu.ac.kr – 152 fungal genomes
• e-Fungi – http://www.cs.man.ac.uk/~cornell/eFungi/
• Seoul National University Genome Browser (SNUGB): [200] – http://genomebrowser.snu.ac.kr/ – A versatile genome browser supporting comparative and functional fungal genomics – 137 fungal genome datasets
• The Genomes OnLine Database (GOLD) [201] – http://www.genomesonline.org/ – Comprehensive database for genomic and metagenomic projects – version 4.0 contains information for more than 11,000 sequencing projects

(continued)

Table 1 (continued)

Databases and tools focusing primarily on functional annotation
• FungiFun version 2 [202] – https://elbe.hki-jena.de/fungifun/fungifun.php – Enrichment analysis of functional annotations including KEGG, GO and FunCAT – Comprises 321 species (functional annotation not available for all)
• FetGOat [139] – www.broadinstitute.org/fetgoat – Gene Ontology Enrichment analysis for ten *Aspergillus* strains from eight species
• Fungal secretome database [101] – http://fsd.snu.ac.kr/index.php?a=view – provides data for 217 fungal species
• Fungal secretome knowledge base [203] – http://proteomics.ysu.edu/secretomes/fungi.php – Prediction of secreted proteins applying different tools including (SignalP, WolfPsort,—Phobius and others) – 54 fungal species including eight *Aspergilli*
• Fungal transcription factor databases [204] – http://ftfd.snu.ac.kr/index.php?a=view – 178 fungal strains
• Fungal cytochrom P450 database [205] – http://passport.riceblast.snu.ac.kr/?t=P450 – 213 fungal genomes
Databases and tools for metabolic network/model analysis
• KEGG pathway database – http://www.genome.jp/kegg/pathway.html – Pathway maps for diverse organisms including seven *Aspergillus* species – 114 pathway maps for *A. niger* ranging from glycolysis to endocytosis
• BioMet toolbox version 2 [206] – http://biomet-toolbox.org/ – Web-based resources for integrative analysis of metabolic models and transcriptome data – Repository contains models for *A. nidulans, A. niger* and *A. oryzae*
• MEMOSys 2.0 [207] – https://memosys.i-med.ac.at/MEMOSys/home.seam – Database for genome-scale models and genomic data – Repository contains models for *A. nidulans, A. niger* and *A. oryzae*
• FAME [208] – http://f-a-m-e.org/ajax/page1.php#top – Flux analysis and modelling framework

(continued)

Table 1 (continued)

• MetaCyc/BioCyc pathway database [209] – http://metacyc.org/ – Exclusively experimentally determined pathways, curation and tight integration of data and references, non-redundant – Covers all domains of life – So far no filamentous fungi included but it provides a resource for network-based annotations
• Aspercyc – www.aspercyc.org.uk – Database for metabolic pathways for eight *Aspergillus* species
Databases and tools for genome-wide expression profiling
• GEO [210] – http://www.ncbi.nlm.nih.gov/geo – Microarray and RNAseq datasets – Web-based analysis of datasets with using the R interface GEO2R
• ArrayExpress [211] – http://www.ebi.ac.uk/arrayexpress/ – Microarray and RNAseq datasets
Tools for predicting secondary metabolism gene clusters
• SMURF [212] – http://jcvi.org/smurf/index.php – Prediction of secondary metabolite biosynthesis genes and pathways in fungal genomes – Based on protein domains and chromosomal position – Precomputed results available for 27 fungal species
• AntiSmash [213] – http://antismash.secondarymetabolites.org/ – Analysis platform for genome mining of SM biosynthetic gene clusters – Combination of protein domain searches and examination of neighbouring regions
Other databases
• Fung-growth – http://www.fung-growth.org/ – Surface growth profiles using 35 different carbon sources including monosaccharides and crude plant biomass
• FANTOM [214] – http://www.sysbio.se/Fantom/ – Functional and taxonomic analysis of metagenomes

project has been initiated for the model fungus *N. crassa* [144] and knockout mutants have been developed for more than two-third of the 10,000 *N. crassa* genes [145]. The basis for this achievement was high-throughput production of gene deletion cassettes, high transformation rates via electroporation of conidia and levels of homologous recombination equal or higher than 90 % [145]. In 2009, an effort from within the *Aspergillus* community was initiated to generate deletion constructs for all *A. nidulans* genes as well. Very recently, gene knockout constructs for about 93 % of the 10,560 predicted *A. nidulans* genes were provided to the community [146]. To prove the utility if these constructs, gene deletion strains were established for 128 predicted protein kinases which uncovered a function for previously uncharacterised kinases in vesicle trafficking, polar growth, septation and secondary metabolism [146].

Similar community-wide efforts for genome-deletion projects for other *Aspergilli* will hopefully become initiated soon as they are one of the most essential and invaluable resources for comprehensively understanding the genetic basis of their lifestyle. In fact, we have all necessary tools in hand for the industrial *Aspergillus* strains: (i) fast molecular cloning methods including fusion PCR, bipartite, in vivo yeast recombination, Golden Gate and Gibson assembly (for review see [147]), (ii) a broad collection of suitable promoters, selection systems and transformation methods including protoplast-mediated transformation, electroporation and shock wave assisted transformation (for details see [1, 4, 148–153]), and (iii) recipient strains with most efficient homologous recombination frequencies [154–158], (iv) a promoter system with which the function of essential genes can be studied [24]. The availability of these tools has made it now possible to lift functional genomics approaches to the higher, i.e. high-throughput level.

7 Mathematical Modelling of *Aspergillus* Cultivations

7.1 What Good Are Models?

As in many other disciplines as well, mathematical modelling of *Aspergillus* cultivations offers a substantial benefit in various aspects. In the most ambitious approach, the huge amount of physiological data obtained could be potentially lumped together to draw a comprehensive picture of morphology elucidating microscopic morphology, colony/pellet growth and interaction with the environment. This would finally allow for a directed control of morphology. Today, this goal seems to be too ambitious not only for *Aspergillus*, but for all fungi. However, less challenging objectives can be and have to be tackled with mathematical models.

By the very nature of living cells, experimental data are obtained from a highly dynamical system where the behaviour of a cell does not only depend on the actual stimuli but what had happened to the cell in the past. On a rational basis, such data can only be interpreted and condensed in the context of a dynamical model. For fungal organisms, interpretation of physiological data is even more challenging. Besides the compartmentalisation of biological functions in distinct organelles, space- and time-dependent distributions of organelles, other compounds of the cytoplasm, and of stimuli in and around a mycelium occur (see Chap. 4). Models may help here in analysing biochemical data which depends on spatial distribution as well. For technical applications, morphology influences culture broth viscosity, productivity and downstream processing. With image analysis, data quantifying morphology from *A. niger* can be linked to biomass-independent rheological data see [159, 160]. In solid-state cultivation, aerial hyphae can give rise to a non-uniform oxygen supply or an increased pressure drop in packed-bed cultivations of *A. oryzae* or *Rhizopus oligosporus* [161]. In this area, mathematical models are needed to predict rheological and other process data from morphology. If this is possible, such rheological data could be estimated at a very early stage of a process development before performing image analysis. Liu et al. [162] propose an interesting approach for *R. oryzae* where they predict the probability of pellet formation with a multiple logistic regression model. This method could be transferred to *Aspergillus*. Details and recent approaches to control morphology and productivity in liquid cultures by the addition of inorganic microparticles, changes of the conidia adhesion capacity, spore concentration, osmolality, pH, mechanical stress, etc., with *A. niger*, *A. terreus* and other strains can be found in [107, 108, 163–165].

In addition to the interpretation and compression of biological data, models can be finally exploited in powerful and versatile methods for process design, optimisation, closed-loop control and supervision. In this respect, cultivation processes have not yet reached the same maturity as traditional chemical processes [166]. This huge area of model-based applications, although very important, will not be considered in this short paragraph as it would deserve a separate review (see [167, 168] for an introduction). Here, we rather focus on the challenges coming from an interpretation of physiological data in light of a complex morphology. To this end, some more general comments concerning modelling of mycelial organisms seem appropriate before we direct our attention to some recent developments and necessary future investigations. The references given in what follows mainly cover publications of the last couple of years. Morphological modelling of fungi, however, goes back, at least, to the year 1967 [169]. Excellent reviews of the period in between can be found, e.g. in [107, 170, 171] or in [172] for filamentous bacteria such as *Streptomyces* species. The cited newer articles, moreover, give credit to other important results from literature which cannot be repeated here for sake of brevity.

7.2 What Can be Modelled?

Mathematical modelling of the morphology of mycelial organisms can focus on scales ranging from nanometers to meters or more [173, 174]. A variety of mechanisms and effects can be elucidated by means of mathematical models. Examples, ordered by a growing length scale, could be the study of exo- and endocytosis for polar growth, of tropism, the onset of new branches, anastomosis, organelle distribution in hyphae, or growth of hyphae over long inert habitats by intrahyphal translocation of nutrients, to name just a few potential applications. Typically, processes involved on a biological level will depend on time- and space-dependent environmental factors as well. A mycelium growing on a solid substrate will deplete it after a while, leading, e.g. to autolysis. A submerged, pellet-forming organism such as *A. niger* will see in its pellet centres a reduced substrate concentration, especially with respect to oxygen, when the size reaches a certain level [159, 172, 175–177]. This may lead to autolysis, enhanced susceptibility to mechanical stress, or improved productivity due to favourable nutrient concentrations inside the pellet. Here, mathematical models have to describe the particle, i.e. pellet size distribution [175] to predict its effect on productivity [178], and, for example, on viscosity, and, hence, on important parameters such as heat and mass transfer coefficients (see, e.g. [163] and references therein).

All morphological details are the result of highly regulated biochemical reactions on a metabolic level. Due to polarised, i.e. apical growth, and apical or subapical branching, metabolic fluxes will be space-dependent. Additionally, in fungi, a coordinated interplay between various organelles inside a hypha has to be considered to determine, e.g. the timing and location of branch initiation. This is far beyond what has been tackled so far with mathematical models for *Aspergillus* or other fungi. Even a comprehensive mathematical model of a subset of this multi-scale problem is still out of reach although more and more biological details on a molecular level are discovered.

7.3 Which Model When?

Before continuing this discussion, a classification of models seems necessary. In contrast to unstructured models, structured approaches do not describe the biotic phase by just one variable but by a number of selected compounds, real or hypothetical compartments, morphological compartments, or by a detailed description, e.g. of metabolic fluxes in the realm of systems biology [13]. For fungal cultivations, as for actinomycetes, growth and production described by unstructured or structured models can be further refined accounting for the space-dependent nutrient situation inside large mycelia and pellets [172, 179], and, equally important, in non-ideally mixed large-scale fermenters [166, 180]. For fungi, rather very few attempts with respect to morphologically structured approaches have been done so far. An old example goes back to Megee et al. [181] who have built up a model for hyphae of *A. awamori* including active, inactive hyphae, tips and substrates. Based on it, refinements were formulated later from different researchers, for example for *P. chrysogenum, Geotrichum candidum* and the actinomycete *S. hygroscopicus.* Examples and more references are given in [182–185].

Micro- or macro-morphological models to describe growth, branching and geometrical appearance of small and large mycelia and of pellets, respectively, can be formulated in a continuous or discrete manner or a combination of both (see [174, 183] for an extended discussion). Continuous models go back to the work of Edelstein and co-workers [184] who combined discrete elements into a continuous description of the evolution of the biomass density by the introduction of a tip density and apical growth. This approach, leading to a set of partial differential equations, is especially suited for studying larger mycelia or pellets in homogenous environments. It was subsequently refined by different researchers [173, 174, 185, 186]. However, the inclusion of -omics data in this kind of continuous approach is hardly possible.

Discrete models, on the other hand, often start from predefined lattices along which growth can occur (lattice-based models) [161, 185], from vector-based approaches [187], or from a geometrically unconstrained description [175]. Discrete models can be readily extended to hybrid ones to include continuous descriptions of nutrient transport outside the mycelium. If stochastic elements are included, appealing images are obtained from simulations that are almost indistinguishable from real microscopic pictures. A first model including deterministic and stochastic parts was already formulated in the year 1992 by Yang et al. [188]. It showed a striking similarity to microscopic images after they had been processed by image analysis algorithms. The model was calibrated with data from *S. tendae* and *G. candidum*. Later, Lejeune and Baron used it as a basis for *T. reesei* [177]. More recent examples describing the geometric form from different fungi or streptomycetes can be found in [179, 183, 189], to name just a few. Typically, the rules and kinetics used in these models do not have to rely on detailed biological knowledge to still produce 'realistic' images, i.e. they do not start from biological data from, e.g. *A. niger*. To give an example, the simulations shown in Fig. 6 just use the information that the apical growth rate in three dimensions is constant and that septa and branches are formed when a critical length of a hyphal compartment is obtained. Including a random growth direction in the simulation produces the results shown. Such basic model structures can now be used to include biological information to get more from the model than just a nice visual impression. Moreover, such models are needed to predict the macroscopic implications with respect to nutrient transport, rheology, etc. when, e.g. the branching rate is changed by a genetic manipulation [179].

A last group of models concentrates on individual hyphae to more accurately describe its growth rate or geometric form. Data from the -omics initiatives are missing here as well. A well-known example goes back to Bartnicki-Garcia et al.

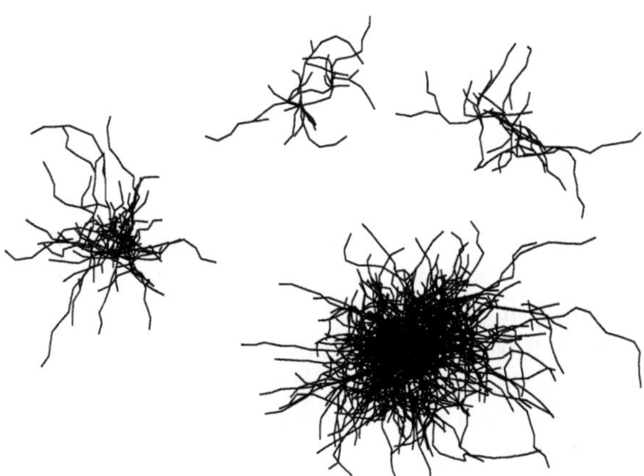

Fig. 6 2D projections of four simulated 3D mycelia using a very simple model. All mycelia start as a spore, grow with the same apical growth rate and only differ in age

[190]. They describe the geometrical form of an apex of various fungi with a simple model. It is based on a set of hypotheses how vesicles are transported from the Spitzenkörper to the wall. The model has been refined later to better account for the three-dimensional shape of a tip and the way vesicles are transported to the wall (see, e.g. [72, 191–193]). Recently, a quantitative model to describe the growth of an apex was introduced by Caballero-Lima et al. [74]. They assume that enzyme carrying vesicles fuse with the plasma membrane at the apex proportional to the local exocyst concentration. Realistic shapes of a hypha, however, are only obtained when these enzymes, e.g. (1,3)-ß-glucan synthase, are removed from the plasma membrane by endocytosis at some distance from the apex. This is one of the very few models which are built upon well-established cell biological processes and are supported by quantitative measurements. For the explanation of another micro-morphological appearance, Sugden et al. [194] argue that the long-range transport of material in hyphae is important for length growth. They explain it by a particle transport along a single, hypothetical microtubule extending over the whole length of a hypha. Monte Carlo simulation reveals an interesting result showing a raising concentration of the particles towards the tip as it is observed by laser scanning microscopy from other groups for organelles in fungi. In contrast, [195] explains length growth of *Phanerochaete velutina* mathematically by a turgor driven intrahyphal flow towards the tip. The model accounts for increasing radii of hyphae when the flow increases which can be shown experimentally.

7.4 What More Is to Come?

As in these last examples, morphological models are mostly based on a very simplified description of the underlying biochemical processes. It is our believe that such unstructured or phenomenologically structured models of small size describing individual hyphae, mycelia or pellets have already reached a high level of maturity. One of the main challenges for future work, in contrast, lies in the exploitation and explanation of Big Data information in view of space-dependent processes in hyphae leading to specific morphologies. A gap exists today between what is known and what can be used in mathematical models. When we proposed the first 'mechanistic' model more than 20 years ago [188], we had to assume hypothetical compounds in a hypha to describe an initial exponential and then constant growth rate of an individual branch as biological details for this process were unknown. Today, for fungi and for actinomycetes, a whelm of data is available. Interpreting this molecular and cell biological data on a mechanistic level and combining it with a morphological description seems to be a promising, though intricate route to follow. It is understood, however, that such a detailed description, for example on an organelle-basis, cannot be used to describe growth of larger mycelia and address process control issues. Therefore, as a next step, detailed models have to be reduced again. This reduction, however, will be based on more detailed, biologically better founded models. In the long run, such an approach might help to replace heuristic rules exploited today to model morphology by more physics- or bio-based expressions and close the circle towards a rational control of morphology.

8 Conclusions

The *Aspergillus* Big Data era creates entirely new opportunities and challenges. In order to make industrial *Aspergillus* strains completely transparent and productive for us, we need to address the challenges on different levels: (i) the Big Data analysis pipeline—data acquisition, extraction, integration, modelling and interpretation [12]—has to be approached entirely not only with a focus on data acquisition. We should not leave -omics data under-analysed but should constantly re-analyse and re-interpret them in view of 'old' and newly incoming -omics data from database repositories. Most importantly, data noisiness and heterogeneity of *Aspergillus* populations have to be taken into account when analysing these data. Likewise, co-regulation analyses are key to identify sub networks, to map transcriptional networks or to identify connections between signalling pathways. (ii) -Omics data have to be combined with in vivo life imaging studies to link time with space. In addition, functional genomics studies and 'good old' enzyme assays are key to ascribe functions to predicted proteins and enzymes. Furthermore, we should not lose the use of electromobility shift assays out of sight, to study and prove binding of predicted transcription factors to predicted promoter binding sites. (iii) And finally, there is a clear bottleneck in the number of fungal scientists working on *Aspergillus* and empowered to ask questions and finding the right experiments to answer them. We thus need to get many more students—biologist, bioinformaticians and engineers—interested and educated in fungal research. What we can offer? Exciting times for gold diggers!

References

1. Meyer V (2008) Genetic engineering of filamentous fungi–progress, obstacles and future trends. Biotechnol Adv 26:177–185
2. Lubertozzi D, Keasling JD (2009) Developing *Aspergillus* as a host for heterologous expression. Biotechnol Adv 27:53–75
3. Ward OP (2011) Production of recombinant proteins by filamentous fungi. Biotechnol Adv 30:1119–1139
4. Fiedler MRM, Nitsche BM, Franziska W, Meyer V (2013) *Aspergillus*: a cell factory with unlimited prospects. In: Gupta VK, Schmoll M, Maki M (eds) Appl Microb Eng. CRC Press, Taylor & Francis Group, Boca Raton, London, pp 1–51
5. Read ND (2011) Exocytosis and growth do not occur only at hyphal tips. Mol Microbiol 81:4–7
6. Hayakawa Y, Ishikawa E, Shoji J-Y, Nakano H, Kitamoto K (2011) Septum-directed secretion in the filamentous fungus *Aspergillus oryzae*. Mol Microbiol 81:40–55

7. Seidl V, Seibel C, Kubicek CP, Schmoll M (2009) Sexual development in the industrial workhorse *Trichoderma reesei*. Proc Natl Acad Sci USA 106:13909–13914
8. Böhm J, Hoff B, O'Gorman CM, Wolfers S, Klix V, Binger D, Zadra I, Kürnsteiner H, Pöggeler S, Dyer PS, Kück U (2013) Sexual reproduction and mating-type-mediated strain development in the penicillin-producing fungus *Penicillium chrysogenum*. Proc Natl Acad Sci USA 110:1476–1481
9. Richter L, Wanka F, Boecker S, Storm D, Kurt T, Vural Ö, Süßmuth R, Meyer V (2014) Engineering of *Aspergillus niger* for the production of secondary metabolites. Fungal Biol Biotechnol 1:4
10. Li A, van Luijk N, ter Beek M, Caspers M, Punt P, van der Werf M (2011) A clone-based transcriptomics approach for the identification of genes relevant for itaconic acid production in *Aspergillus*. Fungal Genet Biol 48:602–611
11. Van der Straat L, Vernooij M, Lammers M, van den Berg W, Schonewille T, Cordewener J, van der Meer I, Koops A, de Graaff LH (2014) Expression of the *Aspergillus terreus* itaconic acid biosynthesis cluster in *Aspergillus niger*. Microb Cell Fact 13:11
12. Challenges and opportunities with big data challenges and opportunities with big data. http://www.cra.org/ccc/files/docs/init/bigdatawhitepaper.pdf
13. Caspeta L, Nielsen J (2013) Toward systems metabolic engineering of *Aspergillus* and *Pichia* species for the production of chemicals and biofuels. Biotechnol J 8:534–544
14. Nitsche BM, Meyer V (2014) Transcriptomics of industrial filamentous fungi: a new view on regulation, physiology, and application. In: Nowrousian M (ed) Fungal genomics, vol 13. Berlin, Springer, pp 209–232
15. Statins: a success story involving FDA, academia and industry. http://www.fda.gov/AboutFDA/WhatWeDo/History/ProductRegulation/SelectionsFromFDLIUpdateSeriesonFDAHistory/ucm082054.htm
16. Gusakov AV (2011) Alternatives to *Trichoderma reesei* in biofuel production. Trends Biotechnol 29:419–425
17. Pensupa N, Jin M, Kokolski M, Archer DB, Du C (2013) A solid state fungal fermentation-based strategy for the hydrolysis of wheat straw. Bioresour Technol 149:261–267
18. Kim S, Dale BE (2004) Global potential bioethanol production from wasted crops and crop residues. Biomass Bioenerg 26:361–375
19. Kolasa M, Ahring BK, Lübeck PS, Lübeck M (2014) Co-cultivation of *Trichoderma reesei* RutC30 with three black *Aspergillus* strains facilitates efficient hydrolysis of pretreated wheat straw and shows promises for on-site enzyme production. Bioresour Technol 169:143–148
20. Yamada R, Yoshie T, Wakai S, Asai-Nakashima N, Okazaki F, Ogino C, Hisada H, Tsutsumi H, Hata Y, Kondo A (2014) *Aspergillus oryzae*-based cell factory for direct kojic acid production from cellulose. Microb Cell Fact 13:71
21. Klement T, Büchs J (2013) Itaconic acid–a biotechnological process in change. Bioresour Technol 135:422–431
22. Li A, Caspers M, Punt P (2013) A systems biology approach for the identification of target genes for the improvement of itaconic acid production in *Aspergillus* species. BMC Res Notes 6:505
23. Van der Straat L, de Graaff LH (2014) Pathway transfer in fungi: transporters are the key to success. Bioengineered 5
24. Meyer V, Wanka F, van Gent J, Arentshorst M, van den Hondel CAMJJ, Ram AFJ (2011) Fungal gene expression on demand: an inducible, tunable, and metabolism-independent expression system for *Aspergillus niger*. Appl Environ Microbiol 77:2975–2983
25. Meyer V, Wu B, Ram AFJ (2011) *Aspergillus* as a multi-purpose cell factory: current status and perspectives. Biotechnol Lett 33:469–476
26. Grigoriev I V, Nikitin R, Haridas S, Kuo A, Ohm R, Otillar R, Riley R, Salamov A, Zhao X, Korzeniewski F, Smirnova T, Nordberg H, Dubchak I, Shabalov I (2014) MycoCosm portal: gearing up for 1000 fungal genomes. Nucleic Acids Res 42(Database issue):D699–D704

27. Andersen MR, Salazar MP, Schaap PJ, van de Vondervoort PJI, Culley D, Thykaer J, Frisvad JC, Nielsen KF, Albang R, Albermann K, Berka RM, Braus GH, Braus-Stromeyer SA, Corrochano LM, Dai Z, van Dijck PWM, Hofmann G, Lasure LL, Magnuson JK, Menke H, Meijer M, Meijer SL, Nielsen JB, Nielsen ML, van Ooyen AJJ, Pel HJ, Poulsen L, Samson RA, Stam H, Tsang A et al (2011) Comparative genomics of citric-acid-producing *Aspergillus niger* ATCC 1015 versus enzyme-producing CBS 513.88. Genome Res 21:885–897
28. Jørgensen TR, Nielsen KF, Arentshorst M, Park J, van den Hondel C a, Frisvad JC, Ram AF (2011) Submerged conidiation and product formation by Aspergillus niger at low specific growth rates are affected in aerial developmental mutants. Appl Environ Microbiol 77:5270–5277
29. Delmas S, Pullan ST, Gaddipati S, Kokolski M, Malla S, Blythe MJ, Ibbett R, Campbell M, Liddell S, Aboobaker A, Tucker GA, Archer DB (2012) Uncovering the genome-wide transcriptional responses of the filamentous fungus *Aspergillus niger* to lignocellulose using RNA sequencing PLoS Genet 8:e1002875
30. Hayer K, Stratford M, Archer DB (2014) Germination of *Aspergillus niger* conidia is triggered by nitrogen compounds related to L-amino acids. Appl Environ Microbiol 80:6046–6053
31. Hayer K, Stratford M, Archer DB (2013) Structural features of sugars that trigger or support conidial germination in the filamentous fungus *Aspergillus niger*. Appl Environ Microbiol 79:6924–6931
32. Novodvorska M, Hayer K, Pullan ST, Wilson R, Blythe MJ, Stam H, Stratford M, Archer DB (2013) Trancriptional landscape of *Aspergillus niger* at breaking of conidial dormancy revealed by RNA-sequencing. BMC Genomics 14:246
33. Van Leeuwen MR, Krijgsheld P, Bleichrodt R, Menke H, Stam H, Stark J, Wösten HAB, Dijksterhuis J (2013) Germination of conidia of *Aspergillus niger* is accompanied by major changes in RNA profiles. Stud Mycol 74:59–70
34. Pel HJ, de Winde JH, Archer DB, Dyer PS, Hofmann G, Schaap PJ, Turner G, de Vries RP, Albang R, Albermann K, Andersen MR, Bendtsen JD, Benen JAE, van den Berg M, Breestraat S, Caddick MX, Contreras R, Cornell M, Coutinho PM, Danchin EGJ, Debets AJM, Dekker P, van Dijck PWM, van Dijk A, Dijkhuizen L, Driessen AJM, D'Enfert C, Geysens S, Goosen C, Groot GSP et al (2007) Genome sequencing and analysis of the versatile cell factory *Aspergillus niger* CBS 513.88. Nat Biotechnol 25:221–231
35. Jedd G, Pieuchot L (2012) Multiple modes for gatekeeping at fungal cell-to-cell channels. Mol Microbiol 86:1291–1294
36. Harris SD, Turner G, Meyer V, Espeso EA, Specht T, Takeshita N, Helmstedt K (2009) Morphology and development in *Aspergillus nidulans*: a complex puzzle. Fungal Genet Biol 46(Suppl 1):S82–S92
37. Levin AM, de Vries RP, Conesa A, de Bekker C, Talon M, Menke HH, van Peij NNME, Wösten HAB (2007) Spatial differentiation in the vegetative mycelium of *Aspergillus niger*. Eukaryot Cell 6:2311–2322
38. Krijgsheld P, Altelaar AFM, Post H, Ringrose JH, Müller WH, Heck AJR, Wösten HAB (2012) Spatially resolving the secretome within the mycelium of the cell factory *Aspergillus niger*. J Proteome Res 11:2807–2818
39. Gordon CL, Khalaj V, Ram a F, Archer DB, Brookman JL, Trinci a P, Jeenes DJ, Doonan JH, Wells B, Punt PJ, van den Hondel CA, Robson GD (2000) Glucoamylase::green fluorescent protein fusions to monitor protein secretion in *Aspergillus niger*. *Microbiology* 146(2):415–426
40. Vinck A, de Bekker C, Ossin A, Ohm RA, de Vries RP, Wösten HAB (2011) Heterogenic expression of genes encoding secreted proteins at the periphery of *Aspergillus niger* colonies. Environ Microbiol 13:216–225
41. De Bekker C, van Veluw GJ, Vinck A, Wiebenga LA, Wösten HAB (2011) Heterogeneity of *Aspergillus niger* microcolonies in liquid shaken cultures. Appl Environ Microbiol 77:1263–1267

42. Van Veluw GJ, Teertstra WR, de Bekker C, Vinck A, van Beek N, Muller WH, Arentshorst M, van der Mei HC, Ram AFJ, Dijksterhuis J, Wösten HAB (2013) Heterogeneity in liquid shaken cultures of *Aspergillus niger* inoculated with melanised conidia or conidia of pigmentation mutants. Stud Mycol 74:47–57
43. Wösten HAB, van Veluw GJ, de Bekker C, Krijgsheld P (2013) Heterogeneity in the mycelium: implications for the use of fungi as cell factories. Biotechnol Lett 35:1155–1164
44. Nitsche BM, Jørgensen TR, Akeroyd M, Meyer V, Ram AFJ (2012) The carbon starvation response of *Aspergillus niger* during submerged cultivation: insights from the transcriptome and secretome. BMC Genomics 13:380
45. Nitsche BM, Burggraaf-van Welzen AM, Lamers G, Meyer V, Ram AFJ (2013) Autophagy promotes survival in aging submerged cultures of the filamentous fungus *Aspergillus niger*. Appl Microbiol Biotechnol 97:8205–8218
46. Shoji J, Craven KD (2011) Autophagy in basal hyphal compartments: A green strategy of great recyclers. Fungal Biol Rev 25:79–83
47. Adams TH, Wieser JK, Yu JH (1998) Asexual sporulation in *Aspergillus nidulans*. Microbiol Mol Biol Rev 62:35–54
48. Krijgsheld P, Bleichrodt R, van Veluw GJ, Wang F, Müller WH, Dijksterhuis J, Wösten HAB (2013) Development in *Aspergillus*. Stud Mycol 74:1–29
49. Harris SD (2012) Evolution of modular conidiophore development in the aspergilli. Ann N Y Acad Sci 1273:1–6
50. Jørgensen TR, Nitsche BM, Lamers GE, Arentshorst M, van den Hondel CA, Ram AF (2010) Transcriptomic insights into the physiology of *Aspergillus niger* approaching a specific growth rate of zero. Appl Environ Microbiol 76:5344–5355
51. Metz B, Seidl-Seiboth V, Haarmann T, Kopchinskiy A, Lorenz P, Seiboth B, Kubicek CP (2011) Expression of biomass-degrading enzymes is a major event during conidium development in *Trichoderma reesei*. Eukaryot Cell 10:1527–1535
52. Van Munster JM, Nitsche BM, Krijgsheld P, van Wijk A, Dijkhuizen L, Wösten HA, Ram AF, van der Maarel MJEC (2013) Chitinases CtcB and CfcI modify the cell wall in sporulating aerial mycelium of *Aspergillus niger*. Microbiology 159(9):1853–1867
53. Brakhage AA, Schroeckh V (2011) Fungal secondary metabolites—strategies to activate silent gene clusters. Fungal Genet Biol 48:15–22
54. Brakhage AA (2013) Regulation of fungal secondary metabolism. Nat Rev Microbiol 11:21–32
55. Wiemann P, Keller NP (2014) Strategies for mining fungal natural products. J Ind Microbiol Biotechnol 41:301–313
56. Van Munster JM, Daly P, Delmas S, Pullan ST, Blythe MJ, Malla S, Kokolski M, Noltorp ECM, Wennberg K, Fetherston R, Beniston R, Yu X, Dupree P, Archer DB (2014) The role of carbon starvation in the induction of enzymes that degrade plant-derived carbohydrates in *Aspergillus niger*. Fungal Genet Biol 72:34–47
57. Andersen MR, Giese M, de Vries RP, Nielsen J (2012) Mapping the polysaccaride degradation potential of *Aspergillus niger*. BMC Genomics 13:313
58. De Souza WR, de Gouvea PF, Savoldi M, Malavazi I, de Souza Bernardes LA, Goldman MHS, de Vries RP, de Castro Oliveira J V, Goldman GH (2011) Transcriptome analysis of *Aspergillus niger* grown on sugarcane bagasse. Biotechnol Biofuels 4:40
59. Banuett F, Herskowitz I (1989) Different a alleles of *Ustilago maydis* are necessary for maintenance of filamentous growth but not for meiosis. Proc Natl Acad Sci USA 86:5878–5882
60. Tey WK, North AJ, Reyes JL, Lu YF, Jedd G (2005) Polarized gene expression determines woronin body formation at the leading edge of the fungal colony. Mol Biol Cell 16:2651–2659
61. Levin AM, de Vries RP, Conesa A, de Bekker C, Talon M, Menke HH, van Peij NNME, Wösten HAB (2007) Spatial differentiation in the vegetative mycelium of *Aspergillus niger*. Eukaryot Cell 6:2311–2322

62. Bleichrodt R-J, van Veluw GJ, Recter B, Maruyama J-I, Kitamoto K, Wösten HAB (2012) Hyphal heterogeneity in *Aspergillus oryzae* is the result of dynamic closure of septa by Woronin bodies. Mol Microbiol 86:1334–1344
63. Taheri-Talesh N, Horio T, Araujo-baza L, Dou X, Espeso EA, Pen MA, Osmani SA, Oakley BR (2008) The tip growth apparatus of *Aspergillus nidulans*. Mol Biol Cell 19:1439–1449
64. Lew RR (2011) How does a hypha grow? The biophysics of pressurized growth in fungi. Nat Rev Microbiol 9:509–518
65. Plamann M (2009) Cytoplasmic streaming in *Neurospora*: disperse the plug to increase the flow? PLoS Genet 5:e1000526
66. Abadeh A, Lew RR (2013) Mass flow and velocity profiles in Neurospora hyphae: partial plug flow dominates intra-hyphal transport. Microbiology 159(11):2386–2394
67. Riquelme M (2013) Tip growth in filamentous fungi: a road trip to the apex. Annu Rev Microbiol 67:587–609
68. Takeshita N, Manck R, Grün N, de Vega SH, Fischer R (2014) Interdependence of the actin and the microtubule cytoskeleton during fungal growth. Curr Opin Microbiol 20C:34–41
69. Riquelme M, Sánchez-León E (2014) The Spitzenkörper: a choreographer of fungal growth and morphogenesis. Curr Opin Microbiol 20C:27–33
70. Free SJ (2013) Fungal cell wall organization and biosynthesis. Adv Genet 81:33–82
71. Hettema EH, Erdmann R, van der Klei I, Veenhuis M (2014) Evolving models for peroxisome biogenesis. Curr Opin Cell Biol 29C:25–30
72. Gierz G, Bartnicki-Garcia S (2001) A three-dimensional model of fungal morphogenesis based on the vesicle supply center concept. J Theor Biol 208:151–164
73. Bartnicki-Garcia S, Bracker CE, Gierz G, López-Franco R, Lu H (2000) Mapping the growth of fungal hyphae: orthogonal cell wall expansion during tip growth and the role of turgor. Biophys J 79:2382–2390
74. Caballero-Lima D, Kaneva IN, Watton SP, Sudbery PE, Craven CJ (2013) The spatial distribution of the exocyst and actin cortical patches is sufficient to organize hyphal tip growth. Eukaryot Cell 12:998–1008
75. Dijksterhuis J, Molenaar D (2013) Vesicle trafficking via the Spitzenkörper during hyphal tip growth in *Rhizoctonia solani*. Antonie Van Leeuwenhoek 103:921–931
76. Kwon MJ, Arentshorst M, Fiedler M, de Groen FLM, Punt PJ, Meyer V, Ram AFJ (2014) Molecular genetic analysis of vesicular transport in *Aspergillus niger* reveals partial conservation of the molecular mechanism of exocytosis in fungi. Microbiology 160 (2):316–329
77. Higuchi Y, Ashwin P, Roger Y, Steinberg G (2014) Early endosome motility spatially organizes polysome distribution. J Cell Biol 204:343–357
78. Steinberg G (2014) Endocytosis and early endosome motility in filamentous fungi. Curr Opin Microbiol 20C:10–18
79. Sudbery P (2011) Fluorescent proteins illuminate the structure and function of the hyphal tip apparatus. Fungal Genet Biol 48:849–857
80. Durand H, Clanet M, Tiraby G (1988) Genetic improvement of *Trichoderma reesei* for large scale cellulase production. Enzyme Microb Technol 10:341–346
81. Finkelstein DB (1987) Improvement of enzyme production in *Aspergillus*. Antonie Van Leeuwenhoek 53:349–352
82. Su X, Schmitz G, Zhang M, Mackie RI, Cann IKO (2012) Heterologous gene expression in filamentous fungi. Adv Appl Microbiol 81:1–61
83. Pakula TM, Laxell M, Huuskonen A, Uusitalo J, Saloheimo M, Penttilä M (2003) The effects of drugs inhibiting protein secretion in the filamentous fungus *Trichoderma reesei*. Evidence for down-regulation of genes that encode secreted proteins in the stressed cells. J Biol Chem 278:45011–45020
84. Kwon MJ, Jørgensen TR, Nitsche BM, Arentshorst M, Park J, Ram AFJ, Meyer V (2012) The transcriptomic fingerprint of glucoamylase over-expression in *Aspergillus niger*. BMC Genomics 13:701

85. Van den Berg BA, Reinders MJT, Hulsman M, Wu L, Pel HJ, Roubos JA, de Ridder D (2012) Exploring sequence characteristics related to high-level production of secreted proteins in *Aspergillus niger*. PLoS One 7:e45869
86. Shoji J-Y, Kikuma T, Kitamoto K (2014) Vesicle trafficking, organelle functions, and unconventional secretion in fungal physiology and pathogenicity. Curr Opin Microbiol 20C:1–9
87. Le Bihan M-C, Bigot A, Jensen SS, Dennis JL, Rogowska-Wrzesinska A, Lainé J, Gache V, Furling D, Jensen ON, Voit T, Mouly V, Coulton GR, Butler-Browne G (2012) In-depth analysis of the secretome identifies three major independent secretory pathways in differentiating human myoblasts. J Proteomics 77:344–356
88. Steringer JP, Müller H-M, Nickel W (2014) Unconventional secretion of fibroblast growth factor 2-a novel type of protein translocation across membranes? J Mol Biol. doi: 10.1016/j.jmb.2014.07.012
89. Ding Y, Robinson DG, Jiang L (2014) Unconventional protein secretion (UPS) pathways in plants. Curr Opin Cell Biol 29:107–115
90. De Marchis F, Bellucci M, Pompa A (2013) Unconventional pathways of secretory plant proteins from the endoplasmic reticulum to the vacuole bypassing the Golgi complex. Plant Signal Behav 8:piie25129
91. Fajardo-Somera RA, Bowman B, Riquelme M (2013) The plasma membrane proton pump PMA-1 is incorporated into distal parts of the hyphae independently of the Spitzenkörper in Neurospora crassa. Eukaryot Cell 12:1097–1105
92. Riquelme M, Bartnicki-García S, González-Prieto JM, Sánchez-León E, Verdín-Ramos JA, Beltrán-Aguilar A, Freitag M (2007) Spitzenkorper localization and intracellular traffic of green fluorescent protein-labeled CHS-3 and CHS-6 chitin synthases in living hyphae of *Neurospora crassa*. Eukaryot Cell 6:1853–1864
93. Stock J, Sarkari P, Kreibich S, Brefort T, Feldbrügge M, Schipper K (2012) Applying unconventional secretion of the endochitinase Cts1 to export heterologous proteins in *Ustilago maydis*. J Biotechnol 161:80–91
94. Arnaud MB, Cerqueira GC, Inglis DO, Skrzypek MS, Binkley J, Chibucos MC, Crabtree J, Howarth C, Orvis J, Shah P, Wymore F, Binkley G, Miyasato SR, Simison M, Sherlock G, Wortman JR (2012) The *Aspergillus* Genome Database (AspGD): recent developments in comprehensive multispecies curation, comparative genomics and community resources. Nucleic Acids Res 40(Database issue):D653–9
95. Cerqueira GC, Arnaud MB, Inglis DO, Skrzypek MS, Binkley G, Simison M, Miyasato SR, Binkley J, Orvis J, Shah P, Wymore F, Sherlock G, Wortman JR (2014) The *Aspergillus* Genome database: multispecies curation and incorporation of RNA-Seq data to improve structural gene annotations. *Nucleic Acids Res* 42(Database issue):D705–10
96. Andersen MR, Nielsen ML, Nielsen J (2008) Metabolic model integration of the bibliome, genome, metabolome and reactome of *Aspergillus niger*. Mol Syst Biol 4:178
97. Vongsangnak W, Olsen P, Hansen K, Krogsgaard S, Nielsen J (2008) Improved annotation through genome-scale metabolic modeling of *Aspergillus oryzae*. BMC Genomics 9:245
98. Andersen MR (2014) Elucidation of primary metabolic pathways in *Aspergillus* species: Orphaned research in characterizing orphan genes. Brief Funct Genomics 13:451–455
99. Battaglia E, Visser L, Nijssen A, van Veluw GJ, Wösten HAB, de Vries RP (2011) Analysis of regulation of pentose utilisation in *Aspergillus niger* reveals evolutionary adaptations in Eurotiales. *Stud Mycol* 69:31–38
100. Liu L, Feizi A, Österlund T, Hjort C, Nielsen J (2014) Genome-scale analysis of the high-efficient protein secretion system of *Aspergillus oryzae*. BMC Syst Biol 8:73
101. Choi J, Park J, Kim D, Jung K, Kang S, Lee Y-H (2010) Fungal secretome database: integrated platform for annotation of fungal secretomes. BMC Genomics 11:105
102. Sanchez JF, Somoza AD, Keller NP, Wang CCC (2012) Advances in *Aspergillus* secondary metabolite research in the post-genomic era. Nat Prod Rep 29:351–371

103. Peñalva MA, Galindo A, Abenza JF, Pinar M, Calcagno-Pizarelli AM, Arst HN, Pantazopoulou A (2012) Searching for gold beyond mitosis: Mining intracellular membrane traffic in *Aspergillus nidulans*. Cell Logist 2:2–14
104. Juvvadi PR, Lamoth F, Steinbach WJ (2014) Calcineurin-mediated regulation of hyphal growth, septation, and virulence in *Aspergillus fumigatus*. Mycopathologia 178:341–348
105. Martín JF (2014) Calcium-containing phosphopeptides pave the secretory pathway for efficient protein traffic and secretion in fungi. Microb Cell Fact 13:117
106. Malavazi I, Goldman GH, Brown NA (2014) The importance of connections between the cell wall integrity pathway and the unfolded protein response in filamentous fungi. Brief Funct Genomics 13:456–470
107. Papagianni M (2004) Fungal morphology and metabolite production in submerged mycelial processes. Biotechnol Adv 22:189–259
108. Colin VL, Baigorí MD, Pera LM (2013) Tailoring fungal morphology of *Aspergillus niger* MYA 135 by altering the hyphal morphology and the conidia adhesion capacity: biotechnological applications. AMB Express 3:27
109. Meyer V, Arentshorst M, Flitter SJ, Nitsche BM, Kwon MJ, Reynaga-Peña CG, Bartnicki-Garcia S, van den Hondel C a MJJ, Ram AFJ (2009) Reconstruction of signaling networks regulating fungal morphogenesis by transcriptomics. *Eukaryot Cell* 8:1677–1691
110. McIntyre M, Müller C, Dynesen J, Nielsen J (2001) Metabolic engineering of the morphology of *Aspergillus*. Adv Biochem Eng Biotechnol 73:103–128
111. Muller C, McIntyre M, Hansen K, Nielsen J (2002) Metabolic Engineering of the Morphology of *Aspergillus oryzae* by altering chitin synthesis. Appl Environ Microbiol 68:1827–1836
112. Te Biesebeke R, Record E, van Biezen N, Heerikhuisen M, Franken A, Punt PJ, van den Hondel CAMJJ (2005) Branching mutants of *Aspergillus oryzae* with improved amylase and protease production on solid substrates. Appl Microbiol Biotechnol 69:44–50
113. Wongwicharn A, McNeil B, Harvey LM (1999) Effect of oxygen enrichment on morphology, growth, and heterologous protein production in chemostat cultures of *Aspergillus niger* B1-D. Biotechnol Bioeng 65:416–424
114. Kwon MJ, Nitsche BM, Arentshorst M, Jørgensen TR, Ram AFJ, Meyer V (2013) The transcriptomic signature of RacA activation and inactivation provides new insights into the morphogenetic network of *Aspergillus niger*. PLoS One 8:e68946
115. Harris SD (2008) Branching of fungal hyphae: regulation, mechanisms and comparison with other branching systems. Mycologia 100:823–832
116. Meyer V, Minkwitz S, Schütze T, Hondel CAMJJ Van Den, Ram AFJ (2010) The *Aspergillus niger* RmsA protein. A node in a genetic network? Commun Integr Biol 3:195–197
117. Kwon MJ, Arentshorst M, Roos ED, van den Hondel C a MJJ, Meyer V, Ram AFJ (2011) Functional characterization of Rho GTPases in *Aspergillus niger* uncovers conserved and diverged roles of Rho proteins within filamentous fungi. Mol Microbiol 79:1151–1167
118. Harris SD (2011) Cdc42/Rho GTPases in fungi: variations on a common theme. Mol Microbiol 79:1123–1127
119. Scazzocchio C (2014) Fungal biology in the post-genomic era. Fungal Biol Biotechnol 2014:1
120. Krishnan K, Ren Z, Losada L, Nierman WC, Lu LJ, Askew DS (2014) Polysome profiling reveals broad translatome remodeling during endoplasmic reticulum (ER) stress in the pathogenic fungus *Aspergillus fumigatus*. BMC Genomics 15:159
121. Ramsubramaniam N, Harris SD, Marten MR (2014) The phosphoproteome of *Aspergillus nidulans* reveals functional association with cellular processes involved in morphology and secretion. Proteomics 14:2454–2459
122. Schachtschabel D, Arentshorst M, Nitsche BM, Morris S, Nielsen KF, van den Hondel CAMJJ, Klis FM, Ram AFJ (2013) The transcriptional repressor TupA in *Aspergillus niger* is involved in controlling gene expression related to cell wall biosynthesis, development, and nitrogen source availability. PLoS One 8:e78102

123. Brown NA, Ries LNA, Goldman GH (2014) How nutritional status signalling coordinates metabolism and lignocellulolytic enzyme secretion. Fungal Genet Biol 72:48–63
124. Conrad M, Schothorst J, Kankipati HN, Van Zeebroeck G, Rubio-Texeira M, Thevelein JM (2014) Nutrient sensing and signaling in the yeast *Saccharomyces cerevisiae. FEMS Microbiol Rev* 38:254–299
125. Paul JA, Barati MT, Cooper M, Perlin MH (2014) Physical and genetic interaction between ammonium transporters and the signaling protein, Rho1, in the plant pathogen *Ustilago maydis. Eukaryot Cell*
126. Goffeau A, Barrell BG, Bussey H, Davis RW, Dujon B, Feldmann H, Galibert F, Hoheisel JD, Jacq C, Johnston M, Louis EJ, Mewes HW, Murakami Y, Philippsen P, Tettelin H, Oliver SG (1996) Life with 6000 genes. Science 274(546):563–567
127. Aguilar-Pontes MV, de Vries RP, Zhou M (2014) (Post-)Genomics approaches in fungal research. Brief Funct Genomics 72:73–81
128. Galagan JE, Calvo SE, Cuomo C, Ma L-J, Wortman JR, Batzoglou S, Lee S-I, Baştürkmen M, Spevak CC, Clutterbuck J, Kapitonov V, Jurka J, Scazzocchio C, Farman M, Butler J, Purcell S, Harris S, Braus GH, Draht O, Busch S, D'Enfert C, Bouchier C, Goldman GH, Bell-Pedersen D, Griffiths-Jones S, Doonan JH, Yu J, Vienken K, Pain A, Freitag M et al (2005) Sequencing of *Aspergillus nidulans* and comparative analysis with *A. fumigatus* and *A. oryzae*. Nature 438:1105–1115
129. Machida M, Asai K, Sano M, Tanaka T, Kumagai T, Terai G, Kusumoto K-I, Arima T, Akita O, Kashiwagi Y, Abe K, Gomi K, Horiuchi H, Kitamoto K, Kobayashi T, Takeuchi M, Denning DW, Galagan JE, Nierman WC, Yu J, Archer DB, Bennett JW, Bhatnagar D, Cleveland TE, Fedorova ND, Gotoh O, Horikawa H, Hosoyama A, Ichinomiya M, Igarashi R et al (2005) Genome sequencing and analysis of *Aspergillus oryzae*. Nature 438:1157–1161
130. Nierman WC, Pain A, Anderson MJ, Wortman JR, Kim HS, Arroyo J, Berriman M, Abe K, Archer DB, Bermejo C, Bennett J, Bowyer P, Chen D, Collins M, Coulsen R, Davies R, Dyer PS, Farman M, Fedorova N, Fedorova N, Feldblyum TV, Fischer R, Fosker N, Fraser A, García JL, García MJ, Goble A, Goldman GH, Gomi K, Griffith-Jones S et al (2005) Genomic sequence of the pathogenic and allergenic filamentous fungus *Aspergillus fumigatus*. Nature 438:1151–1156
131. Hibbett DS, Stajich JE, Spatafora JW (2013) Toward genome-enabled mycology. Mycologia 105:1339–1349
132. Coutinho PM, Andersen MR, Kolenova K, vanKuyk PA, Benoit I, Gruben BS, Trejo-Aguilar B, Visser H, van Solingen P, Pakula T, Seiboth B, Battaglia E, Aguilar-Osorio G, de Jong JF, Ohm RA, Aguilar M, Henrissat B, Nielsen J, Stålbrand H, de Vries RP (2009) Post-genomic insights into the plant polysaccharide degradation potential of *Aspergillus nidulans* and comparison to *Aspergillus niger* and *Aspergillus oryzae*. Fungal Genet Biol 46(1): S161–S169
133. Fernandez-Fueyo E, Ruiz-Dueñas FJ, Ferreira P, Floudas D, Hibbett DS, Canessa P, Larrondo LF, James TY, Seelenfreund D, Lobos S, Polanco R, Tello M, Honda Y, Watanabe T, Watanabe T, Ryu JS, San RJ, Kubicek CP, Schmoll M, Gaskell J, Hammel KE, St John FJ, Vanden Wymelenberg A, Sabat G, Splinter BonDurant S, Syed K, Yadav JS, Doddapaneni H, Subramanian V, Lavín JL, et al (2012) Comparative genomics of *Ceriporiopsis subvermispora* and *Phanerochaete chrysosporium* provide insight into selective ligninolysis. Proc Natl Acad Sci USA 109:5458–5463
134. Wiemann P, Guo C-J, Palmer JM, Sekonyela R, Wang CCC, Keller NP (2013) Prototype of an intertwined secondary-metabolite supercluster. Proc Natl Acad Sci USA 110:17065–17070
135. Takeda I, Umemura M, Koike H, Asai K, Machida M (2014) Motif-independent prediction of a secondary metabolism gene cluster using comparative genomics: application to sequenced genomes of *Aspergillus* and ten other filamentous fungal species. DNA Res 21:447–457
136. Ohm RA, Riley R, Salamov A, Min B, Choi I-G, Grigoriev IV (2014) Genomics of wood-degrading fungi. Fungal Genet Biol 72:82–90

137. Andersen MR, Nielsen JB, Klitgaard A, Petersen LM, Zachariasen M, Hansen TJ, Blicher LH, Gotfredsen CH, Larsen TO, Nielsen KF, Mortensen UH (2013) Accurate prediction of secondary metabolite gene clusters in filamentous fungi. Proc Natl Acad Sci USA 110: E99–107
138. Andersen MR, Lehmann L, Nielsen J (2009) Systemic analysis of the response of *Aspergillus niger* to ambient pH. Genome Biol 10:R47
139. Nitsche BM, Crabtree J, Cerqueira GC, Meyer V, Ram AFJ, Wortman JR, Fj A (2011) New resources for functional analysis of omics data for the genus *Aspergillus*. BMC Genomics 12:486
140. Van den Berg RA, Braaksma M, van der Veen D, van der Werf MJ, Punt PJ, van der Oost J, de Graaff LH (2010) Identification of modules in *Aspergillus niger* by gene co-expression network analysis. Fungal Genet Biol 47:539–550
141. Vongsangnak W, Hansen K, Nielsen J (2011) Integrated analysis of the global transcriptional response to α-amylase over-production by *Aspergillus oryzae*. Biotechnol Bioeng 108:1130–1139
142. Van den Berg BA, Reinders MJT, Hulsman M, Wu L, Pel HJ, Roubos JA, de Ridder D (2012) Exploring sequence characteristics related to high-level production of secreted proteins in *Aspergillus niger*. PLoS One 7:e45869
143. Van den Berg BA, Reinders MJT, van der Laan J-M, Roubos JA, de Ridder D (2014) Protein redesign by learning from data. Protein Eng Des Sel 27:281–288
144. Dunlap JC, Borkovich KA, Henn MR, Turner GE, Sachs MS, Glass NL, McCluskey K, Plamann M, Galagan JE, Birren BW, Weiss RL, Townsend JP, Loros JJ, Nelson MA, Lambreghts R, Colot HV, Park G, Collopy P, Ringelberg C, Crew C, Litvinkova L, DeCaprio D, Hood HM, Curilla S, Shi M, Crawford M, Koerhsen M, Montgomery P, Larson L, Pearson M et al (2007) Enabling a community to dissect an organism: overview of the Neurospora functional genomics project. Adv Genet 57:49–96
145. Park G, Colot H V, Collopy PD, Krystofova S, Crew C, Ringelberg C, Litvinkova L, Altamirano L, Li L, Curilla S, Wang W, Gorrochotegui-Escalante N, Dunlap JC, Borkovich KA (2011) High-throughput production of gene replacement mutants in *Neurospora crassa*. Methods Mol Biol 722:179–189
146. De Souza CP, Hashmi SB, Osmani AH, Andrews P, Ringelberg CS, Dunlap JC, Osmani SA (2013) Functional analysis of the *Aspergillus nidulans* kinome. PLoS One 8:e58008
147. Jiang D, Zhu W, Wang Y, Sun C, Zhang K-Q, Yang J (2013) Molecular tools for functional genomics in filamentous fungi: recent advances and new strategies. Biotechnol Adv 31:1562–1574
148. Meyer V, Ram AFJ, Punt PJ (2010) Genetics, genetic manipulation, and approaches to strain improvement of filamentous fungi. Man Ind Microbiol Biotechnol 1:318–329
149. Schuster A, Bruno KS, Collett JR, Baker SE, Seiboth B, Kubicek CP, Schmoll M (2012) A versatile toolkit for high throughput functional genomics with *Trichoderma reesei*. Biotechnol Biofuels 5:1
150. Delmas S, Llanos A, Parrou J-L, Kokolski M, Pullan ST, Shunburne L, Archer DB (2014) Development of an unmarked gene deletion system for the filamentous fungi *Aspergillus niger* and *Talaromyces versatilis*. Appl Environ Microbiol 80:3484–3487
151. Magaña-Ortíz D, Coconi-Linares N, Ortiz-Vazquez E, Fernández F, Loske AM, Gómez-Lim MA (2013) A novel and highly efficient method for genetic transformation of fungi employing shock waves. Fungal Genet Biol 56:9–16
152. Loske AM, Fernández F, Magaña-Ortíz D, Coconi-Linares N, Ortíz-Vázquez E, Gómez-Lim MA (2014) Tandem shock waves to enhance genetic transformation of *Aspergillus niger*. Ultrasonics 54:1656–1662
153. Blumhoff M, Steiger MG, Marx H, Mattanovich D, Sauer M (2012) Six novel constitutive promoters for metabolic engineering of *Aspergillus niger*. Appl Microbiol Biotechnol 97:253–267

154. Ninomiya Y, Suzuki K, Ishii C, Inoue H (2004) Highly efficient gene replacements in *Neurospora* strains deficient for nonhomologous end-joining. Proc Natl Acad Sci USA 101:12248–12253
155. Krappmann S, Sasse C, Braus GH (2006) Gene targeting in *Aspergillus fumigatus* by homologous recombination is facilitated in a nonhomologous end joining-deficient genetic background. Eukaryot Cell 5:212–215
156. Nayak T, Szewczyk E, Oakley CE, Osmani A, Ukil L, Murray SL, Hynes MJ, Osmani SA, Oakley BR (2006) A versatile and efficient gene-targeting system for *Aspergillus nidulans*. Genetics 172:1557–1566
157. Meyer V, Arentshorst M, El-Ghezal A, Drews A-C, Kooistra R, van den Hondel CAMJJ, Ram AFJ (2007) Highly efficient gene targeting in the *Aspergillus niger* kusA mutant. J Biotechnol 128:770–775
158. Carvalho NDSP, Arentshorst M, Kwon MJ, Meyer V, Ram AFJ (2010) Expanding the ku70 toolbox for filamentous fungi: establishment of complementation vectors and recipient strains for advanced gene analyses. Appl Microbiol Biotechnol 87:1463–1473
159. Krull R, Cordes C, Horn H, Kampen I, Kwade A, Neu TR, Nörtemann B (2010) Morphology of filamentous fungi: linking cellular biology to process engineering using *Aspergillus niger*. Adv Biochem Eng Biotechnol 121:1–21
160. Wucherpfennig T, Lakowitz A, Krull R (2013) Comprehension of viscous morphology—evaluation of fractal and conventional parameters for rheological characterization of *Aspergillus niger* culture broth. J Biotechnol 163:124–132
161. Coradin JH, Braun A, Viccini G, Fernando de Lima Luz L, Krieger N, Mitchell DA (2011) A three-dimensional discrete lattice-based system for modeling the growth of aerial hyphae of filamentous fungi on solid surfaces: A tool for investigating micro-scale phenomena in solid-state fermentation. Biochem Eng J 54:164–171
162. Liu Y, Liao W, Chen S (2008) Study of pellet formation of filamentous fungi *Rhizopus oryzae* using a multiple logistic regression model. Biotechnol Bioeng 99:117–128
163. Wucherpfennig T, Kiep KA, Driouch H, Wittmann C, Krull R (2010) Morphology and rheology in filamentous cultivations. Adv Appl Microbiol 72:89–136
164. Krull R, Wucherpfennig T, Esfandabadi ME, Walisko R, Melzer G, Hempel DC, Kampen I, Kwade A, Wittmann C (2013) Characterization and control of fungal morphology for improved production performance in biotechnology. J Biotechnol 163:112–123
165. Gao Q, Liu J, Liu L (2014) Relationship between morphology and itaconic acid production by *Aspergillus terreus*. J Microbiol Biotechnol 24:168–176
166. Formenti LR, Nørregaard A, Bolic A, Hernandez DQ, Hagemann T, Heins A-L, Larsson H, Mears L, Mauricio-Iglesias M, Krühne U, Gernaey KV (2014) Challenges in industrial fermentation technology research. Biotechnol J 9:727–738
167. Dochain D (2010) Automatic Control of Bioprocesses. vol 28. Wiley, New York
168. Mandenius C-F, Titchener-Hooker NJ (2013) Measurement and control of bioprocesses. Adv Biochem Eng Biotechnol 132:5–8
169. Cohen D (1967) Computer simulation of biological pattern generation processes. Nature 216:246–248
170. Prosser JI (1995) Mathematical modelling of fungal growth. In: Gow NAR, Gadd GM (eds) Grow fungus. Springer, Dordrecht, pp 319–335
171. Davidson FA (2007) Mathematical modelling of mycelia: a question of scale. Fungal Biol Rev 21:30–41
172. Celler K, Picioreanu C, van Loosdrecht MCM, van Wezel GP (2012) Structured morphological modeling as a framework for rational strain design of *Streptomyces* species. Antonie Van Leeuwenhoek 102:409–423
173. Davidson FA, Boswell GP, Fischer MWF, Heaton L, Hofstadler D, Roper M (2011) Mathematical modelling of fungal growth and function. IMA Fungus 2:33–37
174. Hopkins SM (2012) A hybrid mathematical model of fungal mycelia : tropisms, polarised growth and application to colony competition. PhD thesis, University of Glamorgan

175. King R (1998) Mathematical modelling of the morphology of streptomyces species. In: Schügerl K (ed) Advances in biochemical engineering, vol 60. Springer, Berlin, pp 95–119
176. Wittler R, Baumgartl H, Lübbers DW, Schügerl K (1986) Investigations of oxygen transfer into *Penicillium chrysogenum* pellets by microprobe measurements. Biotechnol Bioeng 28:1024–1036
177. Lejeune R, Baron GV (1997) Simulation of growth of a filamentous fungus in 3 dimensions. Biotechnol Bioeng 53:139–150
178. Nieminen L, Webb S, Smith MCM, Hoskisson PA (2013) A flexible mathematical model platform for studying branching networks: experimentally validated using the model actinomycete *Streptomyces coelicolor*. PLoS One 8:e54316
179. Rinas U, El-Enshasy H, Emmler M, Hille A, Hempel DC, Horn H (2005) Model-based prediction of substrate conversion and protein synthesis and excretion in recombinant *Aspergillus niger* biopellets. Chem Eng Sci 60:2729–2739
180. Werner S, Kaiser SC, Kraume M, Eibl D (2014) Computational fluid dynamics as a modern tool for engineering characterization of bioreactors. Pharm Bioprocess 2:85–99
181. Megee RD, Kinoshita S, Fredrickson AG, Tsuchiya HM (1970) Differentiation and product formation in molds. Biotechnol Bioeng 12:771–801
182. Nielsen J (1992) Modelling the growth of filamentous fungi. Adv Biochem Eng Biotechnol 46:187–223
183. Boswell GP, Davidson FA (2012) Modelling hyphal networks. Fungal Biol Rev 26:30–38
184. Edelstein L (1982) The propagation of fungal colonies: a model for tissue growth. J Theor Biol 98:679–701
185. Boswell GP, Jacobs H, Ritz K, Gadd GM, Davidson FA (2007) The development of fungal networks in complex environments. Bull Math Biol 69:605–634
186. Schnepf A, Roose T, Schweiger P (2008) Impact of growth and uptake patterns of arbuscular mycorrhizal fungi on plant phosphorus uptake—a modelling study. Plant Soil 312:85–99
187. Meskauskas A, Fricker MD, Moore D (2004) Simulating colonial growth of fungi with the neighbour-sensing model of hyphal growth. Mycol Res 108(11):1241–1256
188. Yang H, King R, Reichl U, Gilles ED (1992) Mathematical model for apical growth, septation, and branching of mycelial microorganisms. Biotechnol Bioeng 39:49–58
189. Heaton L, Obara B, Grau V, Jones N, Nakagaki T, Boddy L (2012) Fricker MD: analysis of fungal networks. Fungal Biol Rev 26:12–29
190. Bartnicki-Garcia S, Hergert F, Gierz G (1989) Computer simulation of fungal morphogenesis and the mathematical basis for hyphal (tip) growth. Protoplasma 153:46–57
191. Boudaoud A (2003) Growth of walled cells: from shells to vesicles. Phys Rev Lett 91:018104
192. Goriely A, Tabor M (2003) Biomechanical models of hyphal growth in actinomycetes. J Theor Biol 222:211–218
193. Tindemans SH, Kern N, Mulder BM (2006) The diffusive vesicle supply center model for tip growth in fungal hyphae. J Theor Biol 238:937–948
194. Sugden KEP, Evans MR, Poon WCK, Read ND (2007) Model of hyphal tip growth involving microtubule-based transport. Phys Rev E Stat Nonlin Soft Matter Phys 75(3 Pt 1):031909
195. Heaton LLM, López E, Maini PK, Fricker MD, Jones NS (2010) Growth-induced mass flows in fungal networks. Proc Biol Sci 277:3265–3274
196. Mabey Gilsenan J, Cooley J, Bowyer P (2012) CADRE the central *Aspergillus* data REpository 2012. Nucleic Acids Res 40(Database issue):D660–666
197. Kersey PJ, Allen JE, Christensen M, Davis P, Falin LJ, Grabmueller C, Hughes DST, Humphrey J, Kerhornou A, Khobova J, Langridge N, McDowall MD, Maheswari U, Maslen G, Nuhn M, Ong CK, Paulini M, Pedro H, Toneva I, Tuli MA, Walts B, Williams G, Wilson D, Youens-Clark K, Monaco MK, Stein J, Wei X, Ware D, Bolser DM, Howe KL, et al (2014) Ensembl genomes 2013: scaling up access to genome-wide data. Nucleic Acids Res 42(Database issue):D546–552

198. Stajich JE, Harris T, Brunk BP, Brestelli J, Fischer S, Harb OS, Kissinger JC, Li W, Nayak V, Pinney DF, Stoeckert CJ, Roos DS (2012) FungiDB: an integrated functional genomics database for fungi. Nucleic Acids Res 40(Database issue):D675–681
199. Choi J, Cheong K, Jung K, Jeon J, Lee G-W, Kang S, Kim S, Lee Y-W, Lee Y-H (2013) CFGP 2.0: a versatile web-based platform for supporting comparative and evolutionary genomics of fungi and Oomycetes. Nucleic Acids Res 41(Database issue):D714–719
200. Jung K, Park J, Choi J, Park B, Kim S, Ahn K, Choi J, Choi D, Kang S, Lee Y-H (2008) SNUGB: a versatile genome browser supporting comparative and functional fungal genomics. BMC Genomics 9:586
201. Pagani I, Liolios K, Jansson J, Chen I-MA, Smirnova T, Nosrat B, Markowitz VM, Kyrpides NC (2012) The Genomes OnLine Database (GOLD) v.4: status of genomic and metagenomic projects and their associated metadata. Nucleic Acids Res 40(Database issue):D571–579
202. Priebe S, Linde J, Albrecht D, Guthke R, Brakhage AA (2011) FungiFun: a web-based application for functional categorization of fungal genes and proteins. Fungal Genet Biol 48:353–358
203. Lum G, Min XJ (2011) FunSecKB: the fungal secretome knowledgebase. Database (Oxford) 2011:bar001
204. Park J, Park J, Jang S, Kim S, Kong S, Choi J, Ahn K, Kim J, Lee S, Kim S, Park B, Jung K, Kim S, Kang S, Lee Y-H (2008) FTFD: an informatics pipeline supporting phylogenomic analysis of fungal transcription factors. Bioinformatics 24:1024–1025
205. Park J, Lee S, Choi J, Ahn K, Park B, Park J, Kang S, Lee Y-H (2008) Fungal cytochrome P450 database. BMC Genomics 9:402
206. Garcia-Albornoz M, Thankaswamy-Kosalai S, Nilsson A, Väremo L, Nookaew I, Nielsen J (2014) BioMet Toolbox 2.0: genome-wide analysis of metabolism and omics data. Nucleic Acids Res 42(Web Server issue):W175–181
207. Pabinger S, Snajder R, Hardiman T, Willi M, Dander A, Trajanoski Z (2014) MEMOSys 2.0: an update of the bioinformatics database for genome-scale models and genomic data. Database (Oxford) 2014:bau004
208. Boele J, Olivier BG, Teusink B (2012) FAME, the flux analysis and modeling environment. BMC Syst Biol 6:8
209. Caspi R, Foerster H, Fulcher CA, Kaipa P, Krummenacker M, Latendresse M, Paley S, Rhee SY, Shearer AG, Tissier C, Walk TC, Zhang P, Karp PD (2008) The MetaCyc database of metabolic pathways and enzymes and the BioCyc collection of pathway/genome databases. Nucleic Acids Res 36(Database issue):D623–631
210. Barrett T, Troup DB, Wilhite SE, Ledoux P, Evangelista C, Kim IF, Tomashevsky M, Marshall KA, Phillippy KH, Sherman PM, Muertter RN, Holko M, Ayanbule O, Yefanov A, Soboleva A (2011) NCBI GEO: archive for functional genomics data sets–10 years on. Nucleic Acids Res 39(Database issue):D1005–1010
211. Rustici G, Kolesnikov N, Brandizi M, Burdett T, Dylag M, Emam I, Farne A, Hastings E, Ison J, Keays M, Kurbatova N, Malone J, Mani R, Mupo A, Pedro Pereira R, Pilicheva E, Rung J, Sharma A, Tang YA, Ternent T, Tikhonov A, Welter D, Williams E, Brazma A, Parkinson H, Sarkans U (2013) ArrayExpress update–trends in database growth and links to data analysis tools. Nucleic Acids Res 41(Database issue):D987–990
212. Khaldi N, Seifuddin FT, Turner G, Haft D, Nierman WC, Wolfe KH, Fedorova ND (2010) SMURF: genomic mapping of fungal secondary metabolite clusters. Fungal Genet Biol 47:736–741
213. Blin K, Medema MH, Kazempour D, Fischbach MA, Breitling R, Takano E, Weber T (2013) antiSMASH 2.0–a versatile platform for genome mining of secondary metabolite producers. Nucleic Acids Res 41(Web Server issue):W204–212
214. Sanli K, Karlsson FH, Nookaew I, Nielsen J (2013) FANTOM: functional and taxonomic analysis of metagenomes. BMC Bioinform 14:38

Adv Biochem Eng Biotechnol (2015) 149: 133–170
DOI 10.1007/10_2014_302

Published Online: 30 January 2015

Bioprocess Engineering Aspects of the Cultivation of a Lovastatin Producer *Aspergillus terreus*

Marcin Bizukojc and Stanislaw Ledakowicz

Abstract The aim of this work is to review bioprocess engineering aspects of lovastatin (antihypercholesterolemia drug) production by *Aspergillus terreus* in the submerged culture in the bioreactors of various scale presented in the scientific literature since the nineties of the twentieth century. The key factor influencing the cultivation of any filamentous species is fungal morphology and that is why this aspect was treated as the starting point for further considerations. Fungal morphology is known to have an impact on the following issues connected with the cultivation of *A. terreus* reviewed in this article. These are broth viscosity in conjunction with non-Newtonian behaviour of the cultivation broths, and multistage oxygen transfer processes: from gas phase (air) to liquid phase (broth) and diffusion in the fungal agglomerates. The latest achievements concerning the controlling *A. terreus* morphology during lovastatin biosynthesis with the use of morphological engineering techniques were also reviewed. Last but not least, some attention was paid to the type of a bioreactor, its operational mode and kinetic modelling of lovastatin production by *A. terreus*.

Keywords Lovastatin · Bioreactor · Fungal morphology · Kinetics · Broth rheology · Oxygen transfer

Contents

M. Bizukojc (✉) · S. Ledakowicz
Faculty of Process and Environmental Engineering, Department of Bioprocess Engineering, Lodz University of Technology, ul. Wolczanska 213, 90-924 Lodz, Poland
e-mail: marcin.bizukojc@p.lodz.pl

1 Introduction

Lovastatin is a polyketide secondary metabolite from filamentous fungi of high industrial importance as a drug lowering the level of endogenous cholesterol in the human blood serum. It acts as a competitive inhibitor of (S)-3-hydroxy-3-methylglutaryl-CoA reductase. Its chemical (IUPAC) name is [8-[2-(4-hydroxy-6-oxooxan-2-yl)ethyl]-3,7-dimethyl-1,2,3,7,8,8a-hexahydronaphthalen-1-yl] 2-methylbutanoate. Lovastatin is a lactone, water insoluble compound formed during the extraction from the cultivation media of the actual metabolite: mevinolinic acid, which is, from the chemical point of view, a β-hydroxy acid. The biosynthesis of lovastatin and its action in the human organism is visualized in a condensed form for the purpose of introduction in Fig. 1.

There are several fungal species capable of producing of this metabolite such as *Aspergillus terreus*, *Monascus ruber* and *Penicillium citrinum*; nevertheless, it is *A. terreus* that became the industrial producer of this metabolite. *A. terreus* produces lovastatin with good efficiency and, if cultivated under proper conditions, a few other secondary metabolites (by-products) are formed. Much more by-products can be found in *P. citrinum* and *M. ruber*, e.g. mycotoxin citrinin and red pigments, respectively.

The quest to discover the inhibitors of (S)-3-hydroxy-3-methylglutaryl-CoA reductase, later called statins, started in the early 1970s in two international pharmaceutical companies Sankyo in Japan and Merck in the USA. This history was described in detail by Endo [24]. Finally, it was Merck, who succeeded and introduced lovastatin onto the market under the trade name Mevacor® in 1986. At present, *A. terreus* mutant strains are employed to produce this metabolite on the industrial scale and the most efficient mutant, about which some public data are revealed comes from Metkinnen OY (Finland). It is capable of producing over 10 g lovastatin per litre. Not only lovastatin but also the variety of semi-synthetic and fully synthetic statins has been produced [65]. The leading global companies are Pfizer Inc., AstraZeneca PLC, Merck & Co Inc. and Novartis AG. The value of the

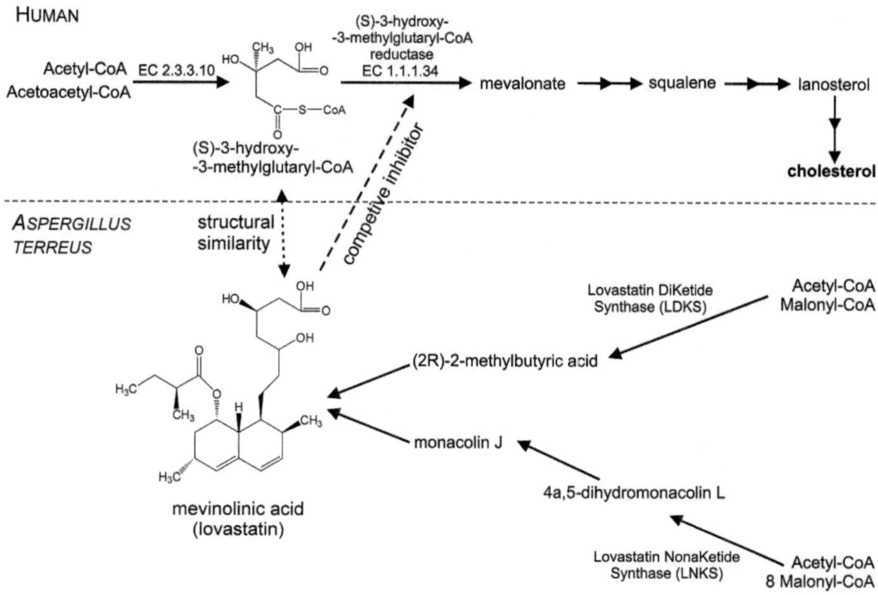

Fig. 1 *Aspergillus terreus* important metabolite lovastatin (mevinolinic acid) and its action in the human organism

global market for statins was estimated to 20.5×10^9 dollars in 2011 with the forecast to decline to 12.2×10^9 dollars in 2018 because of the expiry of the patents (*Source* Research and Markets).

The basic strain of *A. terreus* has been denounced in American Type Culture Collection as ATCC20542. In the scientific literature, often this strain, a recombinant ATCC74135 or other locally (in the lab of the research team) obtained mutants and local strains are the objects of research.

The research on lovastatin production by *A. terreus* was conducted in several universities all over the world. With regard to number of scientific papers published in the last two decades, the team from University of Almeria (Andalusia, Spain) is the leader (inter alia: [15, 62]). Fairly many publications come from Poland, from Lodz University of Technology (inter alia: [8–11, 28, 58]). In Asia, Chaoyang University of Technology in Taiwan leads (inter alia: [39–41]). Older but important papers were published by the Italian team from University of Milan (inter alia: [13, 44, 45]). Few papers were also published in the 1990s by scientists from Technical University of Budapest [66] and University of Ljubljana [52]. Various teams from India (inter alia: [30, 31, 38]) also joined to the research on lovastatin production.

Out of all these teams, the research on lovastatin production is still continued in Poland and recently appeared the Mexican team from two universities of Mexico City specializing in solid-state cultivation for lovastatin production by filamentous fungi (inter alia: 4, 5, 48]).

All the works published so far connected with lovastatin biosynthesis by *A. terreus* can be, in our opinion, subjectively divided into four groups: (1) the earliest works, including patents, focused on the newly discovered substance, i.e. lovastatin (or mevinolin or monacolin *K*, as this metabolite had more names in its history) from the 1980s (inter alia: [1, 2, 49]), (2) the works on genetic and biochemical mechanisms ruling the formation of lovastatin being a polyketide metabolite (inter alia: [3, 36, 68, 71]) (3) microbiologically focused publications mainly connected with media optimization and influence of media composition or any other substances on lovastatin formation (inter alia: [7, 12–16, 32, 39, 40, 44, 45, 66, 67]) and (4) the works with bioprocess engineering approach. The latter group is the subject of this review and the works belonging to it are collected and shortly described in Table 1.

Lovastatin biosynthesis in the submerged culture consists of several steps, which are each time applied by the scientists studying this process. For the sake of clarity and better understanding of the further sections of the review, they are schematically presented in Fig. 2.

2 Fungal Morphology and Lovastatin Formation

Although the issues of fungal morphology do not seem to be, at first glance, closely connected with bioprocess engineering but with mycology or, more generally, microbiology, they are going to be discussed at the beginning. It is a common knowledge that fungal morphology is one of the strongest factors influencing any metabolite formation by filamentous fungi [55], so it is the case of lovastatin production by *A. terreus* too. Furthermore, apart from a strong impact on lovastatin formation, also other parameters and features of the culture, its viscosity and oxygen transfer are extremely influenced by *A. terreus* morphology. Morphology of *A. terreus*, as of any other filamentous fungi, can be affected by the hydrodynamic conditions in the bioreactor and also can be controlled by means of the novel morphological engineering techniques (see Sect. 5). There is no doubt that all aforementioned issues are in the domain of bioprocess engineering.

In the submerged cultures of filamentous fungi, two morphological forms are usually found, i.e. either hyphal aggregates (pellets) or dispersed hyphae (free hyphae). It is connected with the mechanisms and stages of filamentous fungi growth, i.e. spore swelling, hyphal tip extension, branching and finally mycelial aggregation. All these mechanisms were thoroughly described by Papagianni [55]. The aggregation process may occur at different stages of fungal evolution and it results with the formation of fungal aggregates (pellets). Different mechanisms of mycelial aggregation are known. These are the agglomeration of spores in *Aspergillus* sp. and hyphal aggregation, which is characteristic for *Penicillium* sp. Non-agglomerative mechanism (agglomerates formed from an individual spore) is found in prokaryotic *Streptomyces* [47, 51].

Table 1 Publications with bioprocess engineering approach concerning lovastatin production by *A. terreus* in the chronological order

No.	Reference	Major findings
1	Gbewonyo et al. [26]	Significance of *A. terreus* morphology (minimizing its variations) for lovastatin production in conjunction with mass transfer processes
2	Novak et al. [52]	Evidence that fed-batch process is more efficient for lovastatin production than the batch one provided the overfeeding is avoided
3	Kumar et al. [38]	Evidence that fed-batch process is more efficient for lovastatin production than the batch one; determination of k_La during lovastatin production
4	Liu et al. [43]	Morphologically structured model for lovastatin biosynthesis
5	Hajjaj et al. [32]	Starvation (lack of substrate) is required for lovastatin production
6	Lai et al. [41]	Oxygen vector (n-dodecane) increases lovastatin titre in shake flask culture, but not in the bioreactor, and changes fungal morphology
7	Casas Lopez et al. [17]	Increased agitation in a stirred tank bioreactor strongly decreases pellet diameter and lovastatin titre; there is a maximum shear stress that can be tolerated in *A. terreus* culture, independent of better or worse aeration conditions (with air or oxygen-enriched air)
8	Lai et al. [42]	Level of pO_2 influences pellet morphology; optimum pO_2 for lovastatin production is 20 %, initial pH = 6.5 and should not be controlled
9	Rodriguez Porcel et al. [62]	Pellets formed by *A. terreus* in a fluidized bed bioreactor are larger than those in a stirred tank bioreactor; the compactness and fluffiness of pellets is affected by stirring intensity and so is the size of pellets; rheological parameters of the broth are affected by biomass concentration pellets size and pellet fluffiness
10	Rodriguez Porcel et al. [63]	In a fluidized bed bioreactor, aeration with oxygen-enriched air is required to obtain high lovastatin titres at aeration rate equal to 1 L_{air} L^{-1} min^{-1}; mass transfer coefficients k_La depend on the morphological form of the hyphae: dispersed or pelleted
12	Rodriguez Porcel et al. [60]	Lovastatin formation is enhanced in fed-batch and semi-continuous bioreactor (with biomass retention), if fed with non-growth-promoting medium
13	Gupta et al. [30]	Correlation between morphology and rheology of *A. terreus* cultivation broths
14	Bizukojc and Ledakowicz [8]	Unstructured kinetic model for *A. terreus* growth and lovastatin formation on the single carbon substrate (lactose)
15	Bizukojc and Ledakowicz [9]	Determination of lovastatin and (+)-geodin (by-product) formation kinetics in conjunction with carbon substrate uptake

(continued)

Table 1 (continued)

No.	Reference	Major findings
16	Rodriguez Porcel et al. [61]	Organic nitrogen-free media for feeding in fed-batch bioreactors are the most optimum for lovastatin production by *A. terreus*
17	Bizukojc and Ledakowicz [10]	Too intensive aeration of the culture leads to the intensive formation of (+)-geodin, instead of lovastatin; pH control with carbonates is required for better lovastatin titres
18	Gupta et al. [31]	The optimum dilution rate is found for lovastatin production in a continuous bioreactor with free and immobilized biomass
19	Jia et al. [34]	Type of carbon source influences the properties (diameter and structure) of *A. terreus* pellets during lovastatin biosynthesis
20	Bizukojc and Ledakowicz [11]	Lovastatin and (+)-geodin formation depends on the fraction of actively growing filaments in *A. terreus* pellets (the quantitative correlation given); first work on the intrastructure of *A. terreus* pellets
21	Gonciarz et al. [29]	Mineral microparticles (morphological engineering technique) should be added to the preculture in order to increase lovastatin titre; talc powder is better than aluminium oxide microparticles
22	Pawlak et al. [59]	Lovastatin formation strongly depends on the scale of bioreactor
23	Pawlak and Bizukojc [57]	Unstructured kinetic model for *A. terreus* growth and lovastatin formation on two carbon substrates (lactose and glycerol)
24	Pawlak and Bizukojc [58]	Apart from feeding profile, also the level of redox potential and utilization rate of inorganic carbon influence lovastatin formation in the fed-batch bioreactor
25	Gonciarz and Bizukojc [28]	Addition of talc microparticles enhances lovastatin formation in the shake flask culture even by 50 %

With regard to the growth of *A. terreus* mycelium, all researchers dealt with the agglomerates, namely fungal pellets. Thus, at the beginning, the quantitative parameters concerning the pelleted morphology of *A. terreus* are going to be presented. These, who more or less thoroughly referred to *A. terreus* morphology in their works, used above all the simplest parameter to describe it, i.e. pellet diameter. [11, 17, 26, 28, 30, 34, 60, 62, 63]. This selection is fully justified as fungal pellets are the most frequently of spherical shape.

Nevertheless, apart from the shape, a macroscopic structure of *A. terreus* pellets may differ dependent on the conditions, under which they grow. The most often used classification of fungal pellets comes from the review of Metz and Kossen [47]. It distinguishes fluffy loose pellets (with a compact centre and a much looser outer zone), compact smooth pellets (the whole pellet is compact, the surface of the pellet is smooth) and hollow smooth pellets (the centre of the pellet is empty, due to autolysis of hyphae and its surface is smooth). It can be without any doubts applied to *A. terreus*.

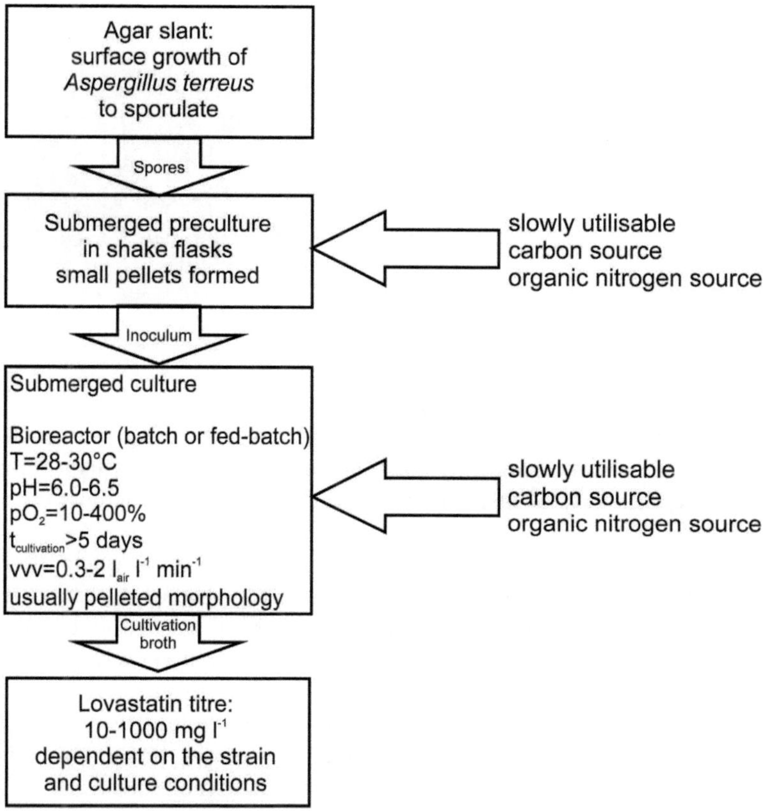

Fig. 2 General flow chart to biosynthesize lovastatin by *A. terreus* in the submerged culture

The quantitative measure of the filamentous growth in the *A. terreus* pellets, describing their hairiness was proposed by Rodriguez Porcel et al. [62] and Casas Lopez et al. [17]. Their concept was based upon the work of Cui et al. [18] and comprised two measures. First, they used the diameter corresponding to a circular area equivalent to the total pellet-projected area, as a one-dimensional measurement of the pellet size. Second, they measured the mean projected area of the whole pellet and the area of the hairy region. Upon this, they introduced the filament ratio that can be upon their description defined as:

$$\text{filament ratio} = \frac{A_Z}{A_Z + A_C} \tag{2.1}$$

where A_Z—projected area of the core of pellet kernel, A_C—projected area of filaments. The definition of filament ratio is presented in Fig. 3 too.

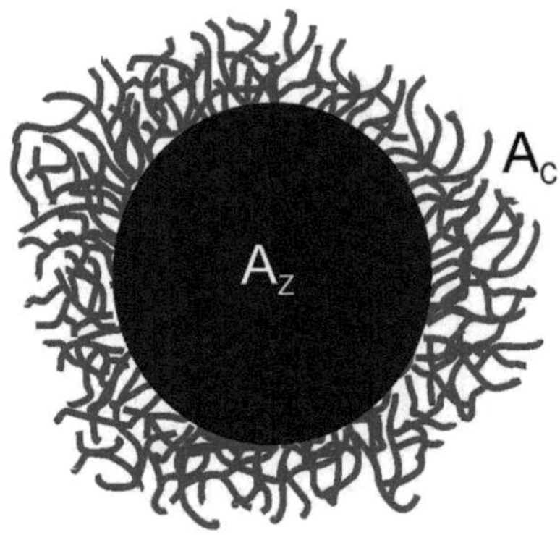

Fig. 3 Definition of filament ratio (upon the description from [62])

All these above-described morphological parameters, being the quantitative measure *of A. terreus* morphology, should be connected with lovastatin formation. Unfortunately, the study of literature revealed that most of the authors did not do this explicitly.

Gbewonyo et al. [26] were the first who noticed the fact of lovastatin production by dispersed (it is not a frequent case) and pelleted hyphae of *A. terreus* without making any detailed morphological measurements. They only wrote that they had pellets of 100–300 μm. No data on lovastatin concentration were given in this paper.

Relatively large amount of data about *A. terreus* morphology were published by the research team from University of Almeria (Andalusia, Spain). Nevertheless, most of them are only the presentation of the data (no doubt that valuable) concerning the change of morphological parameters in time for various process conditions and bioreactors. Upon them hardly can any direct quantitative correlation between *A. terreus* morphology and lovastatin production be found as the quantitation of *A. terreus* morphology was not the main aim of most of these works.

Casas Lopez et al. [17] studied the influence of agitation speed on lovastatin production and thereby showed the difference between pellets formed under high (800 rpm) and low (300 rpm) agitation conditions in a 5-L stirred tank bioreactor (STB). In the first case, pellet diameters were below 1,500 μm and in the second one they achieved even 2,500 μm. Interestingly, they found that filament ratio is not so strongly influenced by the agitation. In all cases it decreased from about 0.9 at the beginning of the process to about 0.4 after 100 h. Casas Lopez et al. [17] did not show any direct mathematical correlation between pellet diameter and lovastatin formation. Nevertheless, they obtained higher titres of lovastatin for lower agitation speed, i.e. in the system with bigger pellets. Quantitative data concerning the mechanical properties of the formed *A. terreus* pellets, for example the correlation between specific energy dissipation rate and fungal morphology, as it had been done for *A. awamori* by Cui et al. [18], were not presented either.

Another work from this team by Rodriguez Porcel et al. [62] did not bring many new data. That time a different bioreactor, i.e. a 17-L fluidized bed bioreactor (FBB), was used. Its application led to obtaining bigger pellets compared to the

STB. Their diameter was in the range from 2,000 to 3,000 μm. But filament ratio remained insensitive to the change of bioreactor. Similar findings were reported in Rodriguez Porcel et al. [63]. Here, the effect of various aeration gas compositions and organic nitrogen source concentration was studied. In the article concerning various feeding strategies for lovastatin production in a FBB [60], no new findings concerning *A. terreus* morphology, compared to the previous articles, can be found again.

There are also some data concerning *A. terreus* morphology from other research teams. Lai et al. [40] showed several distributions of pellet diameter in the optimum media (obtained by means of statistical medium design) for lovastatin production. Lai et al. [42] made the classification of the hyphal objects and presented pellet size distribution in two types of cultures: shake flasks and a 5-L STB. The pellets from shake flask culture were bigger and their diameter ranged from 750 to 1,600 μm with the maximum frequency for 1,200 μm pellets, while in the bioreactor, from 650 to 1,350 μm with the maximum frequency for 1,000 μm pellets. They claimed that stirring velocity is the strong factor influencing pellet size; however, lower shearing stress and consecutively bigger pellets are more useful for lovastatin formation. They also concluded that *A. terreus* pellets with a uniform size distribution (950 μm in average) aided lovastatin production largely. In the earlier work, Lai et al. [39, 41] observed the effect of addition of water-immiscible organic oxygen carrier (n-dodecane) on *A. terreus* morphology. Above all, it decreased pellet diameters (from 1,200 to 800 μm at 2.5 % n-dodecane) and changed their size distribution. Also the structure of pellets was altered. The pellets became less hairy. It increased lovastatin production. In the experiment made in a 5-L STB with the organic solvent added, the star-shape (diffused) pellets were undesirably formed, namely there was a transformation from compact pellets to extremely fluffy loose pellets. To this morphology, together with the elevated oxygen concentration, undesired pH effects and organic solvent droplet formation the aggravation of lovastatin production was attributed [41].

Gupta et al. [30] also traced the changes of *A. terreus* morphology in time in a 2-L STB. Describing methods, they declared the complex quantification of fungal morphology, including the measurement of mean projected area, diameter and circularity. Unfortunately, they did not show these data. They only observed the growth in pellet diameter within the 10-day cultivation from 3,500 μm (on average) in 3rd day up to 5,650 μm (on average) in 10th day. They postulated the deceleration of the process after 8th day to be connected with the breakage of the pellets formed. They also claimed, although they did not show any correlation, that neither very small pellets nor large hollow pellets are suitable for lovastatin production. Only medium-sized pellets 1,800–2,000 μm with the high ratio of filaments to core, actually they can be called fluffy pellets, optimally produced lovastatin at high oxygenation conditions. This study contains the interesting images from the electron microscopy to visualize the destruction of pellets in the late stages of lovastatin production.

Jia et al. [34] characterized *A. terreus* morphology during lovastatin production in shake flasks in terms of pellet core diameter (i.e. the equivalent diameter of the measured core area) and the width of the hairy zone in relation to the used carbon source. In their opinion, regular and compact pellets with slender spongy of outer hyphae and sporangia on the tips of pellet outer hyphae, (the latter is quite unusual opinion, not found in any other sources) were beneficial for *A. terreus* secondary metabolism. Larger pellets were formed on fast utilizable substrates (glucose or sucrose), while smaller ones on lactose, glycerol or starch assuring higher lovastatin titres.

The effect of fungal pellet size and the differentiation of hyphae in *A. terreus* pellets on lovastatin production were studied by Bizukojc and Ledakowicz [11] in the 150-mL shake flask culture. They confirmed the agglomerative mechanism of pellet formation in the 24-h preculture of *A. terreus*. Using various numbers of spores 1.39×10^9 to 2.56×10^{10} L^{-1} in the preculture, and using these precultures for the inoculation of the production medium, they generated smooth pellets of various sizes from about 1,000 to 3,500 μm in the production medium. About 10,400 spores capable of germination were found to form an individual pellet of *A. terreus*. Pellets once formed were not very prone to divide or agglomerate in the later stages of cultivation and their number remained more or less constant with the cultivation [11]. What is the most important, these authors directly found that the smallest pellets occurred to be the most efficient for lovastatin production, while the largest one effectively biosynthesized (+)-geodin, an octaketide by-product of *A. terreus*. Above all, only in this article, the differentiation of hyphae was visualized, quantified and associated with lovastatin formation (Fig. 4). Using microscopic stained slides and, if required, thin cross sections of pellets, two zones of various metabolic activity were distinguished: external active and internal less active or dead.

The specific lovastatin formation rate π_{LOV} was proved to be dependent on the fraction of the active hyphae in the external regions of pellets in accordance with the equation:

$$\pi_{LOV} = 0.117 \cdot Z_1 + 0.047 \tag{2.2}$$

where Z_1 is active external zone fraction and π_{LOV} is specific lovastatin formation rate (related to biomass concentration expressed in mg LOV g X^{-1} h^{-1}).

In Table 2, the most important issues concerning *A. terreus* morphology and lovastatin formation are summarized.

Although there is insufficient amount of data aiming at finding of the direct correlations between *A. terreus* morphology and lovastatin production in the submerged culture, the importance of fungal morphology in these cultivations shall be confirmed in the next sections. Almost all engineering aspects concerning lovastatin production in bioreactors, e.g. broth viscosity and its rheological properties, oxygen transfer in the system, operational parameters of the bioreactor and even its type are stronger or weaker affected and/or correlated with the morphological form of *A. terreus*. Thus, it is not an accident that the articles mentioned in this section will be cited again, as all these engineering aspects were almost always studied and discussed together with the morphology of *A. terreus* mycelium.

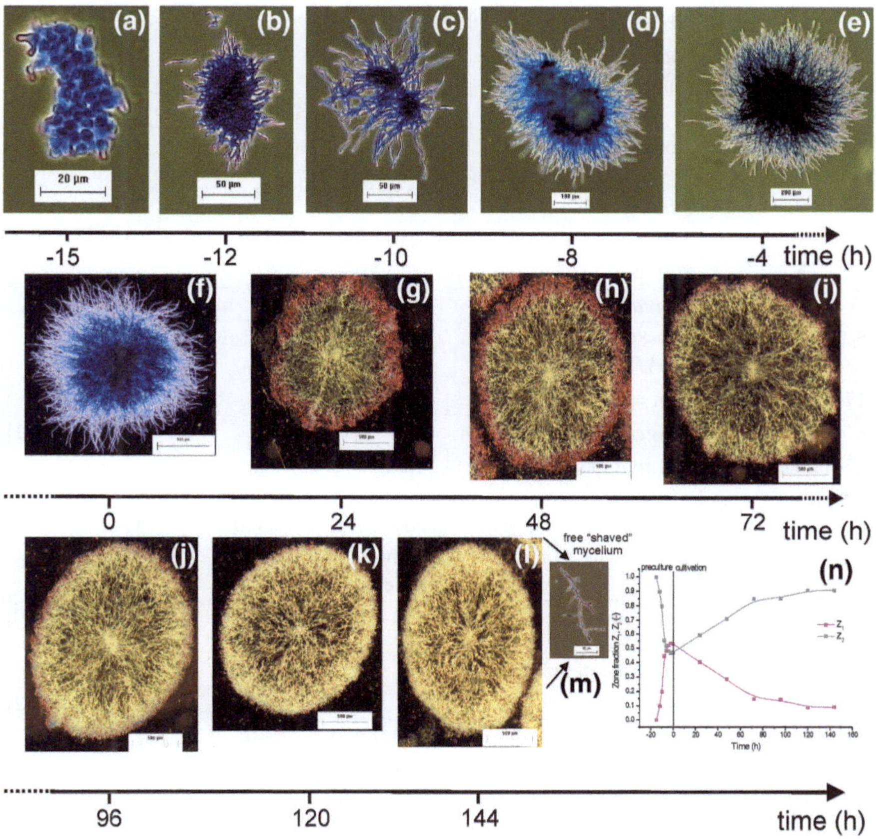

Fig. 4 Visualization of *A. terreus* pellets evolution from spore aggregation to the growth and differentiation of pellets of pellets (from **a** to **m**) and (**n**) changes of fractions of zones Z_1 (active) and Z_2 in time; negative values of time indicate the time of preculture in the reversed order and $t = 0$ h is the inoculation time [11]; reproduced with permission granted from Springer

3 Rheology of *A. terreus* Cultivation Broths

At the beginning, it is worth mentioning that all data which concern rheology of *A. terreus* broths available in literature are the time changes of rheological parameters of the broth in time.

In one of the earliest works containing information on hydrodynamic properties of *A. terreus* broths, Gbewonyo et al. [26] showed the changes of apparent viscosity of the broth in relation to the morphological form of the fungus evolved during the cultivation in a pilot-scale 500-L STB. In the case of filamentous suspension (no pellets formed), apparent viscosity changed from about 50 cP (0.05 Pa s) at the beginning of the cultivation to about 700 cP (0.7 Pa s) for the dispersed morphology and 300 cP (0.3 Pa s) for the pelleted morphology after 100 h of the process. Furthermore, the increase in the apparent viscosity had a permanently increasing

Table 2 *A. terreus* morphology versus lovastatin formation

Bioreactor type and working volume	Factor(s) influencing fungal morphology	Pellet size (μm)	Lovastatin titre (mg L^{-1})	Reference
500-L STB	None (pO_2 controlled)	100–300	Not given	Gbewonyo et al. [26]
5-L STB	*Stirring*			Casas et al. [17]
	300 rpm	2,500	40	
	800 rpm	<1,500	20	
	Bioreactor type			Rodriguez Porcel et al. [62]
5-L STB	300–800 rpm; vvm = 1 L_{air} L^{-1} min^{-1}	500–1,500	Not given	
17-L FBB	vvm = 1 L_{air} L^{-1} min^{-1}	2,000–3,000		
	Oxygen vector			Lai at al. [41]
25-mL shake flask	200 rpm, no $C_{12}H_{26}$	1,200	300	
	200 rpm, 2.5 % $C_{12}H_{26}$	800	>400	
2.5-L STB	2.5 % $C_{12}H_{26}$	Diffused pellets	100	
25-mL shake flask	Response surface methodology (RSM) medium optimization	850–1,050	1158	Lai at al. [40]
	Culture type			Lai at al. [42]
125-mL shake flask		750–1,600 (mean 1,200)	>400	
3-L STB	pO_2 = 20 %	650–1,350 (mean 1,000)	>400	
2-L airlift	None (vvm = 1.5 L_{air} L^{-1} min^{-1})	3,500–5,650	900	Gupta et al. [30]
20-mL shake flask	*Carbon source type*			Jia et al. [34]
	Glucose	400–750	112.5	
	Sucrose		218.8	
	Starch		501.3	
	Lactose		212.5	
	Glycerol		937.5	
150-mL shake flask	*Spore number in the preculture*			Bizukojc and Ledakowicz [11]
	1.39×10^9 to 2.56×10^{10} L^{-1}	1,000–3,500	30–70	

trend for the dispersed morphology, while in the case of pellets it stabilized on the aforementioned level after about 40 h. No rheological model was proposed in this study.

The more detailed research on the rheological properties of *A. terreus* broths was only conducted by the team from University of Almeria, whose works dominate in the scientific literature with regard to this subject. Nevertheless, only a few of these works were more directly aimed at the testing of the rheological properties of *A. terreus* broths. In most of them, the measurements of the rheological parameters were only the additional data included in these works.

Generally, in all these works it was assumed that *A. terreus* broth rheology can be described by the power law, i.e. Ostwald-de Waele equation. Using this law, the apparent viscosity of the broth can be expressed as

$$\mu_{\text{app}} = \frac{\tau}{\dot{\gamma}} = K \cdot \dot{\gamma}^{n-1} \tag{3.1}$$

where K—consistency (N m^{-2} s^n); n—flow behaviour index (−); $\dot{\gamma}$—shear rate (s^{-1}) and τ—shear stress (N m^{-2}).

In the light of the common knowledge on fungal suspensions, the use of the power law was justified as these suspensions satisfy this law being usually shear thinning liquids. These authors did not make any deeper insight into rheology of *A. terreus* cultivation broth than the determination of the parameters for Ostwald-de Waele law throughout the whole duration of the process, namely at various sampling times. No yield stress was described either. They always used the same rheometric measurement device, i.e. rotational viscometer with standard vane spindle of 21.67 mm diameter and 43.33 mm height.

The work from by Casas Lopez et al. [17] was the most detailed and comprised the determination of the rheological parameters satisfying Ostwald-de Waele equation for *A. terreus* broth cultivated in a 5-L STB agitated with Rushton turbines at various speeds, i.e. 300 and 800 rpm at various aeration conditions: with air and oxygen-enriched air. The values of consistency *K* and flow behaviour index *n* were determined within cultivation time and, irrespective of the aeration gas, consistency remained unchanged (0.01 N m^{-2} s^n) within the time of cultivation, if the higher rotation speed of the impeller was used. At the lower rotation speed of the impeller (300 rpm), the increase in consistency from the initial 0.01 to 0.5 N m^{-2} s^n was observed. Changes of flow index were different starting from about 1 (Newtonian fluid): at the beginning of the cultivation, it went towards 1.5 for the higher rotation speed of the impeller (shear thickening fluid) or towards 0.5 (shear thinning fluid) for the lower rotation speed of the impeller. It was also reflected in the apparent viscosities, which were higher for the strongly agitated broth (0.03–0.065 Pa s) and lower for the weaker agitated broth (0.01–0.035 Pa s). Cultivation broths aerated with oxygen-enriched air were usually more viscous. It is the only work, in which the changes of apparent viscosity along the culture time were graphically presented. In all studied runs, the increase of apparent viscosity in time by 0.01–0.03 Pa s within 200 h of the experiment was observed.

The untypical effect of the formation of shear thickening broth the authors explained by the formation of very small pellets (between 500 and 1,500 μm, as shown in the graphs) at the higher rotation speeds of the impeller. Taking the size of the pellets into account, this explanation seems not to be probable as these pellets are not extremely small, compared to other literature data. No correlation between lovastatin formation and broth rheology was explicitly shown. Any positive or negative effects regarding lovastatin yield were rather attributed to the varying process conditions.

Rodriguez Porcel et al. [62] made a comparative study on *A. terreus* broth rheology for two types of bioreactors: a 5-L STB and 17-L FBB. The interesting in this study was that the flow behaviour of the broth in the fluidized bed bioreactor was shear thickening with flow behaviour index value of around 1.2 throughout the whole duration of the cultivation (initially declined from 1.7 to 1.4), despite fairly large pellets (up to 1,500 μm). The reference data for the STB were the same as in Casas Lopez et al. [17]. The consistency index for FBB was more than one magnitude lower (0.005–0.025 Pa s^n) than for STB agitated at 300 rpm (up to 0.5 Pa s^n). In the opinion of the authors, the various values of flow behaviour index were associated to biomass concentration and fungal morphology, namely pellets diameter and, to the lesser extent, fluffiness of pellets. The correlation between lovastatin concentration and broth rheology was here again indirect, via fungal morphology.

Rodriguez Porcel et al. [63] continued their study in the FBB. They again tried to find the correlation between broth rheology and biomass concentration together with pellet size. The experiments were made in the media with varied organic nitrogen concentration. The clear correlation between nitrogen concentration and consistency was then described. Nitrogen-limited conditions (nitrogen initial level 0.15 g N L^{-1}) led to aggravated biomass formation (less than 5 g L^{-1}), which contributed to lower K values around 0.05 Pa s^n. Generally, consistency increased with time of the cultivation until the moment, when pellets ceased growing and retained the same diameter, which happened around 100 h. Consistency was found to be insensitive to the increase in the concentration of pellets of a fixed diameter. As both pellets and filaments were observed in the broths, the correlation between these two morphological forms of biomass and rheological properties was also noted. The filament-rich culture broth was more viscous ($K > 0.5$ Pa s^n) than in the case of pelleted growth ($K = 0.2$–0.4 Pa s^n dependent on the hour of the run). In these cultures, they observed the varying flowing pattern, indicating the change in the properties of the broth from shear thickening ($n = 1.2$–1.8) in the initial stages of the cultivation to strongly shear thinning liquid ($n = 0.5$–0.6) around 100 h of the run. It took place in the nitrogen-rich system, with high amount of biomass. In nitrogen-limited conditions, flow index was usually close to 1. No clear explanations about higher than 1 flow indices were given. Lovastatin yield was to the high extent correlated with high biomass concentration and fungal morphology, but again no direct correlation between the rheological properties of the broth and lovastatin titre was found.

Another work by Rodriguez Porcel et al. [60] was generally focused on establishing a complicated cultivation strategy (batch and semi-continuous operational mode bioreactor in the FBB). The rheological measurements were only the additional data. Here, the broth from the initial shear thickening became shear thinning together with the growth of biomass ($n = 0.4$–0.9) and the consistency index K was of similar level as for the slowly stirred STB as shown in Casas Lopez et al. [17]. Nevertheless, it was lower for the semi-continuous runs (around 0.05). In these runs, 90 % of biomass existed as fluffy small pellets (the rest were filaments) and

Table 3 Rheological parameters of cultivation broth with *A. terreus* in lovastatin biosynthesis

Type of fluid	Rheological parameters		Apparent viscosity Pa s/shear rate (s^{-1})	Varied process conditions	Bioreactor	Literature
	Consistency, K, N m^{-2} s^n	Flow behaviour index, *n*				
Ostwald-de Waele	About 0.01	1–1.5	0.05–0.04/[a]	n = 800 rpm	5-L STB	Casas Lopez et al. [17]
	0.05–0.65	1–0.3	0.0065–0.04/[a]	n = 300 rpm		
	Up to 0.5	0.4–1.6	N/A	Stirred tank	5-L STB; 17-L FBB	Rodriguez Porcel et al. [62]
	0.005–0.025	1.2–1.4	N/A	Fluidized bed		
	0.05–0.55	0.5–1.5	N/A	Initial nitrogen conc. and oxygen conc. in the aeration gas	17-L FBB	Rodriguez Porcel et al. [63]
	0.05–0.35	0.5–1.5	N/A	Various bioreactor operational mode strategy		Rodriguez Porcel et al. [60]
	0.05–0.25	0.4–10.8	N/A			Rodriguez Porcel et al. [61]
	From 0.00978 to 0.06685	0.694–0.48	0.010–0.060/[b]	–	2-L internal loop airlift	Gupta et al. [30]
–	–	–	0.05–0.7/0.17	Dispersed and pelleted mycelium	Pilot-scale 500-L STB	Gbewonyo et al. [26]

[a] Shear rate value not given; only equation from which it was calculated
[b] Apparent viscosity estimated from the power law; shear rate value not given

this value decreased to 70 % in the end of the cultivation. In the work by Rodriguez Porcel et al. [61], a novel bioreactor strategy was proposed and rheological data actually confirmed the previous findings.

The only source, which may allow for the critical comparison of the above-presented data, is the publication of Gupta et al. [30]. They also used Ostwald-de Waele model to describe the rheology of *A. terreus* broths. They showed the changes of apparent viscosity[1] in time. It increased from the value of about 0.01 Pa s at 24 h to 0.06 Pa s at 144 h and subsequently decreased down to 0.035 Pa s. Consistency increased from 0.00978 to 0.06685 Pa s^n (its maximum) and flow behaviour index n changed from 0.694 to 0.48. These authors also correlated the flow behaviour index with biomass concentration introducing an exponent of biomass denounced as α, whose value was actually constant to 2.0. They presented this correlation only in the form of a graph, giving no equation.

All most important data concerning the rheology of *A. terreus* broths for lovastatin production were collected in Table 3.

[1] Upon other literature data, it can be concluded that these authors the most probably made the error in viscosity units attributing Pa s to the values expressed in cP. Here, it was corrected. Also the units of rheological parameters of Ostwald-de Waele equation were incorrect and changed here to be consistent with the data from other authors.

Despite many data available, it would be still useful to seek for the quantitative correlations between lovastatin titre and rheological properties of the broth and fungal morphology. Probably a kind of the multivariate correlation would be useful for this purpose. In our subjective opinion, also the existence of the range, in which *A. terreus* broth is shear thickening, requires the confirmation from the other scientific teams.

4 Role of Oxygen Transfer in Lovastatin Biosynthesis

The main operational parameter that reflects the presence and influence of oxygen in the bioreactor process is dissolved oxygen saturation level, usually denounced as DO or pO_2. Holding its level constant is a convenient and very practical way to assure the required amount of this electron acceptor to the fungal culture. However, the varying composition of the cultivation broth and, above all, the varying fungal morphology in conjunction with broth rheology does not allow for detailed conclusions to be exclusively drawn from its levels. Thereby, the effect of oxygen on fungal culture has to be more thoroughly investigated, especially with regard to its transfer.

Due to the variety of morphological forms in filamentous fungi, the study of mass transfer in the bioreactors, in which filamentous fungi grow, is far more complicated than in the case of such microorganisms as yeasts or bacteria. Its main reason is the formation of macroscopic agglomerates, namely pellets (see Sect. 2), in which several stages of oxygen transfer might be potentially limiting (Fig. 5). Normally, oxygen transfer in biological systems is limited in the liquid film surrounding the air bubbles (stage 3 in Fig. 5). If pellets are formed, both the liquid film surrounding the pellets and diffusion in the pellets alone should be considered. There are three physical parameters to quantify these potentially limiting mass transfer processes, i.e. convective mass transfer coefficients, k_La and k_Sa, and effective diffusion coefficient, D_{eff}. However, not all these steps of oxygen transfer in the fungal cultures were studied with same engagement. It is clearly seen in the detailed review by Garcia-Ochoa and Gomez [25]. At the beginning, they presented all these steps, but further the review dealt with the gas-liquid phase transfer (k_La), about which the literature is really rich.

Similarly, with regard to lovastatin-producing *A. terreus*, the values of k_La were presented in several papers, while the process of oxygen diffusion inside *A. terreus* pellets has been rarely studied. Nowhere can any data concerning the values of convective mass transfer coefficient in the liquid film surrounding the pellet k_Sa be found.

4.1 Impact of Oxygen Saturation

There are few studies concerning directly the impact of oxygen saturation on lovastatin formation by *A. terreus*. The most detailed study was made by Lai et al. [42] in a 5-L STB. They tested four levels of pO_2, i.e. 10, 20, 30 and 40 %

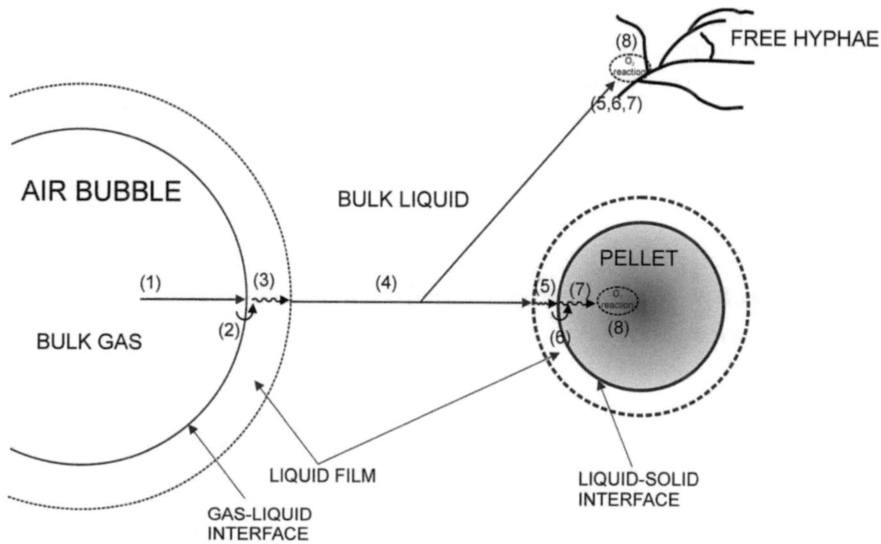

Fig. 5 Oxygen transfer stages in the bioreactor during the cultivation a filamentous fungus: *1* convection in the turbulent bulk gas of the air bubble, *2* diffusion through gas–liquid interface, *3* convection in the liquid film, *4* convection in the turbulent bulk liquid *5* convection through the liquid film surrounding the fungal cells (filaments) or agglomerates (pellets) *6* diffusion through the liquid–solid interface *7* diffusion inside the pellet *8* biological reaction site (inspired and based upon Doran [19])

concluding that the optimum level was 20 %. Higher pO_2 level was in their opinion not favourable for lovastatin formation. This finding is to a certain extent in the contradiction with some works from University of Almeria. These researchers used in their several works the oxygen-enriched (80 % O_2) air for the aeration of *A. terreus* broths. It allowed for holding the oxygen saturation level at 400 % with vvm between 0.5 and 1.5 $L_{air}\,L^{-1}\,min^{-1}$ [17, 60, 63]. On the other hand, Bizukojc and Ledakowicz [10] claimed that high constant aeration rate (above 1 $L_{air}\,L^{-1}min^{-1}$) led to (+)-geodin formation (an octaketide by-product) rather than to lovastatin. Earlier works concerning lovastatin formation were less focused on the issue of oxygen. Only Novak et al. [52], testing the following three levels of oxygen saturation, i.e. 35, 70 and 80 %, found that pO_2 = 70 % assured the highest lovastatin titres (in this publication lovastatin concentrations are not openly given in mg L^{-1} but in "units"). Analysing all these data, it is also important to check what the initial carbon source concentration in the culture was. It is the common biochemical knowledge that more carbon source requires more final electron acceptor for its catabolism. That is why the conclusions about the optimum oxygen saturation level were different. In the works cited above, lactose (C-source concentration) varied from 20 g L^{-1} [10], via 70 g L^{-1} [42] up to 114 g L^{-1} [17, 60, 63]. It certainly influenced the conclusions drawn by all these authors.

The most interesting idea of oxygen supply facilitation in *A. terreus* cultivation was proposed in the earlier mentioned work of Lai et al. [39, 41]. They used a water non-miscible organic oxygen carrier (n-dodecane) to increase lovastatin titre. They succeeded in shake flask culture, while in the bioreactor, the results were not satisfactory mainly due to undesired morphology of mycelium (see Sect. 2) and too high, in their opinion, oxygen saturation level in the broth.

4.2 Convective Mass Transfer Coefficients in Various Bioreactor Systems

The issue of mass transfer from the bubbles into the broth in *A. terreus* cultivation was the object of studies already in the early 1990s. Nevertheless, these data are usually scarce and, as it was with regard to broth rheology, no explicit correlation between k_La and lovastatin production were given. The aforementioned Gbewonyo et al. [26] studied in the limited range the interactions between *A. terreus* morphology (pelleted and dispersed) and mass transfer (k_La) during lovastatin biosynthesis. These measurements concerned the behaviour of their fungal system in response to the changes in agitation (between 100 and 250 rpm) in a STB. Convective mass transfer coefficient was usually lower for the systems with dispersed morphology (Table 4).

Casas Lopez et al. [17] also determined oxygen mass transfer coefficients in their culture. These values are given in Table 4 too. In their STB, pelleted growth of *A. terreus* dominated and they studied the time changes of k_La dependent on the rotation speed of the impeller.

Table 4 Values of convective mass transfer coefficient for *A. terreus* cultivations

k_La values	Fungal morphology	Process conditions	Type of bioreactor	Literature
Below 75 h^{-1} (0.014 s^{-1})	Dispersed	Response to changes in agitation between 50–75 h of the run	Pilot-scale 500-L STB	Gbewonyo et al. [26]
Up to 200 h^{-1} (0.056 s^{-1})	Pelleted			
About 0.04 s^{-1} (144 h^{-1})	Pelleted	O_2-enriched air at 800 rpm	5-L STB	Casas Lopez et al. [17]
0.002–0.003 s^{-1} (7–10 h^{-1})	Pelleted	O_2-enriched air at 400 rpm		
0.004–0.008 s^{-1} (14–28 h^{-1})	Pelleted	Superficial air velocity U_G = 0.014 m/s aerated with O_2-enriched air (80 %)	17-L FBB	Rodriguez Porcel et al. [63]
0.006–0.015 s^{-1} (21–54 h^{-1})	Pelleted	U_G = 0.021 m/s		
280 h^{-1} (0.0778 s^{-1})–100 h^{-1} (0.0278 s^{-1})	Pelleted	Changes of k_La in time without studying the influence of any process condition	1,000-L STB	Kumar et al. [38]

Casas Lopez et al. [17], using a classical chemical engineering approach, showed the experimental correlation for the *A. terreus* lovastatin-producing system between energy dissipation/circulation function EDCF (kW m^{-3} s^{-1}) defined as:

$$\mathrm{EDCF} = \frac{P_g}{k_c D^3 t_c} \tag{4.1}$$

where P_g is gassed power input (kW m^{-3}), k_c—geometric constant, D—impeller diameter (m), t_c—gassed circulation time (s), and convective mass transfer coefficient $k_L a$, achieving the following equation:

$$k_L a\,(s^{-1}) = 7.0 \times 10^{-4} \cdot (\mathrm{ECDF})^{0.76} \tag{4.2}$$

For P_g/V_L ratio (V_L—broth volume), the similar correlation was as follows:

$$k_L a\,(s^{-1}) = 2.24 \times 10^{-2} \cdot \left(P_g/V_L\right)^{0.92} \tag{4.3}$$

Rodriguez Porcel et al. [63] determined $k_L a$ values (Table 4) in a 17-L fluidized bed bioreactor and determined experimentally the exponents and constants of the following correlation for $k_L a$ in the function of broth viscosity and biomass concentration:

$$k_L a = a \cdot U_g^b \cdot \mu_{\mathrm{eff}}^c \cdot c_{\mathrm{b}}^d \tag{4.4}$$

where U_g—superficial gas velocity (m s^{-1}), μ_{eff}—apparent broth viscosity (Pa s) and c_{b}—biomass concentration (kg m^{-3}). As it was in the case of broth rheology, neither direct correlation between $k_L a$ and lovastatin titre was given nor any explicit conclusions drawn. Only general comment confirming that higher aeration is profitable for lovastatin formation and more biomass in the system results in higher lovastatin titre was given.

Kumar et al. [38] determined several values of $k_L a$ in a large 1,000-L bioreactor operating in the discontinuous fed-batch mode. It was maximally 280 h^{-1} (0.0778 s^{-1}) in 24 h of the run. It gradually decreased in time achieving the minimum values below 100 h^{-1} (0.0278 s^{-1}) in 288 h of the run (Table 4).

4.3 *Effective Diffusivities in A. terreus Pellets*

The determination of the effective diffusivities in the immobilized biocatalysts, as fungal pellets can be treated as a form of immobilized biocatalyst, is not an easy task. Precise measurements of oxygen profile in the pellet are required and the tools for this purpose are available since the 1980s [69]. Nevertheless, it must be remembered that these measurements can never be made directly in the bioreactor. Pellets must be withdrawn from the bioreactor or located in the special measuring tube connected to the bioreactor. It is the strong experimental limitation that has an impact on the number and quality of the obtained results.

A very good and detailed analysis how to measure and determine D_{eff} in the individual fungal pellet under various flow conditions was described by Hille et al. [33] for another fungal species *A. niger* AB.1.13 (α-amylase producer). For this purpose, the second Fick's law was used:

$$\frac{dc_{O_2}}{dt} = D_{eff} \cdot \frac{d^2 c_{O_2}}{dr^2} \tag{4.5}$$

where r is pellet radius and D_{eff}-effective diffusivity.

The scarce data, as this paper generally concerned morphological engineering techniques to maximize lovastatin production, on the values of effective diffusivities in *A. terreus* pellets were only obtained by Gonciarz and Bizukojc [28]. Under quiescent flow conditions, the oxygen profiles in at least three *A. terreus* pellets of various diameters in each sampling time from the shake flask culture were made in this work. Effective diffusivities were found from the following oxygen balance (so-called shell model for a biocatalyst) with the assumed Michaelis-Menten kinetics for oxygen utilization r_{O_2} in the pellets.

$$\frac{d^2 c_{O_2}(r)}{dr^2} \cdot r^2 + 2 \cdot r \cdot \frac{dc_{O_2}(r)}{dr} - \frac{1}{D_{eff}} \cdot r_{O_2} \cdot r^2 = 0 \tag{4.6}$$

Some exemplary data, i.e. an oxygen profile and the values of effective diffusivities, are shown in Fig. 6.

The determined values for *A. terreus* pellets from about 1,700 to 500 $\mu m^2\ s^{-1}$ were similar to the ones as in other fungal systems, although they were found as the parameters of model equation and determined in the stationary pellets motionlessly submerged in the cultivation broth. To compare, the relative D_{eff} for *A. niger* pellets ranged from 1,960 to 2,799 $\mu m^2\ s^{-1}$. Dependent on pellet density, it decreased even down to 1,120 $\mu m^2\ s^{-1}$ [33].

5 Controlling Fungal Morphology: Application of Morphological Engineering Tools

5.1 Morphological Engineering with Regard to Filamentous Organisms

There are several techniques to influence fungal morphology in the cultivation of filamentous organisms (both fungi and prokaryotic *Streptomyces*). They can be divided into the classical approach and morphological engineering techniques [37] The classical approach covers the following methods: spore suspension preparation

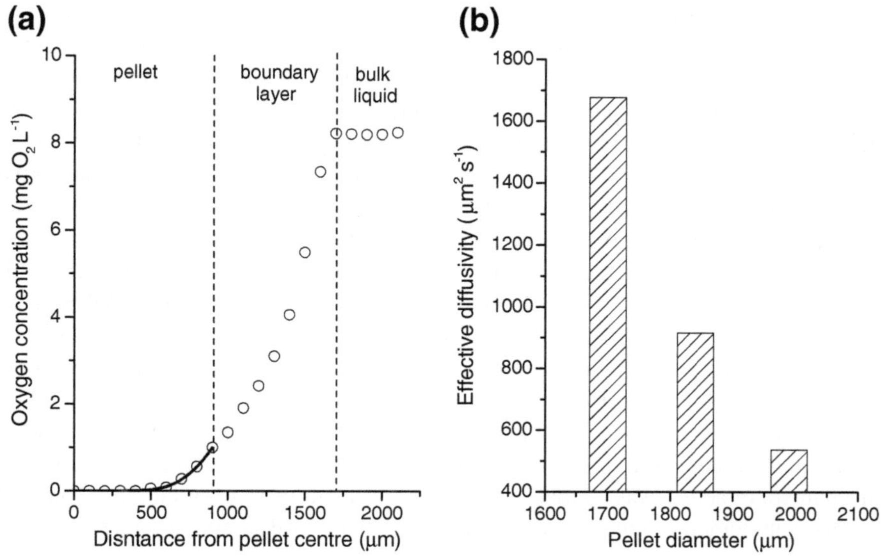

Fig. 6 Diffusion of oxygen inside *A. terreus* pellets: **a** measurement of oxygen profile in an individual pellet; solid line represents the solution of Eq. 4.6 (selected data from [28]) and **b** the effect of pellet diameter on effective diffusivity in *A. terreus* pellets (upon tabularized data from [28])

(manipulating with number of spores), changing of process parameters (e.g. pH and pO_2 shifting), changing of medium compositions (varying carbon and/or nitrogen source) and, above all, changing of hydrodynamic conditions in the bioreactor (shear stress induced by the impeller and aeration). The newer morphological engineering approach seems to be more subtle with regard to microorganisms as it acts on the level of the formation of pellets.

Filamentous fungi (and *Streptomyces*) cultures are evolved from spores. It often leads to obtaining various morphological forms of them. The importance of the morphological form of the filamentous fungi was already discussed in Sects. 2, 3 and 4 and strongly emphasized in the review by MacIntyre et al. [46]. What is the most important, in this work, the phrase "morphological engineering" is used as a keyword for the first time and defined as "tailoring morphologies for specific bioprocesses" [46].

The controversy, which morphological form is more useful for the formation of the given metabolic product, is always the matter of many discussions. Here, a good example was citric acid produced by *A. niger*. Never was it unequivocally said, which form is better [54].

There are two mechanisms of spore agglomeration in filamentous fungi dependent on their genera. In *Penicilli,* one observes the agglomeration of filaments and in *Aspergilli* the agglomeration of spores. In the case of prokaryotic *Streptomyces*, the pellet is formed from the individual spore [47, 51, 55].

In filamentous fungi, the size of pellets to be formed is the issue of the number of spores used. The higher number of spores is introduced to the preculture medium, the smaller pellets are formed. It is also possible that the dispersed mycelium occurs, if the number of spores is high enough [47]. This observation most likely became the foundation of morphological engineering. Various methods can be then applied to influence the process of spores agglomeration and this way to change the fungal morphology. There are several morphological engineering tools that can be used to act on the formation of fungal agglomerates. These are the following: (1) addition of mineral microparticles to the broth at the various stages of cultivation [20–23, 28, 29], called a microparticle-enhanced cultivation (MPEC) by [35], (2) change of broth osmolality [70] and (3) change of broth viscosity [53]. The applications of morphological engineering techniques to the filamentous microorganisms are collected in Table 5.

In all cases, the authors declared better productivity of the desired metabolite owing to the application of the morphological engineering technique. Generally, the mechanism of action was similar in all cases. The agglomerates of biomass obtained in the submerged culture, namely fungal pellets, had their structure changed because of the undertaken actions. The pellets got smaller and looser, what must have facilitated transport of nutrients into the cells and it was reflected in the productivity of a given metabolite.

5.2 Morphological Engineering for Lovastatin Production

The studies concerning the application of morphological engineering techniques for lovastatin production by *A. terreus* are limited (see Table 5). Actually, there are two publications, in which this issue was studied [28, 29] and all the conclusions were drawn upon shake flask culture of *A. terreus* (Table 3). The authors tested the addition of microparticles: talc and aluminium oxide. It was a two-stage cultivation: 24-h preculture and main culture. Talc powder occurred to be more efficient in *A. terreus* cultivation system. Also due to the fact that the microparticles act on the level of spore agglomeration, they had to be added to the preculture. It was the most important finding of these works, i.e. the stage, at which the microparticles should be added to the process. Only then, both the size and structure of *A. terreus* pellets were significantly changed and the positive effect on lovastatin titre was observed. Pellets once formed are not quite sensitive towards the action of microparticles, and in the case of *A. terreus,* the process of pellet formation was actually finished within 24 h in the preculture as earlier claimed by Bizukojc and Ledakowicz [11]. The increase of lovastatin titre with the decrease of fungal pellets diameter in the morphologically engineered culture confirmed the previous findings dealing with the relation of pellet diameter and lovastatin production [11]. The level of 12 g L^{-1} talc powder in the preculture occurred to be optimum for further evolution of the

Table 5 Selected most important approaches to use morphological engineering tools in the cultivation of filamentous organisms

Filamentous species	Metabolite	Morphological engineering technique	Type and working volume of bioreactor	Literature
Caldariomyces fumago	Chloroperoxidase	Aluminium oxide and talc microparticles (<42 μm) added to the preculture (0.05–25 g L^{-1})	100-mL (preculture) and 300-mL shake flasks	Kaup et al. [35]
A. niger AB 1.13 and its mutants	Fructofuranosidase α-amylase	Microparticles (<10 μm) of hydrous magnesium silicate (talc powder) in cultivation medium at conc. from 0.01 to 30 g L^{-1}	50-mL shake flasks and 3-L STB	Driouch et al. [21]
A. niger SKAn1015	Fructofuranosidase	Microparticles (<10 μm) of hydrous magnesium silicate (talc powder) 5 g L^{-1} added to cultivation media prior inoculation	50-mL shake flasks and 2.2-L STB	Driouch et al. [22, 23]
A. niger SKAn1015	Fructofuranosidase	Titanium silicate oxide (8 μm) added to cultivation media prior inoculation	50-mL shake flasks and 2.2-L STB	Driouch et al. [20]
Streptomyces hygroscopicus var. *geldanus*	None	Xanthan gum addition (up to 5 g L^{-1}) to increase medium viscosity and make it more non-Newtonian	100-mL shake flasks	O'Cleireigh et al. [53]
A. niger AB 1.13	Fructofuranosidase α-amylase	Osmolality from 0.35 osmol/kg (standard medium) to 3.6 osmol/kg by sodium chloride addition	3-L STB	Wucherpfennig et al. [70]
A. terreus ATCC20542	Lovastatin	Aluminium oxide and talc microparticles (<10 μm) added to the preculture and growth culture	150-mL shake flask	Gonciarz et al. [29]
A. terreus ATCC20542	Lovastatin	Talc microparticles (<10 μm) added to the preculture (1–20 g L^{-1})	150-mL shake flask	Gonciarz and Bizukojc [28]

mycelium and lovastatin production in the shake flask culture (even a 50 % increase of its titre) [28]. The experiments are being continued by these authors and their preliminary results indicate that the scale-up of lovastatin production by *A. terreus* with the use of talc microparticles to the 5.3-L working volume bioreactor was easy. It is due to the fact that the action of the microparticles takes place in the shake flask preculture step. Thereby, the inoculum was somewhat modified and further evolution of the fungus in the production medium either in shake flask or STB was influenced by its physiological and morphological state. In the bioreactor, the increase of lovastatin titre occurred to be even higher than that in shake flasks. Comparing to the control runs, it was doubled in the batch mode and even 2.5-fold increase was observed in the glycerol-fed fed-batch mode (publication in preparation).

It can be concluded from the statements above that MPEC of the lovastatin producer *A. terreus* is effective and future prospects are very promising.

6 Various Types and Operational Modes of Bioreactors to Produce Lovastatin

A bioreactor is the basic tool for bioprocess engineering research and the equipment without which there would not be any industrial production of useful metabolites. With regard to lovastatin biosynthesis by *A. terreus* in the submerged culture, there are quite many works, in which the experiments were conducted in bioreactors. Referring to the construction, actually type of bioreactor, mainly STBs of various working volume were used. A fluidized bed bioreactor was applied by only one research team [60–63]. One publication deals with an internal loop airlift bioreactor [31].

The majority of experiments were conducted in two operational modes of bioreactors, i.e. batch mode and various fed-batch modes. In the latter ones, higher metabolite titres are usually obtained. As there are several fed-batch modes of bioreactor operation and sometimes various names are used for it, here the classification of Moser [50] is going to be used irrespective of the fact how the authors of the given publication named it. In almost all cases of lovastatin production, two fed-batch modes have been used so far: either discontinuous fed-batch (feeding as an impulse flow in the set intervals) or continuous fed-batch mode, in which the feeding solution is pumped continuously with a constant or not constant rate to the bioreactor.

One of the earliest scientific works, in which lovastatin biosynthesis was studied in the bioreactor, was performed by Novak et al. [52]. They used a 15-L STB and two operational modes of the bioreactor: batch and discontinuous fed-batch at the initial glucose concentration of 100 g L^{-1}. Their lovastatin titres in the batch mode were different dependent on the set oxygen saturation but fast glucose utilization and its deficiency (below 10 g L^{-1}) signalized by dissolved oxygen level made these authors to apply discontinuous feeding with glucose solution at 150 h. The authors called their process as repeated fed-batch, but they did not write, whether during the feeding event a portion of broth was pumped out of bioreactor. Feeding allowed for regaining the activity of the fungus and higher lovastatin titre compared to the batch run. Next feeding step at 250 h occurred to be ineffective.

The work of Hajjaj et al. [32] was only aimed at testing various types of media, complex and synthetic ones, for lovastatin production by *A. terreus* ATCC 74135, but they applied 7- and 15-L STBs in the batch mode. They used constant aeration rate (vvm = 1 L_{air} L^{-1} min^{-1}) and agitation (400 rpm). In the discussion, they did not refer to process conditions in the bioreactor.

The experiments of Kumar et al. [38] were made in the largest bioreactor of all discussed here. They cultivated a mutant originated from *A. terreus* ATCC20541 in an indigenously designed 1,000-L STB. It was equipped with four baffles, and three six-bladed Rushton turbines (tank diameter was 780 mm and impeller diameter 320 mm). Due to the scale of the bioreactor, they had to use a two-stage preculture preparation (the second stage in a smaller bioreactor). Using a highly complex media (containing three carbon sources: glucose, maltodextrin, starch and three nitrogen sources: corn steep liquor, yeast extract and peptonized milk), they carried out the batch and discontinuous fed-batch experiments (they actually named it repeated batch, although no medium was withdrawn from the bioreactor) fed with maltodextrin solution. They claimed the increase in lovastatin titre from 1,270 to 2,000 mg L^{-1} in the fed-batch run.

Two types of bioreactors and various operational modes can be found in the publications from University of Almeria [17, 60, 61, 63].

Casas Lopez et al. [17] applied a 5-L STB in the batch mode for *A. terreus* ATCC20542. They studied the influence of various agitation speed (300, 600, 800 rpm) on the run of lovastatin formation in conjunction with the issues of fungal morphology, broth rheology and aeration of the broth (see Sects. 2, 3 and 4). Various aeration gases were used by them, i.e. air and oxygen-enriched air. They concluded that the highest speed of impeller 800 rpm irrespective of the aeration gas led to smaller pellets and worse lovastatin titre (below 40 mg L^{-1}). In their next work, a 17-L FBB (bubble column) for lovastatin production in the batch mode was applied [63]. Here, they studied the effect of various organic nitrogen levels and aeration for the process. In 2007, the same authors proposed a three-stage operational mode of the same FBB. The process started with the batch mode lasting from 4 to 7 days; next, it was fed with a fresh medium (2.4 L), and after 24 h, the semi-continuous phase occurred, in which the filtered broth was pumped in and out of the bioreactor (biomass was retained in the system). Batch runs were used to get reference data; nevertheless, the positive effect of the new strategy was not very significant. This strategy was repeated in another publication [61], but this time, the fed-batch stage was omitted and the feeding media in the semi-continuous mode were different. These were the complete initial medium, the medium without nitrogen source and the medium without carbon and nitrogen source. Feeding with the complete medium gave the worst results. The positive effects of the other two feeding strategies for lovastatin production were similar. All the works from this team are difficult to evaluate with regard to the positive effects of fed-batch or semi-continuous strategies, as theses authors never supplied data concerning carbon source (lactose) utilization.

Bizukojc and Ledakowicz [10] studied lovastatin formation by *A. terreus* ATCC20542 in a 2-L STB in batch and discontinuous fed-batch mode. What is interesting, apart from lovastatin, they traced for the first time the formation of (+)-geodin in the bioreactor. Their important finding concerning the operation of the bioreactor for lovastatin production in the batch mode was the introduction of pH control, never suggested in the previous works, with concentrated carbonates solution at the levels 7.6 and 7.8. It depressed by-product (+)-geodin formation and enhanced lovastatin production. They also confirmed that feeding of the batch culture with a carbon substrate led to the increase in lovastatin titre as lovastatin is believed to be strongly dependent on carbon substrate availability. There it was the discontinuous fed-batch mode and the bioreactor was fed with concentrated lactose solution. The solution was added when lactose concentration in the broth started to be deficient.

The work of Gupta et al. [31] is to a highest extent different from all the works presented above and below. They used strain *A. terreus* NRLL255. The authors proposed the continuous mode of bioreactor operation for lovastatin production. The biomass was either in the form of fungal pellets or immobilized hyphae on siran particles. They used an internal loop airlift bioreactor of 5-L working volume.

On the contrary to the previous works (Rodriguez Porcel et al. [60, 61]), biomass was not retained in the bioreactor, but continuously withdrawn from the bioreactor. The continuous mode started after 72 h of the batch process, and the effect of varied of dilution rates from 0.01 to 0.05 h^{-1} was studied. The optimum value of dilution rate at about 0.02 h^{-1} and wash-out rate at 0.045 h^{-1} was found. The authors claimed that they obtained steady states at each studied dilution rate. It may be discussible as fungal species differentiate and their cells are never identical. The most important achievement of this work is that these researchers successfully ran their continuous cultivation system for 45 days. Lovastatin productivity (volumetric formation rate) in the continuous mode was equal to 0.022 g L h^{-1} and 0.0255 g L^{-1} h^{-1} (at the most optimum vvm = 0.6 L_{air} L^{-1} min^{-1}) for pellets and immobilized biomass, respectively, and its concentration between 1.1 and 1.2 g L^{-1} [31]. It is clearly seen that immobilized system did not occur much better.

The only work that focused on the impact of bioreactor scale on lovastatin production was reported by Pawlak et al. [59]. The variety of cultivation data is presented there. They come from 150-mL shake flasks, 2-L and 5-L STBs. At first glance, it seemed that shake flasks were generally a better “bioreactor” for lovastatin production by *A. terreus* than a “real” bioreactor, even when the same medium composition was used. Thus, there was a need to tune cautiously the conditions of the cultivation in the bioreactor to achieve similar titres as in shake flasks. To these factors belong the valid aeration to achieve comparable substrate utilization rate and pH control, which was previously applied by Bizukojc and Ledakowicz [10].

The latest work from Pawlak and Bizukojc [58] proved upon the continuous fed-batch experiment in a 5-L STB that feeding profile is not the sole factor that may influence lovastatin formation by *A. terreus*. No direct and clear correlation between the feeding profile, either constant or varied, nor feeding substrate (lactose and/or glycerol) was found and the hypothesis about the other factors influencing lovastatin formation in the bioreactor was proposed. Much attention in this article was for the first time devoted to the issue of redox potential of the culture broth and assimilation of bicarbonate ions from the pH correction solution. The authors concluded that the best lovastatin titres correlated with steep decrease of redox potential in the first hours of the process and high rate of inorganic carbon assimilation. These two process parameters were shown to have the association with the biochemical mechanism of lovastatin production [58].

7 Kinetic Modelling of Lovastatin Biosynthesis

7.1 Lovastatin Formation Kinetics

The issue of product formation kinetics is not an easy task, especially with regard to secondary metabolites as there is a plethora of kinetic expressions in literature to describe the formation of a metabolic product. In the most classical approach, product formation kinetics can be divided into four types upon the correlation

Table 6 Types of association of lovastatin formation with biomass growth

Type of lovastatin formation association	Reference	Comments
Mixed growth association	Szakacs et al. [66]	
Mixed growth association	Kumar et al. [38]	
Mixed growth association	Lai et al. [42]	
Mixed growth association	Jia et al. [34]	
Non-growth association	Hajjaj et al. [32]	Starvation is required for lovastatin formation
Non-growth association	Casas Lopez et al. [15]	Lack of nitrogen and excess carbon is required
Non-growth association	Rodriguez Porcel et al. [64]	Production of lovastatin in the stationary phase (age of spores and its influence on lovastatin formation was discussed here)
Mixed growth association (dominates growth association)	Bizukojc and Ledakowicz [9]	Correlation between biomass growth rate and lovastatin formation rate shown

between product specific formation rate and specific biomass growth rate. These are (1) growth-associated product formation, (2) mixed growth-associated product formation (two phase are then present growth-associated and non-growth-associated phase, described usually by Luedeking-Piret equation, (3) non-growth-associated product formation and (4) negative correlation between product formation and biomass growth [50]. Nevertheless, in some cases metabolite formation mechanism is complicated and none of these cases can be attributed to them.

With regard to lovastatin, this aspect was also considered; however, practically no articles were exclusively devoted to it. Various authors rather expressed their opinions about this fact discussing the other issues connected with lovastatin production by *A. terreus*. Furthermore, the opinions are to a certain extent divided. They are all collected in Table 6 (articles concerning kinetic models are not included).

The opinion that lovastatin formation is mixed associated with biomass growth dominates (Table 6). It is the most difficult to agree with the opinion of Hajjaj et al. [32] that starvation is required for lovastatin formation, as all other authors associate lovastatin production with the sufficient amount of carbon source in the culture (see Sect. 6 too).

Quantitative data concerning lovastatin formation kinetics, e.g. the values of specific lovastatin formation rate, are not frequently met in literature. Hajjaj et al. [32] found that in the late phase of cultivation, when the complex medium was used, it was equal to 0.034 mg LOV g X^{-1} h^{-1}, while in the chemically defined medium (with sodium glutamate as the N-source) it was higher reaching 0.093 mg LOV g X^{-1} h^{-1}). In the articles coming from Lodz University of Technology team several times, lovastatin formation rate was estimated and it reached even 0.25 mg LOV g X^{-1} h^{-1} in the medium with lactose and yeast extract (Bizukojc and Ledakowicz [9, 11]. It was twice higher (up to 0.5 mg LOV g X^{-1} h^{-1}), when the mixture of glycerol and lactose was used as the carbon sources in the discontinuous fed-batch system [56]. In this work, the changes of specific lovastatin formation rate in time were also shown and its highest values were always observed in the trophophase.

7.2 Mathematical Models to Describe Lovastatin Production by A. terreus

In this section, a critical review of the so-far-published mathematical model for lovastatin biosynthesis is going to be presented. The advantages and drawbacks of these models are going to be discussed.

The oldest model for lovastatin formation by *A. terreus* can be found in Liu et al. [43]. This model belongs to the class of morphologically structured models. It means that the hyphae were divided into compartments of various morphological and physiological properties. The differentiation of hyphae was also taken into account. Nevertheless, this model is not fully original as these authors adopted the model for penicillin production by *Penicillium chrysogenum*. At the same time, they also used several model parameters for *P. chrysogenum*, especially the part of the model concerning the differentiation of hyphae was directly taken from *P. chrysogenum* model. It can be regarded as a certain drawback as Liu et al. [43] did not show any evidence that the differentiation process in *Penicilli* and *Aspergilli* is the same. Furthermore, the morphology of these two genera can be different due to mechanism of pellet formation (see Sects. 2 and 5).

Some parameters were estimated upon their own experimental data from a 10-L bioreactor. Finally, they verified it upon the experimental data from an industrial scale 1,000-L bioreactor. The main assumptions and equations of the model are as follows. There were three zones of the mycelium distinguished, i.e. zone M (z_M) to denounce actively growing hyphae, zone N (z_N) to denounce non-growing hyphae and zone D (z_D) of deactivated cells. Zone M could be transformed with rate r_N into zone N in accordance with the equation:

$$r_N = \frac{k_{D1}}{c_S + K_{D1}} \cdot z_M \tag{7.1}$$

where k_{D1} and K_{D1} are the constants, and zone N with rate r_D into zone D in accordance with equation:

$$r_D = k_{D2} \cdot z_N \tag{7.2}$$

where k_{D1} and k_{D2} are rate constants of this process and K_{D1} is the saturation constant.

Carbon substrate (hydrolysed starch denounced as c_S) was assimilated by zone M for growth

$$r_M = \mu_{maxAT} \frac{c_S}{c_S + K_S} \cdot z_M \tag{7.3}$$

where μ_{maxAT}—maximum specific biomass growth rate and K_S is Monod constant.

Nitrogen substrate was not taken into account, which can be regarded as another drawback due to its inhibitive action with regard to lovastatin formation [16].

The most important equation for lovastatin production (c_P) with rate r_P was presented in the form:

$$r_P = k_P \cdot \frac{c_S}{K_P + c_S \cdot \left(1 + \frac{c_S}{K_I}\right)} \cdot (z_N + \gamma \cdot z_M) - k_D \cdot c_P \tag{7.4}$$

where k_P is product (lovastatin) formation rate constant, k_D lovastatin decay rate and γ takes the participation of zone M in product formation into account.

The participation of both growing and non-growing hyphae in lovastatin formation indicates that these authors assumed mixed growth-associated product formation. It is here not understandable why glucose is regarded to be inhibitive in this system, as actually no other authors ever postulated it. The term of lovastatin decay is a good idea to describe frequently observed phenomenon of decreasing lovastatin concentration.

Substrate was assumed to be consumed by the actively growing zone to form biomass (X_{AT}) in accordance with Eq. 7.5, in which also the maintenance term (m) was included.

$$r_{SC} = -\frac{1}{Y_{M/S}} \cdot \mu_{maxAT} \cdot \frac{c_S}{K_S + c_S} \cdot z_M - m \cdot X_{AT} \tag{7.5}$$

where $Y_{M/S}$ is a yield of actively growing zone on substrate.

The fit of this model to the experimental data occurred to be very good.

The other model was proposed by Bizukojc and Ledakowicz [8]. This model was unstructured. It was formulated for the growth of *A. terreus* on an individual carbon substrate, lactose (LAC). Apart from lactose, organic nitrogen (N), lovastatin (LOV) and biomass (X) were balanced. It made four equations to describe both batch and fed-batch bioreactor.

The assumptions of this model were as follows. Lactose was regarded as a sole carbon source in the studied system. Amino acids originated from yeast extract were not utilized as a carbon source irrespective of lactose and yeast extract concentrations. Contois model (saturation constants K_{LAC} and K_N) which takes the amount of biomass into account was used as the limitation term in all rate expressions. Yeast extract was the sole nitrogen source. The excess of nitrogen exerted an inhibitive effect ($K_{I,N}$ and $K_{I,N}^{LOV}$) on lovastatin biosynthesis and lactose uptake. Mevinolinic acid (lovastatin) biosynthesis was mixed growth-associated.

Lactose was both utilized for biomass formation (yield coefficient $Y_{X/LAC}$) and biosynthesis of lovastatin (yield coefficient $Y_{LOV/LAC}$). Therefore, two terms for these phenomena (growth-associated q_{max}^{LOV} and non-growth-associated k_{LOV} product formation and substrate utilization) were applied in lactose and lovastatin balance. All that resulted in the following forms of equations to express the specific rates of lactose uptake (σ_{LAC}), nitrogen uptake (σ_N),, lovastatin formation (π_{LOV}) and biomass growth (μ):

$$\begin{aligned}\sigma_{LAC} = & -\frac{1}{Y_{X/LAC}} \cdot \mu_{max} \cdot \frac{c_{LAC}}{c_{LAC} + K_{LAC} \cdot c_X} \cdot \frac{c_N}{c_N + K_N \cdot c_X} \cdot \frac{K_{1,N}}{K_{1,N} + c_N} \\ & - \frac{1}{Y_{LOV/LAC}} \cdot q_{max}^{LOV} \cdot \frac{c_{LAC}}{c_{LAC} + K_{LAC}^{LOV} \cdot c_X} \cdot \frac{K_{1,N}^{LOV}}{K_{1,N}^{LOV} + c_N}\end{aligned} \tag{7.6}$$

$$\sigma_N = -\frac{1}{Y_{X/N}} \cdot \mu_{max} \cdot \frac{c_{LAC}}{c_{LAC} + K_{LAC} \cdot c_X} \cdot \frac{c_N}{c_N + K_N \cdot c_X} \tag{7.7}$$

$$\pi_{LOV} = q_{max}^{LOV} \cdot \frac{c_{LAC}}{c_{LAC} + K_{LAC}^{LOV} \cdot c_X} \cdot \frac{K_{1,N}^{LOV}}{K_{1,N}^{LOV} + c_N} + k_{LOV} \cdot c_{LAC} \tag{7.8}$$

$$\mu = \mu_{max} \cdot \frac{c_{LAC}}{c_{LAC} + K_{LAC} \cdot c_X} \cdot \frac{c_N}{c_N + K_N \cdot c_X} \tag{7.9}$$

where K_{LAC}, K_N and K_{LAC}^{LOV} are Contois-type saturation constants and c_{LAC}, c_N, c_{LOV} and c_X are concentrations of lactose, organic nitrogen, lovastatin and biomass, respectively.

The parameters of the model were either directly determined from the experimental data, namely μ_{max} and yield coefficients, or identified with the use of the optimization algorithm. The data from shake flask experiments (of 150 mL working volume) performed at various concentrations of carbon and nitrogen were used to find all model parameters. A separate set of data (also from shake flask culture) was used for its verification. Additionally, the model was tested on the set of experimental data coming from the discontinuous fed-batch process (in shake flask). Ultimately, another verification of this model with the use of the experimental data from the bioreactor runs was later published by Bizukojc [6]. The main drawback of this model is its simplicity, which hardly represent the biological phenomenon during *A. terreus* growth, moderately good fit of the simulated curves with experimental data. It especially concerned the character of biomass curve.

This model was later extended by Pawlak and Bizukojc [57] for a two carbon substrates system. Most symbols used in it are the same as above. The assumptions of the model were very similar. The main difference was that two carbon sources were used, i.e. lactose and glycerol (GLC). Due to the experimental observations that these substrates were utilized consecutively, first glycerol (with yield $Y_{X/GLC}$), then lactose, it was assumed that glycerol inhibits lactose uptake. Lovastatin formation was assumed to be mixed growth-associated (expressed by yields $Y_{LOV/X}$,

$Y_{LOV/LAC}$ and rate constant q_{max}^{LOV}). The equations of this model, expressed as above, in the forms of specific rates σ_{LAC}, σ_{GLC}, σ_{N}, π_{LOV} and μ, were as follows:

$$\begin{aligned}\sigma_{LAC} = & -\frac{1}{Y_{X/LAC}}\cdot\mu_{max}\cdot\frac{c_{GLC}}{c_{GLC}+K_{GLY}^{X}\cdot c_X}\cdot\frac{c_N}{c_N+K_N^X\cdot c_X}\cdot\frac{K_{I,GLC,1}}{K_{I,GLC.1}+c_{GLC}}\\ & -\frac{1}{Y_{LOV/LAC}}\cdot q_{max}^{LOV}\cdot\frac{c_{LAC}}{c_{LAC}+K_{LAC}^{LOV}\cdot c_X}\cdot\frac{K_{I,N}^{LOV}}{K_{I,N}^{LOV}+c_N}\cdot\frac{K_{I,GLC,2}}{K_{I,GLC,2}+c_{GLC}}\end{aligned} \quad (7.10)$$

$$\begin{aligned}\sigma_{GLC} = & -\frac{1}{Y_{X/GLC}}\cdot\mu_{max}\cdot\frac{c_{GLC}}{c_{GLC}+K_{GLY}^{X}\cdot c_X}\cdot\frac{c_N}{c_N+K_N^X\cdot c_X}\\ & -\frac{1}{Y_{LOV/GLC}}\cdot q_{max}^{LOV}\cdot\frac{c_{GLC}}{c_{GLC}+K_{GLC}^{LOV}\cdot c_X}+\frac{K_{I,N}^{LOV}}{K_{I,N}^{LOV}+c_N}\end{aligned} \quad (7.11)$$

$$\sigma_N = -\frac{1}{Y_{X/N}}\cdot\mu_{max}\cdot\frac{c_{LAC}}{c_{LAC}+K_{LAC}^{X}\cdot c_X}\cdot\frac{c_N}{c_N+K_N^X\cdot c_X} \quad (7.12)$$

$$\begin{aligned}\pi_{LOV} = & \; Y_{LOV/X}\cdot\mu_{max}\cdot\frac{c_{GLC}}{c_{GLC}+K_{GLY}^{X}\cdot c_X}\cdot\frac{c_N}{c_N+K_N^X\cdot c_X}\\ & +q_{max}^{LOV}\cdot\frac{c_{LAC}}{c_{LAC}+K_{LAC}^{LOV}\cdot c_X}\cdot\frac{K_{I,N}^{LOV}}{K_{I,N}^{LOV}+c_N}\end{aligned} \quad (7.13)$$

$$\mu = \mu_{max}\cdot\frac{c_N}{c_N+K_N^X\cdot c_X} \quad (7.14)$$

where $K_{LAC}^{X}, K_{GLY}^{X}, K_{LAC}^{LOV}$ and K_N^X are Contois-type saturation constants, and $K_{I,GLC,1}$ and $K_{I,GLC,2}$ are inhibition constants referring to glycerol.

It is important to notice here that in this model, several limitation Contois terms were eliminated. It was done for the sake of better robustness of this model [57].

The latest approach to modelling of lovastatin production, but with the participation of another microorganism *A. flavipes* was shown by Gomes et al. [27]. Although it was not *A. terreus,* few comments should be added about it. First of all, this model was formulated for the fed-batch system with glucose, lactose (C-sources) and sodium glutamate (N-source). Also lovastatin, biomass, oxygen were balanced. The expressions for the reaction rates were as follows. Biomass growth on glucose was assumed to be influenced by dissolved oxygen level and expressed as

$$\mu_1 = \mu_M\frac{g}{K_G+g}\cdot\frac{c_L}{c_L^*} \quad (7.15)$$

where μ_M is maximum biomass growth rate, g—glucose concentration, and c_L and c_L^* oxygen concentration, actual and saturation, respectively.

Biomass growth on lactose was expressed by Eq. (7.16)

$$\mu_2 = \mu_{\mathrm{M}} \frac{l}{K_L + l} \cdot (1 - \kappa) \tag{7.16}$$

and product formation on lactose by Eq. (7.17)

$$\pi = \pi_{\mathrm{M}} \frac{l}{K_L + l} \cdot \kappa \tag{7.17}$$

where π_{M}—maximum lovastatin formation rate and l—lactose concentration, and at the same time, dependent on ratio κ

$$\kappa = \frac{n}{n + g} \tag{7.18}$$

which expressed the presence of sodium glutamate n (N-source) in the medium. The interesting thing is that the structure of this model was checked with the use of control theory methods for its correctness. The authors also claimed that it was found to be suitable for process control applications [27]. Its undoubtful advantage is taking of the oxygen level in the culture broth into account.

7.3 Linear Growth of Biomass in A. terreus. Returning to Fungal Morphology

One of the most important issues connected with *A. terreus* growth kinetics is usually its untypical growth curve. It can be found in the variety of publications. In the growth phase, namely trophophase, as it is traditionally named for filamentous fungi, the character of growth is hardly exponential (inter alia: [27, 42]).

Some authors, like Casas Lopez et al. [17], did not show biomass concentration points from the beginning of the process but after 50 h of the cultivation. All in all, it is unlikely to find in literature the exponential growth curve for *A. terreus*. A linear biomass growth observed also by Bizukojc and Ledakowicz [8–10] and Pawlak and Bizukojc [57] seems to be untypical. Lack of typical exponential phase may make it difficult to model *A. terreus* growth as the existence of this phase is often assumed in the kinetic models, including the ones described above. A closer look to this phenomenon was made by Pawlak and Bizukojc [57]. They thoroughly studied the very early stages of *A. terreus* growth in the shake flask culture and found that the exponential growth phase lasted not longer than 12 h since the inoculation with the 24-h preculture (Fig. 7).

Later, it changed into the linear growth phase. This change was easily correlated with the size of pellets formed, as they were usually not bigger that 500 μm in the exponential growth phase. As long as the pellets remained small enough, no

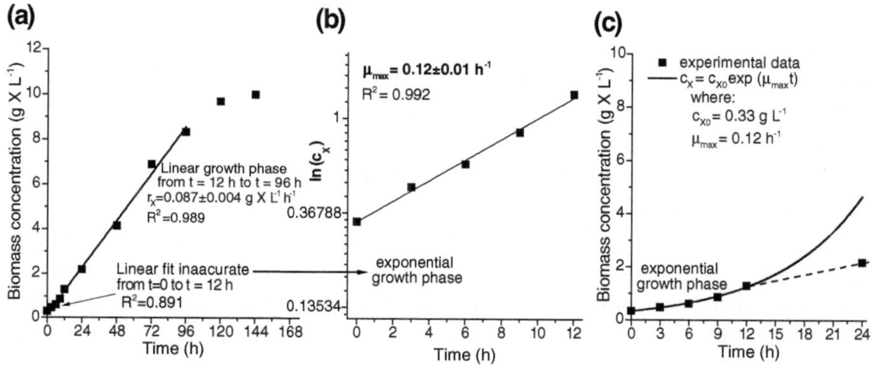

Fig. 7 Untypical biomass growth during lovastatin biosynthesis by *A. terreus* on lactose as a sole carbon source **a** linear growth phase between 12 and 96 h of the experiment, **b** determination of $\mu_{\max}$ upon the data from the first twelve hours of the run **c** experimental data versus simulation made for the 12-h exponential growth (graph prepared upon the experimental data from [57])

additional growth limitations were present in the system and the *sine qua non* condition, evolving from the definition of exponential growth phase about unlimited growth was fulfilled. Further increase in pellet diameter and change of its structure led probably to the decrease of oxygen effective diffusivities (see Sect. 4.2) in the pellets and oxygen limitation. These findings remain also in agreement with the observation of Pawlak and Bizukojc [58]. A strong peak of oxygen utilization (increasing air flow rate in the bioreactor to hold the set 20 % dissolved oxygen level) was observed within the first 12 h of cultivation in the 5-L STB.

To conclude, the linear growth of *A. terreus* shows how important for the investigations of lovastatin biosynthesis is to take the issue of fungal morphology into account.

8 Future Prospects

The bioprocess engineering aspects of lovastatin biosynthesis by *A. terreus* were studied, in our subjective opinion, at the various levels of details. There are many studies in literature concerning lovastatin production in bioreactors of various types, sizes and, what is the most important, of various operational modes (from batch, through the variety of fed-batch, up to continuous). The fairly clear image of the most optimum bioreactor set-up to produce lovastatin by *A. terreus* can be drawn upon them. Also a lot of rheometric and viscosity data for *A. terreus* broths obtained at various process conditions and bioreactors supply the sufficient knowledge to associate the run of lovastatin production process with varying rheological properties of the fungal broth. Also there are numerous data concerning *A. terreus* morphology.

However, there are several engineering issues with regard to lovastatin biosynthesis by *A. terreus* that in our opinion still require thorough studies. These are (1) issue of culture aeration with regard to mass transfer, (2) application of morphological engineering techniques for enhancing of lovastatin production and (3) lovastatin formation kinetics.

The first point should be studied at the following three levels, i.e. oxygen transfer from gas to liquid phase, from liquid phase to solid (mycelium) phase and at the level of oxygen diffusion inside fungal agglomerates (pellets). The issues connected with oxygen transfer from gas to liquid phase seemed to be studied in detail in many papers as the values of convective mass transfer coefficients for *A. terreus* broths were determined and a variety of aeration strategies to hold the desired pO_2 levels were applied. Nevertheless, lack of direct correlations and clear conclusions associating the amount of carbon substrate introduced to the culture with oxygen concentration levels and lack of the correlations between oxygen uptake rate (OUR), k_La and lovastatin formation makes this area of knowledge still incomplete. Also the changes of apparent viscosity and rheological parameters of the cultivation broth may have their contribution in these correlations. Furthermore, due to the strong experimental limitations, the studies on oxygen transfer into fungal pellets (film surrounding the pellet and intrapellet diffusion) were just began. In our opinion, the hypothesis can be propounded that these steps of oxygen transfer in *A. terreus* culture may have the predominant influence on the overall efficiency of the lovastatin biosynthetic system. It awaits the verification.

With regard to the second point, i.e. the application of morphological engineering techniques for enhancing of lovastatin production, there are very limited data. As generally, the application of morphological techniques is extremely promising upon the example of other filamentous fungi species and metabolites and at the same time the application of these techniques is not expensive even in the large scale, the research regarding lovastatin production by *A. terreus* should be directed into this route. Analysing literature data and own experiences the most significant factor that undoubtedly influence lovastatin production to the highest extent is fungal morphology. It cannot but have the effect on oxygen supply to the fungus, which connects this point with the issues above. That is why morphological engineering techniques shall prove successful in increasing the productivity of *A. terreus* for lovastatin biosynthesis.

Referring to the third point, it is known that the association of product formation kinetics with biomass growth can be fairly complicated. But it does not exclude thorough studies on this subject, although the elegant kinetic correlation would not be found. But the discrepancy in the opinions about lovastatin formation kinetics is at least anxious. It can be supposed that it is more dependent on growth conditions than one expects and the changes in medium composition or bioreactor processes conditions influence on it. It should be verified too.

References

1. Albers-Schoenberg G, Joshua H, Lopez MB, Hensens OD, Springer JP, Chen J, Ostrove S, Hoffman CH, Alberts AW, Patchett AA (1981) Dihydromevinolin, a potent hypocholesterolemic metabolite produced by *Aspergillus terreus*. J Antibiot 34:507–512
2. Alberts AW, Chen J, Kuron G, Hunt V, Huff J, Hoffman C, Rothrock J, Lopez M, Joshua H, Harris E, Patchett A, Monaghan R, Currie S, Stapley E, Albers-Schonberg G, Hensens O, Hirshfield J, Hoogsteen K, Liesch J, Springer J (1980) Mevinolin: a highly potent competitive inhibitor of hydroxymethylglutaryl-coenzyme a reductase and a cholesterol-lowering agent. Proc Natl Acad Sci USA 77:3957–3961
3. Auclair K, Kennedy J, Hutchinson R, Vederas JC (2001) Conversion of cyclic nonaketides to lovastatin and compactin by a lovc deficient mutant of *Aspergillus terreus*. Bioorg Med Chem Lett 11:1527–1531
4. Banos JG, Tomasini A, Szakacs G, Barrios-Gonzalez J (2009) High lovastatin production by *Aspergillus terreus* in solid-state fermentation on polyurethane foam: an artificial inert support. J Biosci Bioeng 108:105–110
5. Barrios-González J, Banos JG, Covarrubias AA, Garay-Arroyo A (2008) Lovastatin biosynthetic genes of *Aspergillus terreus* are expressed differentially in solid-state and in liquid submerged fermentation. Appl Microbiol Biotechnol 79:179–186
6. Bizukojc M (2009) Biosynthesis of polyketide compounds by *Aspergillus terreus* (in Polish). DSc thesis, Lodz University of Technology, Poland
7. Bizukojc M, Ledakowicz S (2005) The influence of environmental factors on mycelial growth and biosynthesis of lovastatin by *Aspergillus terreus* (in Polish). Biotechnologia-monografie 2(2):25–36
8. Bizukojc M, Ledakowicz S (2007) A macrokinetic modelling of the biosynthesis of lovastatin by *Aspergillus terreus*. J Biotechnol 130:422–435
9. Bizukojc M, Ledakowicz S (2007) Simultaneous biosynthesis of (+)-geodin by a lovastatin-producing fungus *Aspergillus terreus*. J Biotechnol 132:453–460
10. Bizukojc M, Ledakowicz S (2008) Biosynthesis of lovastatin and (+)-geodin by *Aspergillus terreus* in batch and fed-batch culture in the stirred tank bioreactor. Biochem Eng J 42:198–207
11. Bizukojc M, Ledakowicz S (2010) The morphological and physiological evolution of *Aspergillus terreus* mycelium in the submerged culture and its relation to the formation of secondary metabolites. World J Microbiol Biotechnol 26:41–54
12. Bizukojc M, Pawlowska B, Ledakowicz S (2007) Supplementation of the cultivation media with B-group vitamins enhances lovastatin biosynthesis by *Aspergillus terreus*. J Biotechnol 127:258–268
13. Bradamante S, Barenghi L, Beretta G, Bonfa M, Rollini M, Manzoni M (2002) Production of lovastatin examined by an integrated approach based on chemometrics and DOSY-NMR. Biotechnol Bioeng 80:589–593
14. Casas Lopez JL, Rodriguez Porcel EM, Vilches Ferron EM, Sanchez Perez JA, Fernandez Sevilla JM, Chisti Y (2004) Lovastatin inhibits its own synthesis in *Aspergillus terreus*. J Ind Microbiol Biotechnol 31:48–50
15. Casas Lopez JL, Sanchez Perez JA, Fernandez Sevilla JM, Acien Fernandez FG, Molina Grima E, Chisti Y (2003) Production of lovastatin by Aspergillus terreus: effects of the C:N ratio and the principal nutrients on growth and metabolite production. Enzyme Microb Technol 33:270–277
16. Casas Lopez JL, Sanchez Perez JA, Fernandez Sevilla JM, Acien Fernandez FG, Molina Grima EM, Chisti Y (2004) Fermentation optimization for the production of lovastatin by *Aspergillus terreus*: use of response surface methodology. J Chem Technol Biotechnol 79:1119–1126
17. Casas Lopez JL, Sanchez Perez JA, Fernandez Sevilla JM, Rodriguez Porcel EM, Chisti Y (2005) Pellet morphology, culture rheology and lovastatin production in cultures of *Aspergillus terreus*. J Biotechnol 116:61–77

18. Cui YQ, van der Lans RGJM, Luyben KCAM (1997) Effects of agitation intensities on fungal morphology of submerged fermentation. Biotechnol Bioeng 55:715–726
19. Doran PM (1995) Bioprocess engineering principles. Elsevier, Oxford
20. Driouch H, Hansch R, Wucherpfennig T, Krull R, Wittmann C (2012) Improved enzyme production by superior bio-pellets of *Aspergillus niger*—targeted morphology engineering using titanate micro particles. Biotechnol Bioeng 109:462–471
21. Driouch H, Roth A, Dersch P, Wittmann C (2010) Optimized bioprocess for production of fructofuranosidase by recombinant *Aspergillus niger*. Appl Microbiol Biotechnol 87:2011–2024
22. Driouch H, Roth A, Dersch P, Wittmann C (2011) Filamentous fungi in good shape: microparticles for tailor-made fungal morphology and enhanced enzyme production. Bioeng Bugs 2:100–104
23. Driouch H, Sommer B, Wittmann C (2010) Morphology engineering of *Aspergillus niger* for improved enzyme production. Biotechnol Bioeng 105:1058–1068
24. Endo A (2004) The origin of statins. Int Congr Ser 1262:3–8
25. Garcia-Ochoa F, Gomez E (2009) Bioreactor scale-up and oxygen transfer rate in microbial processes: an overview. Biotechnol Adv 27:153–176
26. Gbewonyo K, Hunt G, Buckland B (1992) Interactions of cell morphology and transport processes in the lovastatin fermentation. Bioproc Eng 8:1–7
27. Gomes J, Pahwa J, Kumar S, Gupta BS (2014) Lovastatin biosynthesis depends on the carbon-nitrogen proportion: model development and controller design. Eng Life Sci 14:201–210
28. Gonciarz J, Bizukojc M (2014) Adding talc microparticles to *Aspergillus terreus* ATCC 20542 preculture decreases fungal pellet size and improves lovastatin production. Eng Life Sci 14:190–200
29. Gonciarz J, Pawlak M, Bizukojc M (2012) Forcing of morphological changes of *Aspergillus terreus* mycelium by the addition of inorganic microparticles (in Polish). Inz Ap Chem 51:123–124
30. Gupta K, Mishra PK, Srivastava P (2007) A correlative evaluation of morphology and rheology of *Aspergillus terreus* during lovastatin fermentation. Biotechnol Bioproc Eng 12:140–146
31. Gupta K, Mishra PK, Srivastava P (2009) Enhanced continuous production of lovastatin using pellets and siran supported growth of *Aspergillus terreus* in an airlift reactor. Biotechnol Bioproc Eng 14:207–212
32. Hajjaj H, Niederberger P, Duboc P (2001) Lovastatin biosynthesis by *Aspergillus terreus* in a chemically defined medium. Appl Environ Microbiol 67:2596–2602
33. Hille A, Neu TR, Hempel DC, Horn H (2009) Effective diffusivities and mass fluxes in fungal biopellets. Biotechnol Bioeng 103:1202–1213
34. Jia Z, Zhang X, Cao X (2009) Effects of carbon sources on fungal morphology and lovastatin biosynthesis by submerged cultivation of *Aspergillus terreus*. Asia-Pacific J Chem Eng 4:672–677
35. Kaup BA, Ehrich K, Pescheck M, Schrader J (2008) Microparticle-enhanced cultivation of filamentous microorganisms: increased chloroperoxidase formation by *Caldariomyces fumago* as an example. Biotechnol Bioeng 99:491–498
36. Kennedy J, Auclair K, Kendrew SG, Park C, Vederas J, Hutchinson CR (1999) Modulation of polyketide synthase activity by accessory proteins during lovastatin biosynthesis. Science 284:1368–1372
37. Krull R, Wucherpfennig T, Esfandabadi ME, Walisko R, Melzer G, Hempel DC, Kampen I, Kwade A, Wittmann C (2013) Characterization and control of fungal morphology for improved production performance in biotechnology. J Biotechnol 163:112–123
38. Kumar MS, Jana SK, Senthil V, Shashanka V, Kumar SV, Sadhukhan AK (2000) Repeated fed-batch process for improving lovastatin production. Proc Biochem 36:363–368
39. Lai L-ST, Pan CC, Tzeng BK (2002) Medium optimization for lovastatin production by *Aspergillus terreus* in submerged cultures. J Chin Inst Chem Eng 33:517–527

40. Lai L-ST, Pan CC, Tzeng BK (2003) The influence of medium design on lovastatin production and pellet formation with a high producing mutant of *Aspergillus terreus* in submerged cultures. Proc Biochem 38:1317–1326
41. Lai L-ST, Tsai T-H, Wang TC (2002) Application of oxygen vectors to *Aspergillus terreus* cultivation. J Biosci Bioeng 94:453–459
42. Lai L-ST, Tsai T-H, Wang TC, Cheng TY (2005) The influence of culturing environments on lovastatin production by *Aspergillus terreus* in submerged cultures. Enzyme Microb Technol 36:737–748
43. Liu G, Xu Z, Cen P (2000) A morphologically structured model for mycelial growth and secondary metabolite formation. Chin J Chem Eng 8:46–51
44. Manzoni M, Rollini M, Bergomi S, Cavazzoni V (1998) Production and purification of statins from *Aspergillus terreus* strains. Biotechnol Techn 12:529–532
45. Manzoni M, Bergomi S, Rollini M, Cavazzoni V (1999) Production of statins by filamentous fungi. Biotechnol Lett 21:253–257
46. McIntyre M, Müller Ch, Dynesen J, Nielsen J (2001) Metabolic engineering of the morphology of *Aspergillus*. Adv Biochem Eng/Biotechnol 73:104–128
47. Metz B, Kossen NWF (1977) The growth of molds in the form of pellets—a literature review. Biotechnol Bioeng 19:781–799
48. Miranda RU, Gomez-Quiroz LE, Mejía A, Barrios-Gonzalez J (2013) Oxidative state in idiophase links reactive oxygen species (ROS) and lovastatin biosynthesis: Differences and similarities in submerged-and solid-state fermentations. Fungal Biol 117:85–93
49. Monaghan RL, Alberts AW, Hoffman CH, Albers-Schonberg G (1980) Hypocholesterolemic fermentation products and process of preparation. US Patent 4(231):938
50. Moser A (1988) Bioprocess technology. Kinetics and reactors. Springer, New York
51. Nielsen J (1996) Modelling the morphology of filamentous microorganism. Trends Biotechnol 14:438–443
52. Novak N, Gerdin S, Berovic M (1997) Increased lovastatin formation by *Aspergillus terreus* using repeated fed-batch process. Biotechnol Lett 19:947–948
53. O'Cleirigh C, Casey JT, Walsh PK, O'Shea DG (2005) Morphological engineering of *Streptomyces hygroscopicus* var. *geldanus*: regulation of pellet morphology through manipulation of broth viscosity. Appl Microbiol Biotechnol 68:305–310
54. Papagianni M (1999) Fungal morphology In: Kristiansen B, Mattey M, Linden J (eds) Citric Acid Biotechnology. Taylor & Francis, London
55. Papagianni M (2004) Fungal morphology and metabolite production in submerged mycelial processes. Biotechnol Adv 22:189–259
56. Pecyna M, Bizukojc M (2011) Lovastatin biosynthesis by *Aspergillus terreus* with the simultaneous use of lactose and glycerol in a discontinuous fed-batch culture. J Biotechnol 151:77–86
57. Pawlak M, Bizukojc M (2012) Kinetic modelling of lovastatin biosynthesis by *Aspergillus terreus* cultivated on lactose and glycerol as carbon sources. Chem Proc Eng 33:651–665
58. Pawlak M, Bizukojc M (2013) Feeding profile is not the sole factor influencing lovastatin production by *Aspergillus terreus* ATCC20542 in a continuous fed-batch stirred tank bioreactor. Biochem Eng J 81:80–89
59. Pawlak M, Bizukojc M, Ledakowicz S (2012) Impact of bioreactor scale on lovastatin biosynthesis by *Aspergillus terreus* ATCC 20542 in a batch culture. Chem Proc Eng 33:71–84
60. Rodriguez Porcel EM, Casas Lopez JL, Sanchez Perez JA, Chisti Y (2007) Enhanced production of lovastatin in a bubble column by *Aspergillus terreus* using a two-stage feeding strategy. J Chem Technol Biotechnol 82:58–64
61. Rodriguez Porcel EM, Casas Lopez JL, Sanchez Perez JA, Chisti Y (2008) Lovastatin production in a two-staged feeding operation. J Chem Technol Biotechnol 83:1236–1243
62. Rodriguez Porcel EM, Casas Lopez JL, Sanchez Perez JA, Fernandez Sevilla JM, Chisti Y (2005) Effects of pellet morphology on broth rheology in fermentations of *Aspergillus terreus*. Biochem Eng J 26:139–144

63. Rodriguez Porcel EM, Casas Lopez JL, Sanchez Perez JA, Fernandez Sevilla JM, Garcia Sanchez JL, Chisti Y (2006) *Aspergillus terreus* broth rheology, oxygen transfer, and lovastatin production in a gas-agitated slurry reactor. Ind Eng Chem Res 45:4837–4843
64. Rodriguez Porcel EM, Casas Lopez JL, Vilches Ferron MA, Sanchez Perez JA, Garcia Sanchez JL, Chisti Y (2006) Effects of the sporulation conditions on the lovastatin production by *Aspergillus terreus*. Bioproc Biosys Eng 29:1–5
65. Sutherland A, Auclair K, Vederas JC (2001) Recent advances in the biosynthetic studies of lovastatin. Curr Opin Drug Discov Develop 4:229–236
66. Szakacs G, Morovjan G, Tengerdy RP (1998) Production of lovastatin by a wild strain of *Aspergillus terreus*. Biotechnol Lett 20:411–415
67. Vinci VA, Hoerner TD, Coffman AD, Schimmel TG, Dabora RL, Kirkepar AC, Ruby CL, Stieber RW (1991) Mutants of a lovastatin-hyperproducing *Aspergillus terreus* deficient in the production of sulochrin. J Ind Microbiol 8:113–120
68. Witter DJ, Vederas JC (1996) Putative Diels-Alder-catalyzed cyclization during the biosynthesis of lovastatin. J Org Chem 61:2613–2623
69. Wittler R, Baumgärtl H, Lubbers DW, Schügerl K (1986) Investigations of oxygen transfer into *Penicillium chrysogenum* pellets by microprobe measurements. Biotechnol Bioeng 28:1024–1036
70. Wucherpfennig T, Hestler T, Krull R (2011) Morphology engineering—osmolality and its effect on *Aspergillus niger* morphology and productivity. Microb Cell Fact 10:58
71. Yoshizawa Y, Witter DJ, Liu Y, Vederas JC (1994) Revision of the biosynthetic origin of oxygens in mevinolin (lovastatin), a hypocholesterolemic drug from *Aspergillus terreus* MF 4845. J Am Chem Soc 116:2693–2694

Adv Biochem Eng Biotechnol (2015) 149: 171–222
DOI 10.1007/10_2014_299

Published Online: 22 January 2015

Modeling the Growth of Filamentous Fungi at the Particle Scale in Solid-State Fermentation Systems

Maura Harumi Sugai-Guérios, Wellington Balmant, Agenor Furigo Jr, Nadia Krieger and David Alexander Mitchell

Abstract Solid-state fermentation (SSF) with filamentous fungi is a promising technique for the production of a range of biotechnological products and has the potential to play an important role in future biorefineries. The performance of such processes is intimately linked with the mycelial mode of growth of these fungi: Not only is the production of extracellular enzymes related to morphological characteristics, but also the mycelium can affect bed properties and, consequently, the efficiency of heat and mass transfer within the bed. A mathematical model that describes the development of the fungal mycelium in SSF systems at the particle scale would be a useful tool for investigating these phenomena, but, as yet, a sufficiently complete model has not been proposed. This review presents the biological and mass transfer phenomena that should be included in such a model and then evaluates how these phenomena have been modeled previously in the SSF and related literature. We conclude that a discrete lattice-based model that uses differential equations to describe the mass balances of the components within the system would be most appropriate and that mathematical expressions for describing the

M.H. Sugai-Guérios · A. Furigo Jr
Departamento de Engenharia Química e Engenharia de Alimentos,
Universidade Federal de Santa Catarina, Centro Tecnológico, Cx.P. 476,
Florianópolis 88040-900, Santa Catarina, Brazil

W. Balmant · D.A. Mitchell (✉)
Departamento de Bioquímica e Biologia Molecular, Universidade Federal do Paraná,
Centro Politécnico Cx.P. 19046, Curitiba 81531-980, Paraná, Brazil
e-mail: davidmitchell@ufpr.br

N. Krieger
Departamento de Química, Universidade Federal do Paraná, Centro Politécnico,
Cx.P. 19081, Curitiba 81531-980, Paraná, Brazil

Present Address:
W. Balmant
Núcleo de Pesquisa e Desenvolvimento em Energia Autossustentável,
Departamento de Engenharia Mecânica, Universidade Federal do Paraná,
Centro Politécnico, Cx.P. 19011, Curitiba 81531-980, Paraná, Brazil

individual phenomena are available in the literature. It remains for these phenomena to be integrated into a complete model describing the development of fungal mycelia in SSF systems.

Keywords Hyphae · Mycelium · Mass transfer · Discrete models · Lattice-based models

Abbreviations

A	Cross-sectional area of the hypha (μm^2)
A_L	Area of contact of the plasma membrane with the extracellular medium (μm^2)
c_1	Proportionality constant (L^3 s^{-1} g-nutrient^{-1} g-biomass^{-1})
c_2	Proportionality constant (L g-transporter g-biomass^{-1})
c_3	Proportionality constant (g-nutrient g-transporter-substrate-complex^{-1} s^{-1})
$C_{O_2}\big\|_j$	O_2 concentration around tank j (g-O_2 L^{-1})
$C^f_{O_2}$	O_2 concentration in the biofilm (g-O_2 L^{-1})
D_E	Effective diffusivity of the enzyme in the solid particle (μm^2 s^{-1})
$D^f_{O_2}$	Effective diffusivity of O_2 in the biofilm layer (μm^2 s^{-1})
D_S	Diffusivity of maltose inside the hypha (μm^2 s^{-1})
D^e_S	Effective diffusivity of the soluble nutrient or hydrolysis product in the solid particle (μm^2 s^{-1})
E	Enzyme concentration (g-enzyme L^{-1})
H(x)	Heaviside function
j	Number of the tank
$J_E\big\|_{z=0}$	Flux of enzyme across the surface of the particle (g-enzyme μm^{-2} s^{-1})
k_c	Maximum rate of vesicle consumption (g-vesicles s^{-1})
K_C	Saturation constant for vesicle consumption (g-vesicles L^{-1})
k_{cat}	Catalytic constant of the enzyme (g-polymeric-nutrient g-enzyme^{-1} s^{-1})
K_m	Saturation constant for the hydrolysis of the polymeric carbon and energy source (g-polymeric-nutrient L^{-1})
k_{max}	Maximum specific transport rate of soluble nutrient or hydrolysis product across the plasma membrane (g-nutrient g-transporter^{-1} s^{-1})
K_{O_2}	Saturation constant for O_2 (g-O_2 L^{-1})
k_p	Maximum rate of vesicle production (g-vesicles L^{-1} s^{-1})
K_P	Saturation constant for vesicle production (g-nutrient L^{-1})
K_s	Saturation constant for glucose in growth rate expression (g-nutrient L^{-1})
K_t	Saturation constant for the absorption of soluble nutrient or hydrolysis product across the membrane (g-nutrient L^{-1})
L	Length of the tip-tank (μm)
m	Maintenance coefficient (g-nutrient g-biomass^{-1} s^{-1})
n	Tank number of the tip-tank
r	Radial position in the biofilm (μm)
r_a	Rate of absorption across the plasma membrane (g-nutrient s^{-1})

r_E Rate of secretion of enzyme (g-enzyme μm^{-2} s^{-1})
r_{EP} Rate of production of extracellular products (g-extracellular-products s^{-1})
r_{IP} Rate of production of intracellular products (g-intracellular-products s^{-1})
r_{LI} Rate of production of lipids (g-lipids s^{-1})
r_N Rate of consumption of alanine (g-alanine s^{-1})
r_{O_2} Rate of consumption of O_2 (g-O_2 s^{-1})
r_s Rate of consumption of glucose (g-glucose s^{-1})
r_X Rate of biomass growth [g-biomass L^{-1} s^{-1} for Eq. (12); g-biomass s^{-1} for Eq. (15)]
S^f Concentration of soluble nutrient or hydrolysis product in the biofilm (g L^{-1})
S_e Concentration of extracellular soluble nutrient or hydrolysis product (g L^{-1})
S_i Concentration of intracellular soluble nutrient or hydrolysis product (g L^{-1})
$S_i|_j$ Concentration of maltose in tank j (g-maltose L^{-1})
S_p Concentration of polymeric nutrient (g-polymeric-nutrient L^{-1})
SSF Solid-state fermentation
t Cultivation time (s)
T Transporter concentration per area of the plasma membrane (g-transporter μm^{-2})
t_E Time when the secretion of enzyme ceases (s)
v Velocity of cytoplasmic flow (μm s^{-1})
X Concentration of biomass (g-biomass L^{-1})
Y_E Yield of glucose from starch (g-glucose g-polymeric-nutrient^{-1})
Y_L Extension of hyphal length per mass of vesicle consumed (μm g-vesicle^{-1})
Y_{X/O_2} Yield of biomass O_2 (g-biomass g-O_2^{-1})
Y_ϕ Yield coefficient for production of vesicles from maltose (g-vesicles g-nutrient^{-1})
z Depth within the solid particle (μm)
η Membrane coordinate (μm)
Δz Length of a "normal" tank (μm)
θ Concentration of the transporter–substrate complex (g-transporter-substrate-complex L^{-1})
μ_{max} Maximum specific growth rate constant (s^{-1})
ρ_X Biomass dry weight per volume (g-biomass L^{-1})
$\phi|_j$ Concentration of vesicles in tank j (g-vesicles L^{-1})
ψ Velocity of active transport of vesicles (μm s^{-1})

Contents

1 Introduction

Solid-state fermentation (SSF) processes involve the growth of microorganisms on moist solid substrate particles, in situations where the spaces between the particles are filled with a continuous gas phase. Although there are some SSF processes that involve bacteria and yeasts, most SSF processes involve filamentous fungi [32]. For example, the filamentous fungi *Rhizopus oligosporus, Aspergillus oryzae*, and *Monascus purpureus* are used to produce *tempe*, soy sauce *koji*, and red rice *koji*, respectively, in SSF processes that have been practiced for many centuries in Asia. Filamentous fungi are also involved in several commercial processes of SSF that have been developed over the last few decades and also in many potential applications that are being studied [39, 44]. Additionally, SSF processes involving filamentous fungi will be important processing steps in biorefineries [11].

Mathematical models are useful tools for guiding the design, scale-up, and operation of bioreactors and various models that describe heat and mass transfer within SSF bioreactors have been developed with this purpose [35, 36]. Models can also be used as tools for gaining insights into how phenomena that occur at microscopic scales control system performance. This review evaluates the state of the art in the development of such models for SSF processes that involve filamentous fungi. We address key issues that are important during the initial steps of developing any mathematical model. Firstly, it is important to know what one wants to achieve with a model, as this will determine the type of model to be developed. Secondly, with the intended use in mind, it is necessary to express ones understanding of the functioning of the system, at an appropriate level of detail, detailing subsystems and phenomena that occur within and between these subsystems. Thirdly, before writing the equations and building the model, it is appropriate to study previously published models, as these can give insights into how the system might be represented and how specific phenomena might be described mathematically. By addressing these issues, we aim to provide a basis for the continued development of mathematical models of fungal growth at the particle scale in SSF systems.

2 The Appropriate Scale for Describing Growth of Filamentous Fungi in SSF Processes

This review deals with models that can provide insights into how the growth of filamentous fungi at the particle scale affects SSF bioreactor performance. This section shows that a model with this purpose must describe the mycelial mode of growth, since not only can the characteristics of the mycelium influence the performance of the bioreactor, but also the manner in which the bioreactor is operated can affect the development of the fungal mycelium. It then concludes that, among the models that have been used to describe the growth of filamentous fungi, "mesoscale" models are the most appropriate to describe how the mycelium develops to produce a complex spatial arrangement in the heterogeneous bed of solid particles.

2.1 The Importance of Mycelial Growth in SSF Processes

The performance of an SSF process that involves a filamentous fungus is determined by a complex interplay between the morphological and physiological characteristics of the fungus and the factors that determine the conditions in the local environment that the fungus experiences. These factors include the design and operating variables of the bioreactor that is used, the transport phenomena that operate across the whole substrate bed, and the transport phenomena that operate at the level of a single particle and the inter-particle spaces around it (Fig. 1). Details of the design and operating variables that are available for the various SSF bioreactor types and the heat and mass transfer phenomena that occur across the substrate bed in these bioreactors are discussed in depth elsewhere [35].

Before considering the interaction between fungal mycelia and SSF bioreactors, it is useful to classify the hyphae present in SSF processes into surface hyphae, biofilm hyphae, penetrative hyphae, and aerial hyphae, according to the specific physical environment in which they are located [50] (Fig. 2). Surface hyphae are located at the air–solid interface and grow horizontally on top of the surface of the solid particle; they are in intimate contact with both the air phase and the solid phase. Biofilm hyphae represent those hyphae in the part of the mycelium that is above the substrate surface but is bathed in a liquid film. The biofilm may be several hundred micrometers in depth [43]. Penetrative hyphae grow below the surface of the solid particle and are therefore surrounded by the moist solid matrix of the particle. Aerial hyphae are those hyphae that are located in the inter-particle spaces or at the exposed surface of the substrate bed and are surrounded on all sides by a gas phase.

This review deals with aerobic SSF processes. The various bioreactor designs that have been proposed for such processes can be classified on the basis of the strategies that are used with respect to the aeration and the agitation of the substrate bed [32] (Fig. 3). In SSF processes that involve filamentous fungi, the agitation

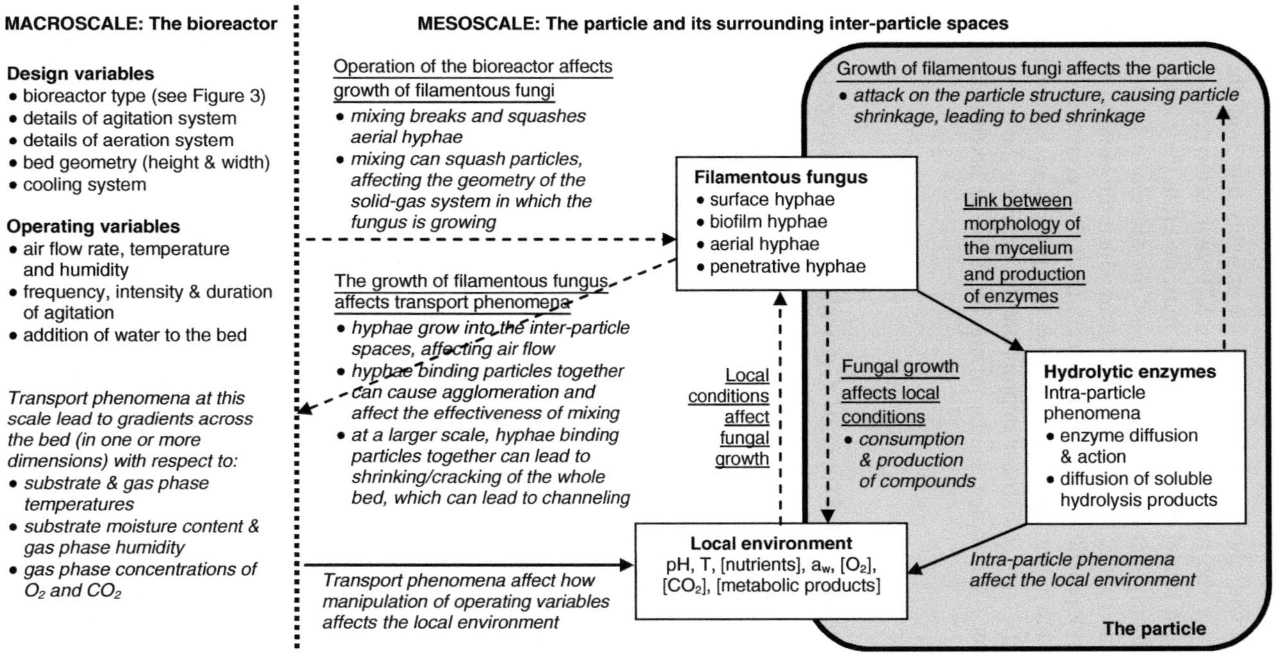

Fig. 1 The performance of an SSF process that involves a filamentous fungus depends on a complex interplay between the morphological and physiological characteristics of the fungus and the factors that determine the conditions of the local environment to which the fungus is subjected. *Solid arrows* represent the production, consumption, or flow of system components. *Dashed arrows* represent effects of the fungus or on the fungus, with these effects being identified with underlined text

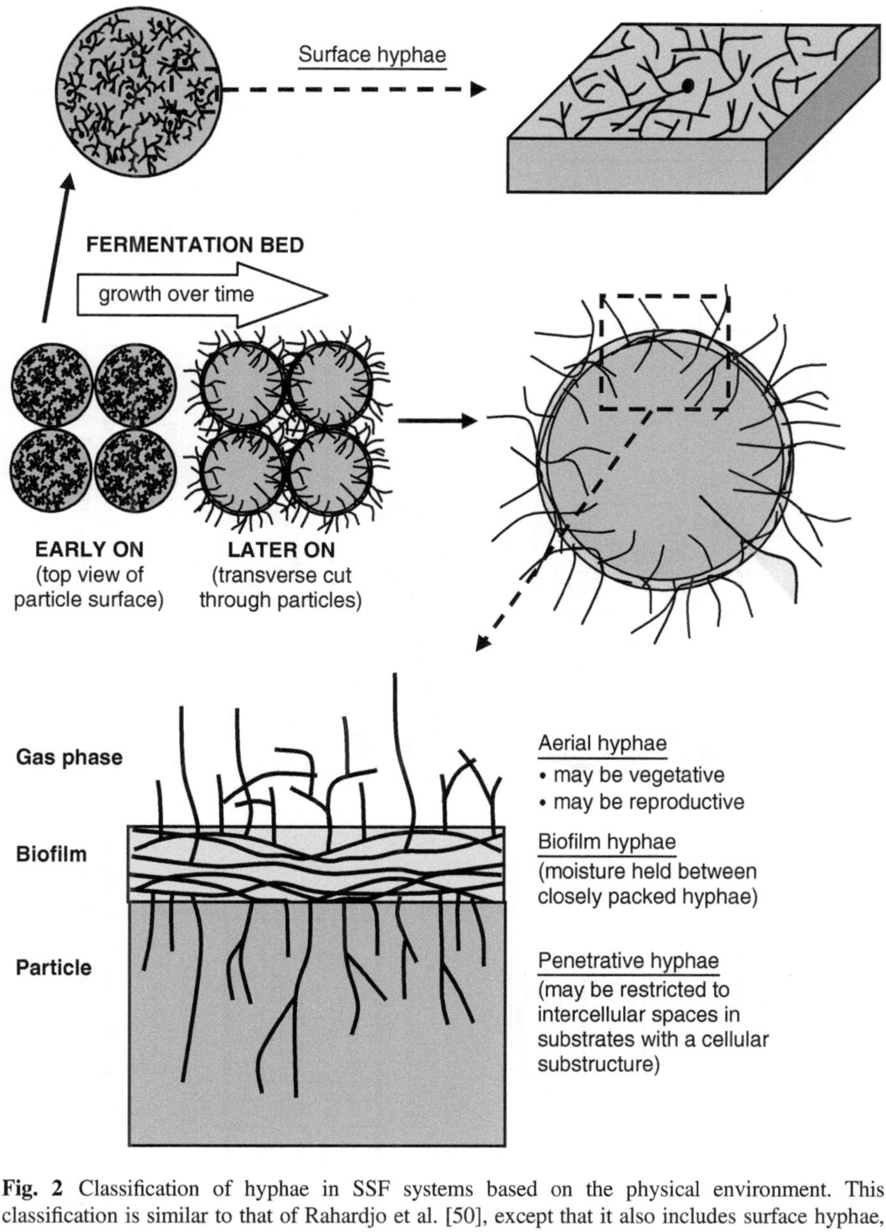

Fig. 2 Classification of hyphae in SSF systems based on the physical environment. This classification is similar to that of Rahardjo et al. [50], except that it also includes surface hyphae, which grow across the surface of the substrate particle early on in the process

strategy that is used will affect the spatial arrangement of the hyphae in the system. In some processes, the bed needs to remain essentially static, such as is the case in the production of fungal spores for use as biopesticides, since agitation may damage aerial reproductive structures or even prevent them from forming. If the bed is left

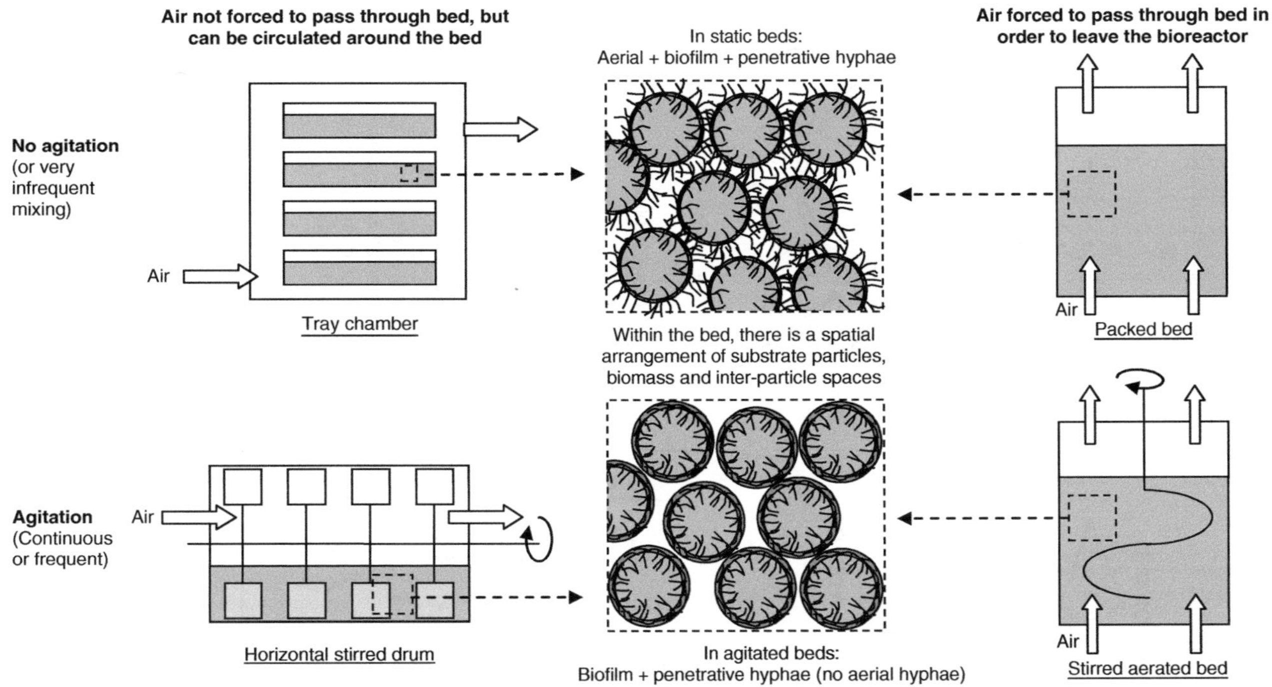

Fig. 3 Classification of SSF bioreactors into four groups on the basis of how they are agitated and aerated [32]. Agitation has significant consequences for the spatial distribution of fungal hyphae in the system, preventing the formation of aerial hyphae

completely static or mixed very infrequently, then hyphae can extend into the inter-particle spaces, becoming aerial hyphae. Aerial hyphae originating from several different particles can intertwine, thereby binding the particles into agglomerates. If the bed is left static for a long time and then mixed, then this mixing may tear the agglomerates apart and squash the aerial hyphae onto the particle surface, forming a biofilm. On the other hand, if the bed is mixed frequently or continuously, then aerial hyphae will simply not develop, as any hyphae that begin to extend into the air spaces will quickly be squashed onto the particle surface.

The aeration strategy that is used in the bioreactor is also important, because it will affect the delivery of O_2 to the inter-particle spaces within the substrate bed. The bed may not be forcefully aerated, but rather placed in an environment in which air is circulated around the bed. In this case, if the bed is static, then gas transport in the bed will be limited to diffusion, although some natural convection may occur if there are significant temperature gradients across the bed [56]. If the bed is agitated, as occurs for horizontal stirred drums and rotating drums, then the effectiveness with which the gas phase within the inter-particle spaces is replenished with O_2 depends on the effectiveness of the mixing. Since the mixing will squash any aerial hyphae into the biofilm on the surface of the particle, then the supply of O_2 to the fungus itself will depend on diffusion of O_2 in this biofilm.

On the other hand, a static bed may be forcefully aerated, meaning that the air is forced to flow through the inter-particle spaces of the bed in order to leave the bioreactor. In order to obtain uniform aeration over the whole bed, it is important that the air is not able to flow preferentially through regions of lower resistance. The growth of filamentous fungi can have important consequences with respect to the airflow in the bed. Firstly, the growth of aerial hyphae into the inter-particle spaces increases the resistance to airflow through the bed, affecting the efficiency of the aeration system [1]. Secondly, the tensile forces exerted by the network of aerial hyphae can contribute to contraction of the bed, causing it to crack or pull away from the bioreactor walls, leading to the phenomenon of channeling, where the air flows preferentially around the bed or through the cracks and not through the inter-particle spaces. If this occurs, then the bed can be agitated in an attempt to reseat it. However, some particles may be bound together strongly by the aerial hyphae, such that large agglomerates of particles resist being broken up. In fact, it has been suggested that the bed should be agitated early during the process to disrupt the aerial hyphae and prevent the formation of agglomerates in the first place, at least for fungi that produce abundant aerial hyphae [55].

Although there are many studies of the relationship between the morphology and productivity of filamentous fungi in submerged culture systems [20, 45], this topic has received much less attention in SSF systems, with only two studies. In the first study, the production of acid protease and acid carboxypeptidase in rice *koji* by *A. oryzae* was correlated with the degree of mycelial penetration into the grain: At low degrees of penetration, the activities of these two enzymes were high [26]. In the second study, mutant strains of *A. oryzae* with increased branching frequency in their aerial hyphae produced significantly higher amylase and protease levels than the wild-type strain, despite producing similar overall biomass levels [61].

2.2 Appropriate Models for Mycelial Growth at the Particle Scale in SSF

It is clear from the above discussion that for a mathematical model to be a useful tool for elucidating how particle-scale phenomena influence the performance of SSF bioreactors in processes that involve filamentous fungi, it must describe the growth of the mycelium at an appropriate scale. Current models of fungal growth in SSF can be classified as being microscale models, mesoscale models, or macroscale models, based on the scale at which the fungal mycelium is described (Table 1). This classification is similar to that proposed by Davidson [10] and Boswell and Davidson [5], but applied to models for SSF.

Microscale growth models are those concerned with phenomena occurring in a single hypha and often do not describe an entire hypha. They are most commonly used to study growth and morphology of the hyphal tip and usually include a description of the functioning of the *Spitzenkörper* [19].

Mesoscale growth models describe a network of hyphae, although, as will be discussed later (Sect. 4.1), they may or may not describe these hyphae as distinct entities. Some models focus on describing how a colony arises from a single spore or single inoculation point; such models are often referred to as single-colony models. Other models describe how a mycelium develops from many points of inoculation over the substrate surface, a situation that is more common in SSF systems and which has been referred to as "overculture" [34]. Mesoscale models may include some of the phenomena that are described in microscale models, but at a lesser degree of detail, and they include other phenomena related to the formation of the mycelium, such as branching. When mesoscale models are used to describe growth in SSF systems, they may recognize the substrate particles, but, if so, they will only describe one or a few particles. They may describe the presence of concentration gradients for key components within these particles.

Macroscale growth models describe the growth of the fungus in a bed containing a very large number of particles and are typically used to describe the growth kinetics within models of heat and mass transfer within bioreactors. These models rarely describe the gradients of components inside particles. Rather, they focus on temperature and gas concentration gradients across the fermentation bed. No attempt is made to describe individual hyphae [37]. Rather, empirical equations are fitted to data of biomass concentrations (often expressed as "grams of biomass per gram of total dry solids"), which will typically be obtained from measurements made on homogenized preparations obtained from samples containing hundreds or even thousands of substrate particles [67].

Mesoscale growth models are most appropriate for the purpose of elucidating how the mycelial mode of growth of filamentous fungi influences the performance of SSF systems. This review focuses on this type of model.

Table 1 Classification of mathematical models of fungal growth on solid particles according to the scale at which the system is described

Category name	Microscale	Mesoscale	Macroscale
Scale of growth	μm	mm to cm	cm to m
Number of particles involved	None or only one	One to a few	A large number (almost uncountable)
The manner of describing the mycelium	Individual hyphae	Individual hyphae or average concentration of biomass	Average concentration of biomass
Main feature	These models focus on a specific region of the hypha (e.g., a few μm behind the tip), and they do not attempt to describe a whole hypha or a mycelium	These models describe the growth of one or a few colonies interacting with each other	These models calculate concentration gradients of biomass/temperature/O_2 across the substrate bed, but not gradients within particles
General purpose	To study intracellular phenomena that determine tip extension rate and tip morphology	To study the phenomena that determine the rate of colonization of solid particles	To study SSF bioreactor performance
What can be investigated with the models	• The mechanisms involved in the *Spitzenkörper* • Patterns of formation of new cell wall • Tip swelling • Tip extension against frictional forces • Molecular motors and microtubule transport	• Fungal morphology in SSF • Biomass distribution inside and around a solid particle • How nutrient/O_2 distribution in the extracellular environment affects development of the mycelium • Shrinkage of solid particles	• Overall growth and product formation in the bed • Local temperature and moisture content and how they affect growth • Pressure drop • Shrinkage of the bed

(continued)

Table 1 (continued)

Category name	Microscale	Mesoscale	Macroscale
Scale of growth	μm	mm to cm	cm to m
Most important phenomena to be described by the model	• Vesicle transport to the *Spitzenkörper* and from the *Spitzenkörper* to the apical cell membrane • Fusion of vesicles into cell membrane and formation of new cell wall • Water uptake and intracellular water transport	• Transport phenomena inside the solid particle and the biofilm layer • Nutrient uptake, intracellular transport, and consumption • Vesicle production, transport, and fusion at the tip • Branching and changes in growth direction	• Mass and heat transfer between particles • Biomass growth rate • Particle shrinkage
Examples	[19, 58, 62]	[9, 31, 48]	[36]
Illustration of the scale of growth	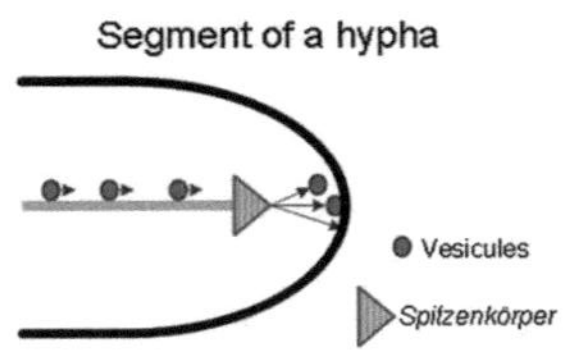	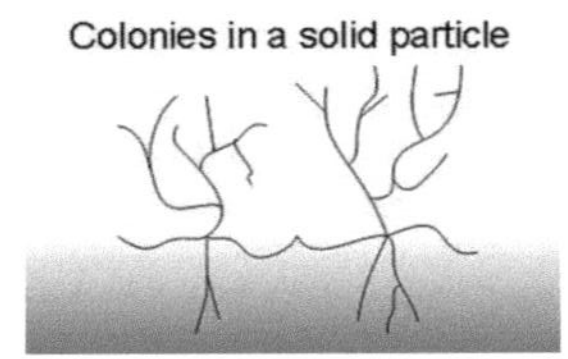	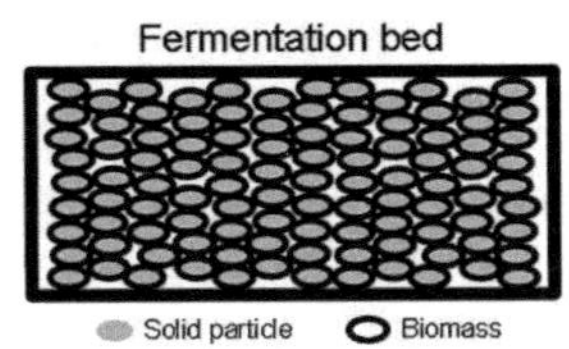

3 How the Fungal Mycelium Grows and Interacts with Its Environment in SSF

In the development of any mathematical model, before writing the equations, it is essential to understand the various phenomena that occur within the system. This section provides an overview of our knowledge about how the fungal mycelium grows and interacts with its environment in SSF, at a level of detail that would be appropriate for the purposes of modeling mycelial development in SSF processes. A mesoscale model of SSF would concern itself with transport phenomena, but would not attempt to describe phenomena that occur at the cellular and molecular levels in detail. Detailed descriptions of these cellular and molecular processes can be found elsewhere [3, 46, 53, 57, 59].

In this section, it is assumed that the filamentous fungus is growing within a bed of solid particles that is initially static. Section 3.1 focuses on what the hyphae themselves do; Sect. 3.2 explores the intracellular and extracellular reactions and transport phenomena that are involved.

3.1 Development of the Mycelium in an SSF System

It is not necessarily the case that all four types of hyphae identified in Sect. 2.1, namely surface, biofilm, penetrative, and aerial hyphae, will be present at the same time in an SSF process. Rather, there is a developmental sequence [25].

Typically, the substrate will be inoculated by mixing a spore suspension in with the substrate, although mycelial fragments are sometimes used. If spores are used, initially they will be distributed over the surface of each substrate particle (Fig. 4). The spores swell and produce a germ tube, in a process that may take several hours. This process is not completely synchronized, so different spores can germinate at different times. Also, some spores may fail to germinate. After extending a certain length, typically 60–200 μm, across the surface of the substrate, this germ tube will branch to produce two daughter hyphae [64]. In turn, these daughter hyphae will extend across the surface and branch, with this extension and branching process being repeated several times. The hyphal tips tend to grow into unoccupied space, in other words, away from other extending hyphae and already established hyphae. This leads to the development of microcolonies around each germinated spore, with these microcolonies being composed essentially of surface hyphae, although penetrative hyphae may also form if the O_2 level in the medium is high [50]. At this stage, it may be difficult to see the fungus with the naked eye, although the texture of the surface of the substrate particle may have changed. Additionally, the biomass level may be so low that it is difficult to measure either components of the biomass (such as ergosterol or glucosamine) or the respiratory activity of the biomass.

As the microcolonies expand, the surface hyphae from different microcolonies begin to approach one another. As they do so, they may change direction to avoid

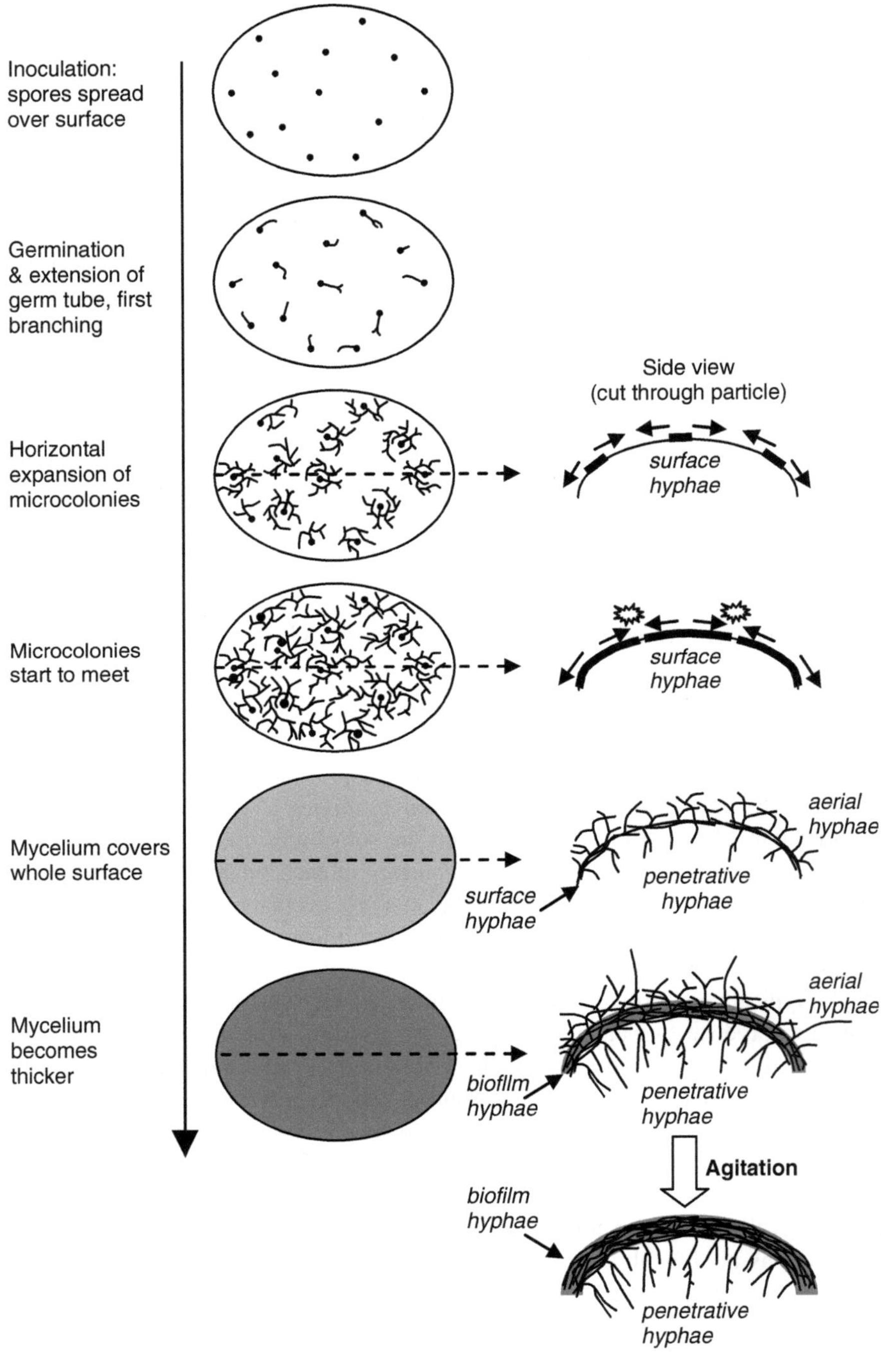

Fig. 4 How the presence of the various types of hyphae varies during the development of the mycelium in an SSF system that is initially left static and then agitated later in the process

one another, cross one another, or stop extending [24]. Once all the available space on the substrate surface is occupied by surface hyphae, the density of the mycelium increases with the production of penetrative and aerial hyphae.

Penetrative hyphae tend to move perpendicularly away from the substrate surface if the substrate matrix is homogeneous and does not contain significant physical barriers [41, 42]. However, many substrates derived from plants will have a cellular substructure. In this case, the penetrative hyphae will tend to follow the path of least resistance, which means that they often grow in the extracellular matrix between the cell walls within the substrate particle [27]. However, if the fungus has the ability to digest the cell walls within the substrate, it may penetrate into the interior of the cells, through the combined action of mechanical force and enzymatic degradation, obtaining access to the nutrients in the cell lumen [7].

Aerial hyphae can be of two types, vegetative or reproductive. Vegetative aerial hyphae typically appear before reproductive aerial hyphae. Reproductive aerial hyphae are normally specialized structures, such as sporangiophores or conidiophores. There are significant differences between different fungi with respect to the degree to which vegetative and reproductive aerial hyphae are produced and the growth patterns of these hyphae. As mentioned in Sect. 2.1, these aerial hyphae occupy part of the inter-particle space that was present in the original substrate bed.

The physical environment of a hypha, and therefore its classification, can change over the course of an SSF process. For example, if the aerial hyphae form a sufficiently dense mat above the substrate surface, a film of water may be drawn up from the substrate by capillary action into the spaces between these aerial hyphae, thereby changing the physical environment, such that these aerial hyphae and the original surface hyphae would now be classified as biofilm hyphae [50]. Biofilm hyphae will also be formed if the substrate bed is mixed, since this will squash aerial hyphae, forming a mat on the substrate surface. Also, as the mycelium becomes denser, it may become difficult to distinguish particular types of hyphae. For example, if the solid material at the substrate surface is consumed during the process, then it may be very difficult to identify where the biofilm ends and the penetrative hyphae begin.

3.2 Phenomena Occurring During Growth of a Filamentous Fungus on a Solid Substrate

In this section, it is assumed that the carbon and energy source is a polymer. This is commonly the case in SSF processes, but not always. Also, in different SSF systems, polymeric carbon and energy sources that are consumed by the microorganism may or may not contribute directly to the structural integrity of the particle. If they do, then the particle can shrink significantly during the process, causing changes in the bed structure and properties. However, this aspect will not be explored in this review. Further, it will be assumed that the bed is forcefully aerated, such that O_2 is freely available in the inter-particle spaces. However, it should be noted that, even with forced aeration, there will be a stagnant gas film attached to any surface present in the system.

3.2.1 Intracellular Transport and Other Phenomena Involved in Hyphal Extension

Hyphae grow by apical extension in a process involving the fusion of vesicles at the tip. These vesicles are produced in a subapical zone, which will be referred to here as the "vesicle-producing zone" and transported to the tip along microtubules within the cytoplasm (Fig. 5) [3, 46, 53, 57, 59]. In order to produce these vesicles, the cytoplasm in the vesicle-producing zone needs to receive nutrients. If the vesicle-producing zone of the hypha is not in contact with an external source of nutrients, then these nutrients must be supplied from more distal parts of the hypha.

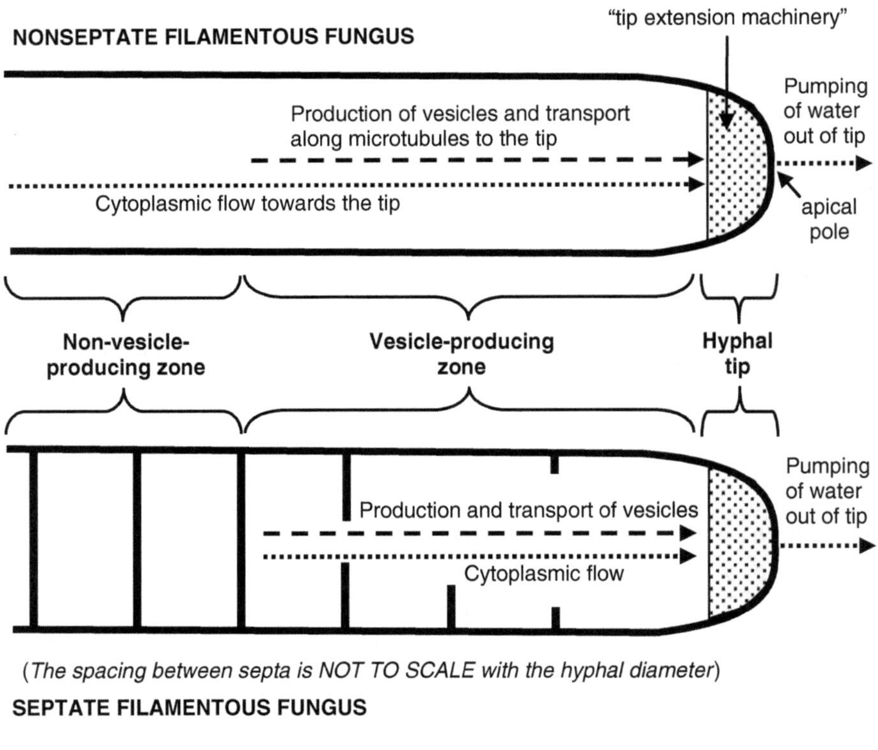

Fig. 5 A simplified representation of the intracellular phenomena involved in extension of the hyphal tip. Riquelme and Bartnicki-García [53] report that according to the classical model for hyphal tip growth, the tip growth machinery is within 1–5 μm of the "apical pole" and the vesicle-producing zone is up to 20 μm long. However, the lengths of these regions are likely to vary between different fungi

The vesicles are surrounded by a membrane and contain enzymes involved in cell wall synthesis, both cell-wall-lysing enzymes and cell-wall-synthesizing enzymes. As the vesicles fuse with the plasma membrane at the tip, their membranes contribute to the extension of the plasma membrane. Cell wall precursors are probably supplied in the cytosol [54]. In fact, the cytosol flows from more distal parts of the hyphae to the extending tip, in a phenomenon called "cytoplasmic streaming" or "cytoplasmic flow." This cytoplasmic flow is driven by the high osmotic pressure in the tip due to the high concentrations of calcium that are maintained there [28]. The velocity of cytoplasmic flow can be higher than the rate of extension of the tip, which is only possible if there is active efflux of water at the hyphal tip. In older regions of septate hypha, the septa may close, preventing cytoplasmic flow in these regions. In some fungi, the septum does not close completely; rather, a central pore is maintained. This pore has a Woronin body beside it, which plugs the pore if the adjacent hyphal segment is damaged [30].

As the hypha extends, the length of the vesicle-producing zone remains essentially constant [63]. In other words, the length of the fungal hypha that contributes directly to the extension of the tip remains constant. Consequently, the physiology of a particular section of a hypha at a certain location in space changes over time. The particular section is "created" when that location is first occupied by the extending tip. As the tip extends ever onward, this section becomes part of the vesicle-producing zone, then loses its ability to produce vesicles, and later undergoes an aging process in which the proportion of the cytoplasm occupied by vacuoles increases significantly [47]. Later still, this section may undergo autolysis [47].

Branching of hyphae can be either apical or lateral or both, depending on the fungus and hyphal type. In apical branching, the tip itself divides. In lateral branching, a new branch appears some distance behind the tip. Apical branching occurs when the rate of supply of vesicles to the tip is greater than the maximum rate at which vesicles can fuse at the tip [21]. Lateral branches often emerge where vesicles accumulate behind septa in septate hyphae, but aseptate hyphae can also branch laterally and the role of vesicle accumulation in triggering lateral branching is not totally clear [21]. It is also possible for the hyphae of some fungi to fuse by anastomosis. This can be tip to tip or tip to hypha [17]. It is not clear to what degree anastomosis occurs in SSF processes.

3.2.2 Extracellular Phenomena Involved in the Growth of Surface Hyphae

Surface hyphae are present during the early stages of expansion of a microcolony. The germination of the spore and the initial stages of extension of the germ tube are supported by the internal reserves contained in the spores, but soon it is necessary for the extending germ tube and the new hyphae to obtain nutrients from the solid particle.

In order to utilize a polymeric carbon and energy source, the hyphae must secrete hydrolytic enzymes (Fig. 6). Once secreted, these enzymes may remain attached to

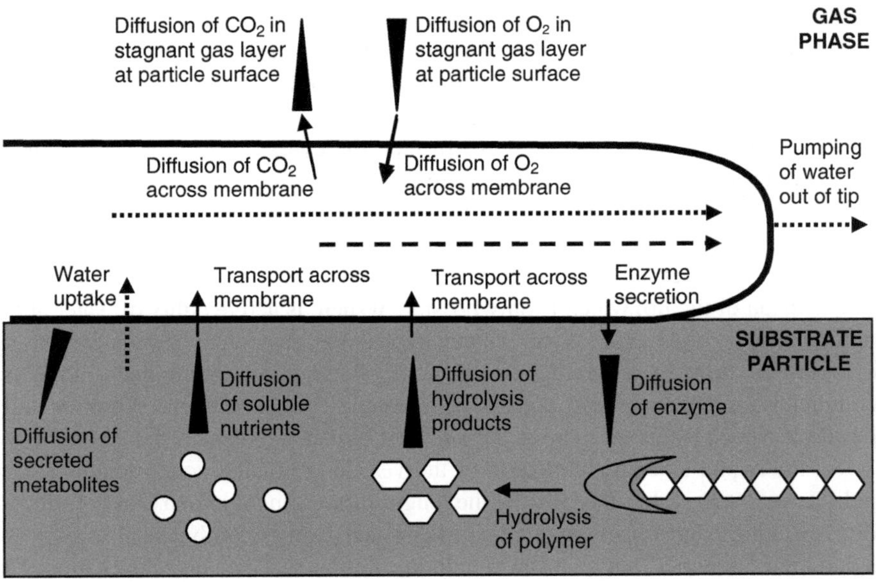

Fig. 6 Phenomena involved in the growth of surface hyphae during the initial stages of the process. *Long triangles* represent diffusion down concentration gradients. *Dotted arrows* represent flow of water or cytoplasm. *Dashed arrows* represent transport of vesicles

the hypha or may be liberated into the surrounding medium. If liberated, then their diffusion in the matrix of the substrate particle will be influenced by the microscopic porosity of the matrix and by the presence of impenetrable barriers, such as intact cell walls. Once they encounter and hydrolyze polymers, then the resulting soluble hydrolysis products will diffuse through the substrate matrix. As the soluble hydrolysis products are taken up at the plasma membrane, their concentration next to the hypha decreases, leading to the formation of concentration gradients in the extracellular medium. Similar diffusion phenomena will occur for any soluble nutrients that are taken up by the hypha. In the early stages of the process, when the microcolonies are expanding across the particle surface, it is likely that the polymeric carbon and energy source will be present very close to the substrate surface, such that the distances over which the extracellular enzyme and soluble hydrolysis products need to diffuse are quite small. In fact, if hydrolysis is faster than uptake, the concentration of the soluble hydrolysis product at the surface can rise significantly [33]. In a forcefully aerated system, at this stage, the supply of O_2 to the surface hyphae will not be limiting. CO_2 will be produced and diffuse away from the hypha into the gas phase and also, in dissolved form, into the substrate particle. Any secreted metabolites will also diffuse away from the hypha. These secreted metabolites can be important in mediating interactions between hyphae.

3.2.3 Extracellular Phenomena Involved in the Growth of Hyphae Within Biofilms

By the time that a biofilm has been produced in an SSF process, there are also likely to be aerial and penetrative hyphae in the system. The processes influencing the growth of aerial and penetrative hyphae are described below; here, we focus specifically on what happens in the biofilm.

As the fermentation process advances, it is quite likely that the polymeric carbon and energy source will be exhausted in the region near the surface (Fig. 7). Any enzyme that has not diffused out of this region will not contribute to the liberation of soluble hydrolysis products. Since the diffusion of enzymes in the matrix of the substrate can be quite slow, this can lead to a significant decrease in the rate of liberation of soluble hydrolysis products within the substrate particle [33]. This, in conjunction with high uptake rates by the hyphae in the biofilm at the surface, can lead to steep concentration gradients of these soluble hydrolysis products within the particle, with very low concentrations at the interface between the biofilm and the particle. As the soluble hydrolysis products and other soluble nutrients diffuse into the biofilm, they are taken up by the hyphae. If the rate of uptake is sufficiently high and the biofilm is sufficiently thick, then these soluble hydrolysis products and nutrients can be totally consumed by the hyphae in the parts of the biofilm close to the particle–biofilm interface, with the parts of the biofilm that are further from this interface being starved of essential nutrients [31, 52].

Additionally, even if high O_2 concentrations are maintained in the gas phase in the inter-particle spaces, the fact that O_2 transport within the biofilm is limited to diffusion means that the respiration of the hyphae in the biofilm can lead to the dissolved O_2 concentration falling to essentially zero within 100 μm depth below the interface between the biofilm and the air phase [43]. In other words, the parts of the biofilm closer to the surface of the substrate particle can become anaerobic.

3.2.4 Extracellular Phenomena Involved in the Growth of Penetrative Hyphae

Penetrative hyphae are in intimate contact with the solid substrate. The matrix into which they grow will typically be moist, but with high concentrations of solids and solutes, such that much of the water will be adsorbed or complexed. These hyphae therefore need to exert turgor pressure in order to penetrate into the matrix [38]. There may also be physical barriers, such as the cell walls of plant materials from which the solid particles were derived. Not all fungi are capable of penetrating such barriers.

It is quite probable that the penetrative hyphae contribute to the release of enzymes into the substrate matrix [66] (Fig. 7). Since penetrative hyphae are located in deeper regions within the substrate particle, when compared to biofilm hyphae, they will experience higher concentrations of soluble hydrolysis products and nutrients, but lower concentrations of O_2, at their plasma membranes. In fact, once

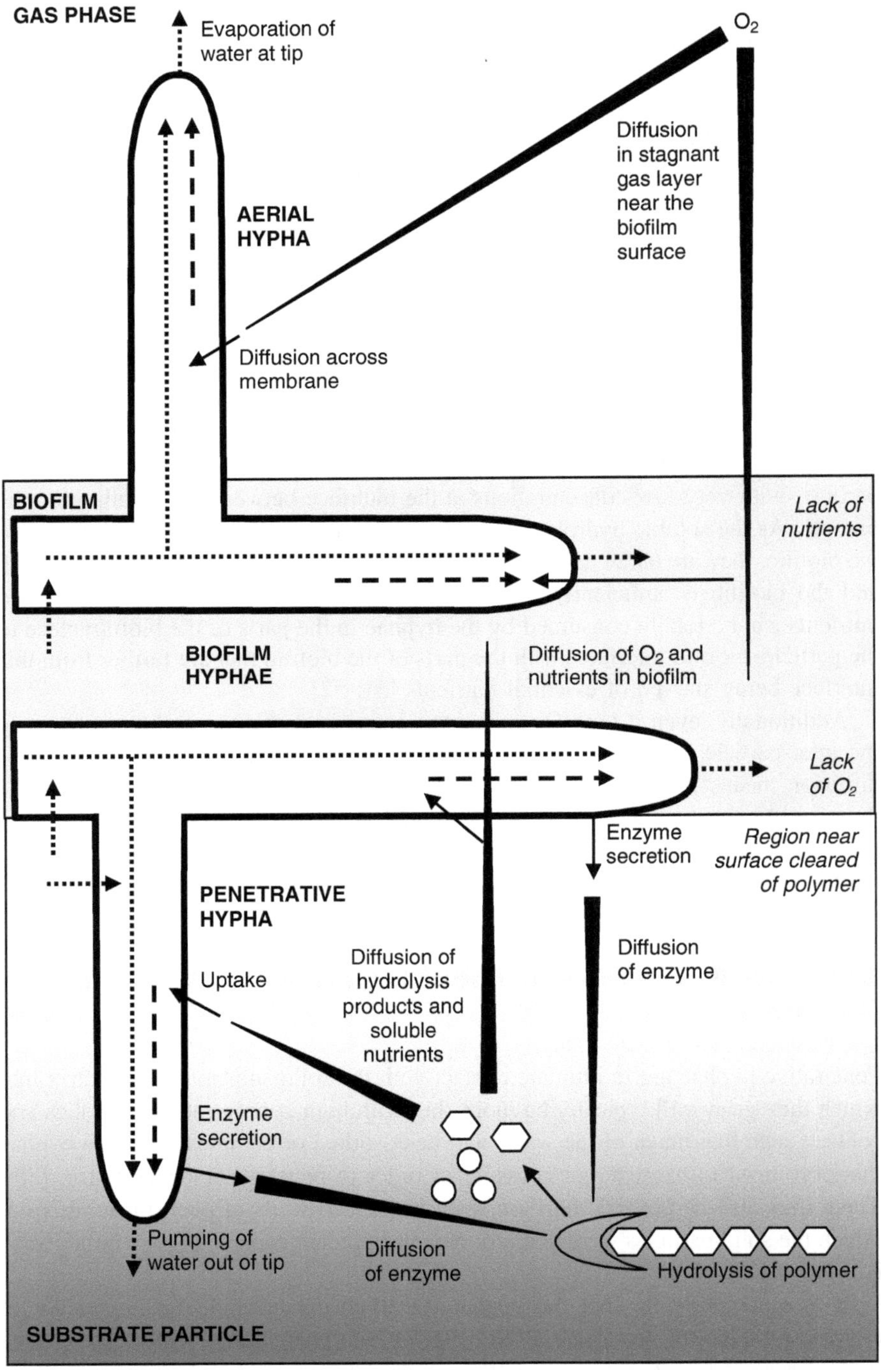

Fig. 7 Phenomena involved in the growth of biofilm, penetrative, and aerial hyphae. "*Long triangles*" represent diffusion down concentration gradients. "*Dotted arrows*" represent flow of water or cytoplasm. "*Dashed arrows*" represent transport of vesicles

an actively metabolizing biofilm is established at the substrate surface, the penetrative hyphae can be left in an anaerobic environment.

3.2.5 Transport Phenomena Involved in the Growth of Aerial Hyphae

Aerial hyphae are in intimate contact with the air phase and can take up O_2 directly from the air (Fig. 7). However, the degree to which they do this varies significantly among different fungi [51]. On the other hand, the vesicle-producing region behind the extending tip of an aerial hypha must receive material from the distal part of the hypha. Since aerial hyphae can be several millimeters or even several centimeters long, depending on the fungus and the growth conditions, material for tip extension may need to be transported within the aerial hyphae over large distances. This transport probably involves cytoplasmic flow. In fact, although studies are limited, in *Aspergillus niger*, cytoplasmic flow occurs from the biofilm hyphae into the aerial hyphae, with cytoplasmic flow rates in the aerial hyphae being of the order of 10–15 $\mu m\ s^{-1}$ [4].

4 Mathematical Models of Phenomena Involved in the Growth of Fungal Hyphae on Solid Surfaces

Based on the qualitative description presented in Sect. 3, a "complete" mathematical model that proposes to describe the development of fungal mycelia in SSF systems should recognize the four different types of hyphae, namely penetrative, aerial, surface, and biofilm, enabling prediction not only of growth over the particle surface, but also of the biomass distribution above and below the particle surface. It should also describe the following phenomena: physiological differentiation within the mycelium; tip extension based on the production, transport, and incorporation of vesicles; nutrient and O_2 uptake and also mass transfer outside and inside the hyphae; the secretion and the hydrolytic activity of extracellular enzymes; apical and lateral branch formation, with angles similar to those found in real mycelia; and septation, in the case of septate fungi.

At present, there is no mathematical model of the growth of filamentous fungi in SSF systems that incorporates, simultaneously, all of these biomass types and growth-related phenomena. On the other hand, most of these phenomena have already been described in various fungal growth models, and the mathematical approaches used in these models could be integrated, in a modified form if necessary, into a complete model. This section describes the most appropriate mathematical approaches, from a selection of mesoscale and microscale fungal growth models. Key features of these selected models are summarized in Table 2.

Table 2 Overview of the mathematical models selected for the review

Model	Components described[a]	Spatial treatment of biomass[b]	Type of lattice (for discrete models)[c]	Number of dimensions used to describe biomass	Treatment of temporal domain	Type of growth	Type of hyphae	Purpose of the model	Type of validation	Type of data used for validation/calibration
Balmant [2]	B+V	D	LB	1	Smooth	Overculture	Reproductive	For the study of intracellular phenomena in a single hypha	Calibrated	Hyphal length over time
Boswell et al. [6]	B+N	D	LB	2	Stepwise	Single colony	Surface	For bioremediation studies	Calibrated	Rate of extension of the tip and rate of branching
Coradin et al. [9]	B	D	LB	3	Stepwise	Overculture	Aerial and surface	For SSF studies	Quantitative	Biomass density over height above solid surface
Edelstein and Segel [13]	B+N	P	–	1	Smooth	Single colony	Surface	For single-colony studies	Qualitative	Biomass and tip densities over the radius of the colony (done in [12])
Fuhr et al. [15]	B+N	D	LF	3	Stepwise	Overculture	Penetrative	For wood degradation studies	Quantitative	Permeability of the wood particle
Georgiou and Shuler [16]	B+N	P	–	1[d]	Smooth	Single colony	Reproductive and surface	For single-colony studies	None	–
Hutchinson et al. [24]	B	D	LB	2	Stepwise	Single colony	Surface	For single-colony studies	Qualitative	Morphology of surface hyphae of a colony

(continued)

Table 2 (continued)

Model	Components described[a]	Spatial treatment of biomass[b]	Type of lattice (for discrete models)[c]	Number of dimensions used to describe biomass	Treatment of temporal domain	Type of growth	Type of hyphae	Purpose of the model	Type of validation	Type of data used for validation/calibration
López-Isunza et al. [29]	B+V+N	M	–	3	Smooth	Single colony	Germ tube	For the study of intracellular phenomena in a single hypha	Calibrated	Hyphal length over time
Meeuwse et al. [31]	B+O+N	P	–	1	Smooth	Overculture	Biofilm	For SSF studies	Calibrated	Overall concentration of biomass, substrates, and products over time
Mitchell et al. [33]	B+N+E	P	–	0[e]	Smooth	Overculture	Aerial and surface	For SSF studies	Calibrated	Overall biomass concentration and glucoamylase activity over time
Prosser and Trinci [48]	B+V	D	LF	2	Stepwise	Single colony	Surface	For single-colony studies	Quantitative	Various measures of growth rates based on number of tips and total mycelial length
Rajagopalan and Modak [52]	B+O+N+E	P	–	1[d]	Smooth	Overculture	Biofilm	For SSF studies	None	–
Tunbridge and Jones [65]	B+V	D	LF	2	Stepwise	Single colony	Surface	For single-colony studies	Quantitative	The simulation results from Prosser and Trinci [48] for number of tips and total mycelial length

(continued)

Table 2 (continued)

Model	Components described[a]	Spatial treatment of biomass[b]	Type of lattice (for discrete models)[c]	Number of dimensions used to describe biomass	Treatment of temporal domain	Type of growth	Type of hyphae	Purpose of the model	Type of validation	Type of data used for validation/calibration
Yang et al. [70]	B+V[f]	D	LF	3	Smooth	Pellet	Pellet	For studying pellet formation	Quantitative	Length and extension rate of the germ tube and a branch, total mycelium length, and number of tips over time

[a] *B* biomass; *V* vesicles; *O* O_2; *N* nutrient; *E* enzymes

[b] *D* discrete; *P* pseudohomogeneous; *M* microscale

[c] *LB* lattice bound; *LF* lattice free

[d] In the model of Georgiou and Shuler [16], the biomass concentration varies along the radius of a circularly symmetrical colony. In the model of Rajagopalan and Modak [52], the dimension described by the model is the radius though a spherical particle covered with a biofilm; the biofilm thickness increases during the simulation

[e] In the model of Mitchell et al. [33], biomass is represented by an overall concentration at the solid surface, without any mention of its distribution in space; thus, biomass is described without any dimension

[f] Some authors use the term "precursors," representing material that is produced inside the hypha and transported to the tip. We have classified these "precursors" as being equivalent to vesicles

4.1 Classification of Mesoscale Mathematical Models

It is useful to classify mesoscale models that describe the growth of filamentous fungi on the basis of which components they describe and how they treat the distribution of biomass within the spatial domain (Table 2). With respect to the components that are described, some models only describe the development of the fungal biomass itself, such as hyphae and tips, while other models, in addition to describing fungal biomass, recognize the contribution of other components to the growth process, such as nutrients, enzymes, and vesicles. With respect to the distribution of biomass within the spatial domain, models may be classified as being either "pseudohomogeneous models" or "discrete models" (Fig. 8). Pseudohomogeneous models, which have also been referred to as "continuum models" [5, 10], "reaction–diffusion models" [50], and "whole-population models" [14], do not recognize fungal hyphae individually and therefore do not specify their exact position in space; rather, they describe hyphae in terms of an average biomass concentration that may vary continuously along one or more dimensions within the spatial domain (Fig. 8a). Conversely, discrete models, which have also been referred to as "individual-based models" [14], specify a particular location in the spatial domain for each hyphal element and tip and distinguish this hyphal biomass from its surroundings in space; in other words, the biomass is "spatially discrete" (Fig. 8b, c). Obviously, if the intention is to produce a model that can generate multidimensional images of fungal networks, similar to those that would be seen in a micrograph, this can only be achieved with discrete models and this review focuses on this type of model. However, where relevant, approaches used to describe particular phenomena in pseudohomogeneous models will be mentioned.

Discrete models often treat the hyphae in a stochastic manner. For example, in the case of an extending tip that is capable of changing growth direction or branching apically, probabilities are assigned to these actions. At certain times during the simulation, a random number is generated to see whether the tip will undertake one of these actions; the action being evaluated will happen for the tip only if the random number is lower than the probability assigned to that action. This feature confers a degree of variability on the results of the simulation, such that two simulations that start with the same initial conditions give different final mycelial morphologies.

Discrete models can be further classified according to two criteria. Firstly, they can be classified as "smooth" or "stepwise," based on how time is treated in the model: Time progresses continuously in a smooth model but in discrete steps in a stepwise model, with these time steps normally being of the same size, although this is not essential. Secondly, they can be classified as being "lattice based" or "lattice free," on the basis of whether the hyphae are restricted to specific geometries or are free to grow in any position and direction in space [23]. These two classifications are discussed in the following subsections.

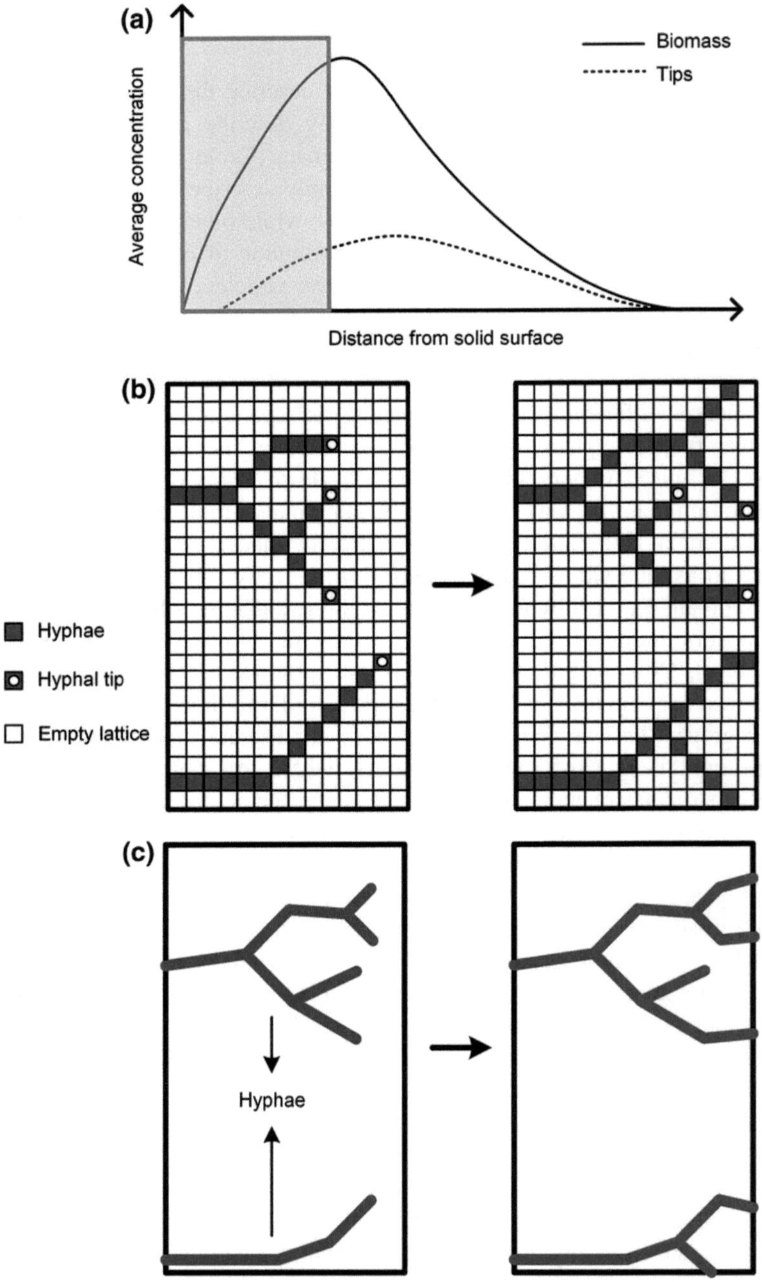

Fig. 8 Typical results that can be obtained with mesoscale models. **a** Plots of the concentrations of biomass and tips as a function of distance from the solid surface, generated by a pseudohomogeneous model. **b** Images generated by a discrete 2D lattice-based model at different times of growth. **c** Images generated by a discrete 2D lattice-free model at different times of growth. The images used to illustrate the discrete models correspond to the marked area of the graph for the pseudohomogeneous model

4.1.1 Smooth and Stepwise Models

In a smooth model, state variables change continuously with time. The rates of change in these state variables, such as hyphal length or vesicle concentration, are typically expressed in the form of continuous differential equations. Conversely, in a stepwise model, state variables change abruptly with each time step, with the changes being expressed in the form of algebraic equations. In the specific case of hyphal extension, in smooth models, the hyphal length increases in a continuous manner over time. In stepwise models, new "units" of biomass are incremented in each time step. These "units" are often of a defined length, but it is also possible to add units of varying lengths.

There are three important considerations with respect to this classification. Firstly, these model types are often referred to in the general mathematical modeling literature as "continuous change and discrete change models" or "continuous time and discrete time models." In the context of the current review, where we have used the word "discrete" in relation to the spatial domain, we prefer to use a different pair of words with relation to the temporal domain. Secondly, the distinction between smooth and stepwise models is based on the formulation of the model equations, not on the mathematical methods used to solve them. This consideration is important, since the set of differential equations that is written in a smooth model is usually too complex to be solved analytically and is therefore solved using numerical methods. These methods may involve discretization of the temporal domain; however, the use of discretization as a mathematical tool to obtain an estimate of the solution of a smooth function does not make the function itself stepwise. Thirdly, even though the use of a very small time step may allow a stepwise model to make predictions that resemble smooth functions, the equations, as formulated, are not smooth.

4.1.2 Lattice-Based and Lattice-Free Models

In lattice-based models (Fig. 8b), hyphae are only allowed to occupy positions within a previously defined lattice, which may consist of a network of segments or of a grid of "cells" (the latter being the case of cellular automata models, where the "cells" do not necessarily represent cells in the biological sense). The status of each segment or "cell" can change during the course of the simulation. For example, in a cellular automata model of fungal growth, when the simulation is initiated, a "cell" in a given location may have the status of "active tip." As the hyphal tip extends away from this location, the status of the "cell" (which now represents a part of the hypha behind the tip) changes to "active hypha" and a previously "empty" neighbor "cell" receives the status "active tip" (Fig. 8b). Since the hyphae are constrained to be within these "cells," the images generated are restricted by the geometry of the lattice, which might not allow the model to represent the true geometry of a fungal mycelium. On the other hand, since the position of each hypha is registered, it is a simple matter to detect when a hyphal tip tries to occupy a "cell" that is already occupied by an existing hypha and thereby describe avoidance reactions or anastomosis.

In lattice-free models (Fig. 8c), hyphae are described as line segments that can have any position and orientation within the spatial domain. Hyphal extension is represented by adding a new line segment at the tip of a previous segment, while branching is represented by adding two line segments at the tip (apical branching) or a new line segment at some position along an already existing segment (lateral branching). The added segments may have variable lengths and directions. In this manner, it is possible to generate images in which there are varying angles between hyphae and branches. The main disadvantages of lattice-free models are, firstly, that they do not describe the volume or area occupied by the hyphae, and secondly, it is computationally much more difficult, compared to lattice-based models, to check whether an extending hypha will try to occupy a space that is already occupied by an existing hypha.

The degree to which these two model types generate images that resemble a real mycelium depends on the size of the lattice for a lattice-based model and the rules that are used, for a lattice-free model. For example, a lattice-based model that uses lattice divisions of the same scale as the diameter of the hypha and allows various angles of growth and branching may result in a more realistic image than a lattice-free model with branching angles being restricted always to 90°.

4.2 The Representation of the Physical System

4.2.1 The Representation of the Hyphae

For a mesoscale model to be useful for describing the morphology of a fungal mycelium in SSF, the manner of representing hyphae must comply with three requirements. Firstly, it must be possible to reproduce the morphology of the mycelium in a three-dimensional environment with branching distances and angles that are similar to those seen under the microscope. Secondly, it must be possible to combine the physical representation with the mathematical equations that determine the rates of growth and other internal processes. Thirdly, it must be possible to represent different physiological states, not only in different regions of the hyphae at the same time, but also in the same hyphal segment, but at different times during the simulation. Table 3 lists various manners in which the hyphae have been physically represented in the selected models.

The representation used by Coradin et al. [9] provides the basis for a model that can comply with all three requirements (Table 3). These authors used a stochastic and discrete lattice-based model to describe the distribution, in three dimensions, of vegetative and reproductive aerial hyphae within the space above a solid surface. This space was divided into cubes, with each cube having the same cross-sectional area as a hypha. A cube could be empty or occupied by a hyphal segment, a hyphal tip, or a spore. Active hyphae extended through the addition of new cubes at their tips. The hyphae grew upward, downward, horizontally, or diagonally. Vegetative hyphae could branch apically, with the daughter hyphae choosing directions of

Table 3 Approaches used in previous models to represent hyphae and their environment

Model	Type of lattice	Size of hyphal lattice	How extension occurs	Type of environment
Ideal approach	Any geometry with volume (i.e., 3D)	At the scale of the hyphal diameter	Based on the rate of vesicle fusion at the tip	Gas phase, biofilm layer, and solid particle
Coradin et al. [9]	Cubic	10 μm	–	Gas phase (as empty cubes)
Balmant [2]	Cubic	10 μm	Uses mass balance equations for intracellular maltose and vesicles, with the extension rate being proportional to the rate of fusion of vesicles at the tip	Gas phase (as empty cubes)
Boswell et al. [6]	Triangular for biomass and hexagonal for environment	100 μm	Extension depends on the probability of the hypha extending to an adjacent part of the lattice; this probability is proportional to the intracellular nutrient concentration	Solid particle, without penetration of hyphae
Tunbridge and Jones [65]	Lattice free	–	The size of the mycelium is calculated first, without defining its distribution in space, with the extension rate being proportional to the rate of fusion of vesicles at the tip	No interactions with or descriptions of the surroundings
Fuhr et al. [15]	Segments positioned randomly for biomass and rectangular parallelepiped for environment	50 μm	Extension depends on probabilities of the hypha extending to another pit, which is related to the nutrient concentration at the pit	Solid particle with penetration of hyphae

growth stochastically, among the available empty spaces around the branching tip. The number of cubes added per iteration was chosen so that the average distance between branches was of the same order as that reported by Trinci [64]. After a certain period of growth, all vegetative hyphae stopped growing and some active tips differentiated into reproductive hyphae. The reproductive hyphae also branched, but contrary to vegetative hyphae, they only grew upward.

The main advantage of the model of Coradin et al. [9] is that the small lattice size allows a realistic representation of hyphal morphology and biomass distribution in space, while facilitating the localization of hyphae. Further, although the model does not contain expressions for mass balances on intracellular components, it was designed for later incorporation of these expressions. This was done by Balmant [2], who treated the series of cubes that represented a hypha in the model of Coradin et al. [9] as a series of well-mixed tanks, with transport of intracellular material between adjacent tanks. Balance equations were written for two components, maltose and vesicles, over each tank. In this well-mixed tank approach, these balance equations have the form of ordinary differential equations, unlike the partial differential equations that arise when balance equations are written over a hypha as a single continuous entity. The tanks forming the hypha were divided into three types: a "tip-tank" at the end of the hypha, vesicle-producing tanks, and non-vesicle-producing tanks. A different set of equations was used for each type of tank, representing the physiological differences between them (more details on the equations used are provided in Sect. 4.3.1). Growth occurred through the addition of new tanks at the tip, with this addition depending on the rate of fusion of vesicles to the membrane within the tip-tank.

Although Balmant [2] only modeled the extension of a single aerial hypha, with maltose being supplied at the base of the hypha, the well-mixed tank approach has the flexibility to be extended to describe phenomena involved in the development of a hyphal network in an SSF system, such as branching, nutrient absorption across the sides of the hypha (i.e., across the plasma membrane), and enzymatic hydrolysis of polymeric carbon and energy sources within the solid matrix, as will be discussed in later subsections. For convenience, from this point on, this approach will be referred to as the "Coradin–Balmant approach."

A model that can produce a realistic mycelial morphology, while describing internal processes, is that of Tunbridge and Jones [65] (Table 3). It is a discrete, lattice-free model describing the growth of surface hyphae from a single spore. It incorporates the earlier model of Prosser and Trinci [48], in which the hyphae are divided into sections. Vesicles are transported between adjacent sections, with the extension rate being proportional to the rate at which vesicles fuse at the tip (more details on the internal processes are provided in Sect. 4.3.1). Since Prosser and Trinci [48] did not aim to reproduce the morphology of the mycelium, they used 90° angles between branches and did not include changes in the direction of growth of extending hyphae. Tunbridge and Jones [65] added random changes in growth directions and different angles between branches, by dividing the simulation into two parts. The first part uses the model of Prosser and Trinci [48] to determine the final size of the mycelium, namely the position of each branch and the overall number of tips. In the second part, hyphal segments and branches are drawn in two-dimensional space, with random variations in growth direction and branch angles.

There would be two main problems with any attempt to adapt the Tunbridge and Jones [65] model for the modeling of SSF systems. Firstly, it is lattice free; therefore, the hyphae have no volume. Consequently, if this approach were to be extended to describe growth in a three-dimensional environment, it would be difficult to track the position of each hypha and thereby avoid two hyphae occupying the same position in space. Secondly, since growth is calculated before the spatial distribution of the mycelium is determined, it would not be possible to include inactivation of hyphae due to high local biomass concentrations or to include the effects of the variation in the concentrations of nutrients in the local surroundings.

Another approach is that of Boswell et al. [6] (Table 3). In this discrete model of the colonization of soil particles by a fungus, both the mycelium and the medium are represented by two-dimensional lattices. The medium is divided into a hexagonal lattice, with homogeneous concentration of nutrient within each hexagon. This is overlaid with a triangular lattice of 100-μm line segments that connect the centers of adjacent hexagons. Hyphae grow along these line segments, such that angles between adjacent hyphal segments are restricted to either 120° or 180°. The advantage of this model is that it already considers the interactions with the extracellular environment, while this still needs to be incorporated into the Coradin–Balmant approach. Also, the superposition of hyphae is prevented, since only a single hypha is allowed to enter each hexagon of the substrate lattice. On the other hand, the 100-μm-long segments would not be appropriate for SSF systems, since a hyphal segment of this length could contain significant gradients of internal compounds and experience significant gradients of external compounds. Further, the two-dimensional triangular lattice only allows representation of surface hyphae, and even then, it does not even allow a proper representation of observed morphologies of these hyphae.

4.2.2 The Representation of the Surroundings

In SSF, a hypha can be in contact with one or more of three phases: the gas phase, the liquid of the biofilm layer, and the matrix of the solid particle. The processes occurring in these phases should be described by mass balance equations that include mass transfer and the consumption and release of key compounds, such as O_2, soluble nutrients, and enzymes.

In discrete models, the extracellular environment is also divided into a regular lattice. In the Coradin–Balmant approach, it would be divided into well-mixed tanks of the same size and shape as the ones used for the hyphae. This would facilitate the representation of the hyphae extending into their surroundings: As the hyphal tip extends into a tank, the status of that tank would change from "surrounding phase" to "hyphal tip." Further, appropriate mass balances for enzymes, their hydrolysis products, and other nutrients would be written over those tanks that represent the matrix of the substrate particle.

The two-dimensional hexagonal lattice that is used to represent the substrate in the model of Boswell et al. [6] would not be appropriate for describing SSF systems: It would not be possible to describe how part of the volume within the solid particle is occupied by penetrative hyphae, nor to represent properly the concentration gradients that occur with depth below the particle surface.

Various models for growth of fungi over solid surfaces have treated the solid matrix as a homogeneous environment, without any physical barriers to diffusion, such as is the case when an artificial medium composed of a nutrient solution and a solidifying agent (e.g., agar or κ-carrageenan) is used [6, 33]. However, in SSF, the substrate particle is often composed of plant tissues, with cellular divisions, and the cell wall structure must be degraded in order for the hyphae to reach the nutrients contained within. The only model that describes penetrative growth of fungal hyphae in a plant tissue was developed by Fuhr et al. [15]. It is a three-dimensional discrete lattice-based and stochastic model for the growth of penetrative hyphae of a white-rot fungus in wood. The wood cells are represented by rectangular parallelepipeds (Table 3), the walls of which contain randomly distributed pits. These pits are locations of the cell wall that are easily degraded by the fungus, allowing hyphae to invade adjacent cells. In the model, hyphae extend from one pit to another, forming straight lines connecting the pits. All nutrients necessary for growth are considered to be concentrated at these pits. This model represents cell wall penetration well but presents some limitations; for example, it represents hyphae as segments without volume and it does not describe the occupation of the cell lumen by the hyphae nor does it describe the consumption of the nutrients inside the lumen.

The division of the extracellular space into cubes, as was done in the model of Coradin et al. [9], allows for the representation of cellular substructures in particles. Agglomerations of cubes could be designated as substrate particles, with particular divisions between cubes representing cell walls. These particular divisions would be barriers to the diffusion of components, and it would be made difficult for the hyphal tip to cross them.

4.3 Intracellular Phenomena Supporting Growth at the Tip

The production of new fungal biomass requires the supply of several different nutrients. However, for simplicity, it is common to assume that the carbon source is limiting and that all other nutrients are in excess; this section will focus on models that make this assumption. However, it should be noted that the models of Georgiou and Shuler [16] and Meeuwse et al. [31] describe the metabolism of both a carbon source and a nitrogen source, which enabled them to relate key events to the absence of nitrogen, such as the formation of reproductive structures or the production of intracellular carbohydrate reserves. The supply of molecular oxygen (O_2) can also be limiting; this aspect is discussed in Sect. 4.6.

4.3.1 Appropriate Mass Balance Equations for Nutrients and Vesicles

Given the key roles of vesicles in tip extension and in initiation of branching (Sect. 3.2.1), a morphological model should describe their production, translocation, and consumption within the hyphae. This should be done in a simplified manner, but the model should have at least three features. Firstly, vesicle production should be described as being limited to a specific region of the hypha (the vesicle-producing zone) and as depending on the concentration of a limiting nutrient. Secondly, translocation of vesicles to the tip along the cytoskeleton should be at a rate different from cytoplasmic flow. Thirdly, the tip extension rate should be proportional to the rate of vesicle consumption at the tip. Also, since the extension rate of a new hyphal branch increases exponentially with time until it reaches a maximum value and then remains constant, even if the vesicle concentration increases behind the tip membrane [48], the rate of vesicle consumption at the tip should be described using saturation kinetics. Saturation kinetics would also be appropriate to describe vesicle production. Finally, a single type of vesicle should be sufficient for mesoscale models, instead of trying to describe subpopulations of different types of vesicles. Among the mesoscale models for fungal growth, the only model to incorporate all of these characteristics is that of Balmant [2], and as a result, it is the only one to have predicted the vesicle concentration peak at the tip that occurs experimentally: Of all the vesicles in the 50 μm nearest to the tip, about 60 % are present in the first 10 μm behind the tip, decreasing to 17 % in the region between 10 and 20 μm behind the tip and to less than 11 % in each of the remaining 10-μm segments [8, 18].

The model of Balmant [2] describes the extension of a single unbranched aerial hypha, using a series of well-mixed tanks (Fig. 9). Mass balance equations for maltose and vesicles were written over each of the three different types of tanks. In non-vesicle-producing tanks and vesicle-producing tanks, maltose is transferred between adjacent tanks by convective flow of the cytoplasm and diffusion, while vesicles are actively transported along the cytoskeleton toward the tip. Maltose is consumed for maintenance and, in the vesicle-producing tanks, also for vesicle production. The mass balance for maltose in the vesicle-producing tanks is given by

$$\frac{\mathrm{d}S_i|_j}{\mathrm{d}t} = \underbrace{\frac{v}{\Delta z}\left(S_i|_{j-1} - S_i|_j\right)}_{\text{convection}} + \underbrace{\frac{D_S}{\Delta z^2}\left(S_i|_{j-1} - 2S_i|_j + S_i|_{j+1}\right)}_{\text{diffusion}} - \underbrace{m\rho_X}_{\text{maintenance}} - \underbrace{\frac{1}{Y_\phi}\frac{k_p S_i|_j}{K_P + S_i|_j}}_{\text{vesicle production}} \tag{1}$$

where $S_i|_j$ is the concentration of maltose in tank j, t is the cultivation time, v is the velocity of flow of the cytoplasm, Δz is the length of a tank, D_S is the diffusivity of maltose inside the hypha, m is the maintenance coefficient, ρ_X is the mass of dry cell material per unit of volume occupied by the hypha, Y_ϕ is the yield coefficient for production of vesicles from maltose, k_p is the maximum rate of vesicle production, and K_P is the saturation constant for vesicle production.

The balance for vesicles in the vesicle-producing tanks is given by

$$\frac{\mathrm{d}\phi|_j}{\mathrm{d}t} = \underbrace{\frac{\psi}{\Delta z}\left(\phi|_{j-1} - \phi|_j\right)}_{\text{activetransport}} + \underbrace{\frac{k_p S_i|_j}{K_P + S_i|_j}}_{\text{vesicle production}} \tag{2}$$

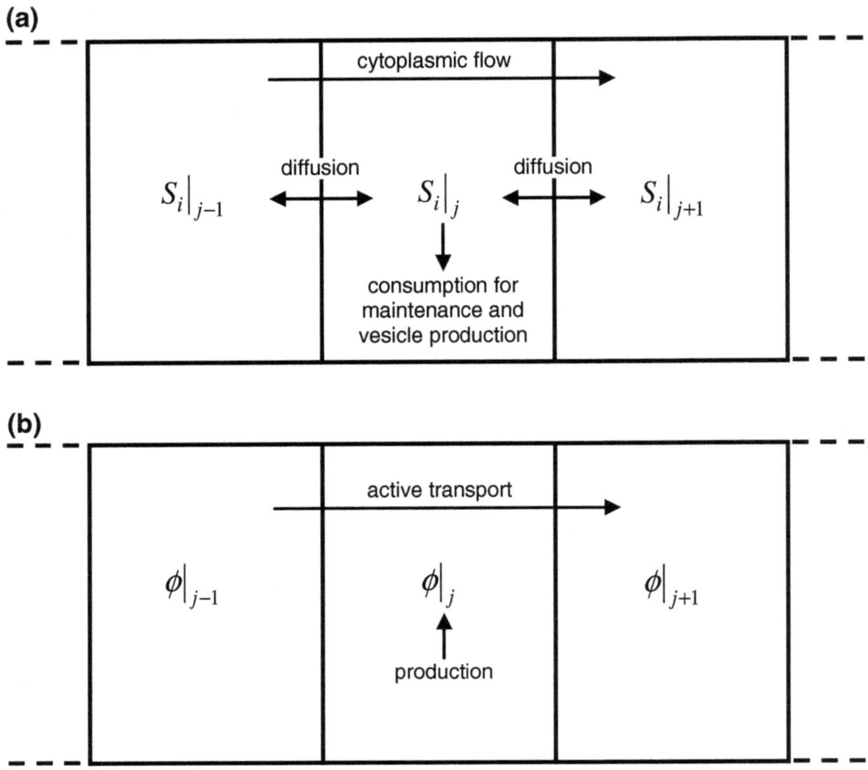

Fig. 9 Mass balances over vesicle-producing tank j in the model of Balmant [2]. **a** balance for maltose and **b** balance for vesicles

where $\phi|_j$ is concentration of vesicles in tank j and ψ is the velocity of active transport of vesicles.

As a new tank is formed at the tip, the vesicle-producing tank that is furthest from the tip is converted into a non-vesicle-producing tank. The balances on the non-vesicle-producing tanks are similar to Eqs. (1) and (2), but without the terms related to vesicle production.

Vesicle consumption occurs only in the tip-tank, increasing the length (L) of this tank at a rate that is proportional to the rate of vesicle consumption. Thus, the velocity of extension of the tip-tank is given by

$$\frac{\mathrm{d}L}{\mathrm{d}t} = Y_L \frac{k_\mathrm{c}\phi|_n}{K_\mathrm{C} + \phi|_n} \tag{3}$$

where Y_L is the extension of hyphal length per mass of vesicles consumed, k_c is the maximum rate of vesicle consumption, and K_C is the saturation constant for vesicle consumption. When the length of the tip-tank reaches double that of a normal tank (i.e., when $L = 2\Delta z$), it divides. The two new normal-sized tanks (i.e., each with $L = \Delta z$) each have concentrations of maltose and vesicles equal to the values for the double-sized tip-tank immediately before division. This approach to tip extension results in a continuous increase in hyphal length. This contrasts with various stepwise models, in which a new segment of hypha of a fixed size is added per growth step, thus resulting in instantaneous production of new segments.

The tip-tank only consumes vesicles; it does not produce them. Thus, the balances on maltose and vesicles in this tank are, respectively:

$$\frac{\mathrm{d}S_i|_n}{\mathrm{d}t} = -\underbrace{\left(\frac{S_i|_n}{L}\right)\frac{\mathrm{d}L}{\mathrm{d}t}}_{\text{tankextension}} + \underbrace{\frac{v}{L}S_i|_{n-1}}_{\text{convection}} + \underbrace{\frac{D_S}{L\left(\frac{L+\Delta z}{2}\right)}(S_i|_{n-1} - S_i|_n)}_{\text{diffusion}} - \underbrace{m\rho_X}_{\text{maintenance}} \tag{4}$$

$$\frac{\mathrm{d}\phi|_n}{\mathrm{d}t} = -\underbrace{\left(\frac{\phi|_n}{L}\right)\frac{\mathrm{d}L}{\mathrm{d}t}}_{\text{tank extension}} + \underbrace{\frac{\psi}{L}\phi|_{n-1}}_{\text{active transport}} - \underbrace{\left(\frac{1}{A_L}\right)\frac{k_C\phi|_n}{K_C + \phi|_n}}_{\text{consumption}} \tag{5}$$

where A is the cross-sectional area of the hypha. The first term on the right-hand side of each equation describes the dilution that occurs due to the increase in the tank volume caused by tip extension.

Two other models that describe vesicles are those of Prosser and Trinci [48] and López-Isunza et al. [29]. In the model of growth of surface hyphae of Prosser and Trinci [48], processes involving nutrients are not described. The vesicle production rate is constant, and vesicle production occurs throughout the entire mycelium, instead of being limited to the vesicle-producing zone. The rate of tip extension is proportional to the rate of vesicle fusion at the tip, which is described by saturation kinetics. The model of López-Isunza et al. [29] describes the extension of a single germ tube, across the substrate surface, before the formation of the first branch. Two types of vesicles are described: Macrovesicles are produced throughout the germ tube, while microvesicles are produced at the tip from macrovesicles; production rates for both vesicles follow Michaelis–Menten-type kinetics. Partial differential equations describe the diffusion of nutrients and macrovesicles in both the axial and radial dimensions of the tube. Tip extension depends on the rate of vesicle fusion at the tip, which is directly proportional to both the concentration of microvesicles and the length of the germ tube and follows saturation kinetics with respect to the concentration of macrovesicles. However, this model would not be appropriate for SSF, since the description of both radial and axial diffusion would be unnecessarily complex for the modeling of a whole mycelium. Another major drawback of this model is that it is not predictive, as the final length of the germ tube is used, at the start of the simulation, to calculate key parameters, namely the Damköhler numbers for vesicle production and cell wall synthesis at the tip.

4.3.2 Nutrient Absorption Across the Membrane

The absorption of soluble nutrients across the plasma membrane of the hypha is often included in mathematical models for fungal growth. Since the transport of polar molecules across this membrane is mediated by carriers, it exhibits saturation kinetics and should be described by an expression similar to the following:

$$r_a = \frac{k_{\max} S_e}{K_t + S_e} T A_L \tag{6}$$

where r_a is the rate of transport, S_e is the concentration of nutrient in the extracellular medium, T is the transporter concentration per area of the plasma membrane, A_L is the area of contact of the plasma membrane with the extracellular medium, $k_{\max}$ is the maximum specific transport rate, and K_t is the saturation constant. The term T could be removed from the equation if the transporters were assumed to be evenly distributed across the area of contact along the hypha, as done by Edelstein and Segel [13], who also considered that the area of contact was proportional to the biomass concentration. Transport could be included in the model of Balmant [2] by adding an expression similar to Eq. (6) to the balance equations for intracellular maltose for those tanks in contact with extracellular nutrients [i.e., to Eqs. (1) and (4)].

Some kinetic expressions that have been used in models to date do not allow for saturation, for example, that of Boswell et al. [6], in which the uptake rate is first order in the external substrate concentration (Table 4). On the other hand, López-Isunza et al. [29] described, individually, the extracellular uptake by the transport protein, the diffusion of the transporter–substrate complex across the membrane, and the release of substrate into the cytosol. However, this added complexity does not bring any tangible advantages over Eq. (6) for mesoscale models.

4.3.3 Consumption of Intracellular Substrate for Cellular Maintenance

During the development of the mycelium, intracellular substrate is required to produce energy for other processes besides the production of vesicles, such as for active transport of nutrients across the membrane, transportation of vesicles along the cytoskeleton, and enzyme production. It is not necessary for a model to describe the rates of intracellular substrate consumption for all of these processes individually; they can be combined in a single maintenance term. This maintenance term is usually expressed as the product of the biomass concentration and a constant maintenance coefficient, as in the models of Balmant [2] and Meeuwse et al. [31].

On the other hand, Boswell et al. [6] explicitly described the substrate consumption associated with active transport across the membrane (Table 4). In their model, the substrate absorption rate (r_a) was expressed as follows:

Table 4 Approaches used by previous models in the description of intracellular substrates

Model	Consumption of substrate	Intracellular transport	Absorption rate
Ideal approach	For vesicle production and cellular maintenance	Axial diffusion and cytoplasmic transport toward the tip	A function of the concentration of extracellular substrate through saturation kinetics and proportional to the absorption area [see Eq. (6)]
Balmant [2]	For vesicle production and cellular maintenance	Axial diffusion and cytoplasmic transport toward the tip	No absorption, rather supply of maltose from a source tank representing vegetative mycelium in the biofilm, from which the single unbranched aerial hypha that was modeled extended
Boswell et al. [6]	For biomass production, cellular maintenance and absorption and active transport costs	Axial diffusion and active transport toward the highest concentration of tips	$r_a = c_1 S_\mathrm{i} S_\mathrm{e} X$
Edelstein and Segel [13]	For biomass production and cellular maintenance	Axial diffusion and cytoplasmic transport toward the highest concentration of tips	$r_a = c_2 \frac{k_{\max} S_\mathrm{e}}{K_t + S_\mathrm{e}} X$
López-Isunza et al. [29]	For microvesicle and macrovesicle production	Axial and radial diffusion and axial convection toward the tip	Rate for nutrient uptake by the transporter: $\frac{k_{\max} S_\mathrm{e} T}{K_t + S_\mathrm{e}}$ Rate for diffusion of the transporter–substrate complex in the membrane: $\propto \frac{\partial^2 \theta}{\partial \eta^2}$ Rate for intracellular release: $c_3 \theta$
Meeuwse et al. [31]	Glucose: for biomass production, cellular maintenance, and product formation. Alanine: for biomass production	Diffusion[a]	Free diffusion[a]

[a] In the model of Meeuwse et al. [31], the biofilm layer is described as a homogeneous layer, without distinguishing the biomass and the liquid film individually

$$r_a = c_1 S_\mathrm{i} S_\mathrm{e} X \tag{7}$$

where c_1 is a proportionality constant, S_i is the concentration of intracellular substrate, S_e is the concentration of extracellular substrate, and X is the concentration of biomass participating in the absorption process. This expression is included in the balance equations for both the intracellular and the extracellular substrate, but with different values for the proportionality constant (c_1), such that more nutrients are removed from the external medium than are added to the intracellular space. The difference between the rates therefore represents the amount of substrate needed to provide the energy for the active transport across the membrane. This approach requires an additional parameter to be determined, related to the energy cost of nutrient transport across the plasma membrane, and, at the moment, it is unclear whether these energy costs are so high that they should be accounted for separately from the maintenance coefficient. For simplicity, it seems better to subsume all the processes for nutrient consumption that are not associated directly with growth within a general maintenance coefficient with a constant value.

4.3.4 Considerations About Cytoplasmic Flow and Transport of Vesicles

Once inside the hypha, nutrients diffuse and are transported toward the tip by the cytoplasmic flow, while vesicles are transported actively along the cytoskeleton, at a rate that is different from the rate of cytoplasmic flow. Although intracellular diffusion of nutrients is commonly described in models, cytoplasmic flow has only been included in the models of López-Isunza et al. [29] and Balmant [2]. Further, only the latter model allowed for different rates of cytoplasmic flow and vesicle transport (Table 4).

The models of López-Isunza et al. [29] and Balmant [2] describe a single non-septate hypha and, hence, do not allow for changes in the cytoplasmic flow or the rate of vesicle transport when septa or branches are formed. In fact, adequate data about how these phenomena affect intracellular transport are not available, although some considerations can be made. For example, unplugged septa, such as those in *Aspergillus*, would not prevent cytoplasmic flow, but might reduce the flow rate. Additionally, the formation of new branches would increase the total cross-sectional area of flow after a branch; thus, if the flow rate in the mother hypha were to remain constant, the flow rate in each daughter branch would decrease. On the other hand, branching increases the number of tips, which is where the driving force of the cytoplasmic flow is generated [28]; thus, the flow rate per tip might remain constant, with an increase in the flow rate in the mother hypha. At the very least, the cytoplasmic flow needs to be high enough to supply cytoplasm to the growing tips: The cytoplasmic flow in a given hyphal segment should be equal to or higher than the sum of the extension rates of the tips that are connected to it [22].

For vesicles, the rate of transport depends on the cytoskeleton and on motor proteins; hence, it is influenced by the formation or interruption of these structures. Therefore, the rate of transport of vesicles may not be affected by an unplugged septum if its formation does not interrupt the cytoskeleton. On the other hand, it will be influenced by the formation of apical branches, which divides the vesicle flow between the two daughter hyphae.

Once an adequate understanding of these phenomena is obtained, descriptions can easily be incorporated into discrete models. One possible approach has already been implemented by Prosser and Trinci [48] for vesicle transport: During the formation of an unplugged septum, the flow rate of vesicles decreased, linearly, to 10 % of the flow rate that was present before septation began. For lateral branches, 90 % of the vesicles continued in the direction of the main hypha, while 10 % entered the lateral branch. On the other hand, the consequences of branching for cytoplasmic flow were not described in their model.

4.3.5 Septation

In septate hyphae, it is important to model septation because it influences intracellular transport of vesicles and nutrients and may be related to the formation of lateral branches. The only two models to have described septation have related it to the duplication of nuclei in the apical compartment. The first model, of Prosser and Trinci [48], was applied to the growth of surface hyphae of *Aspergillus nidulans* and *Geotrichum candidum*. In this model, the apical compartment initially has four nuclei. This compartment extends, in very small steps; when the volume-to-nuclei ratio within it reaches a critical value, the number of nuclei starts to increase exponentially. When there are eight nuclei, a septum is formed in the middle of the apical compartment, dividing the nuclei equally on the two sides of the septum. The second model, of Yang et al. [70], describes the formation of pellets of *Streptomyces tendae* and *G. candidum* in submerged fermentation. This model also starts with four nuclei in the apical compartment. New nuclei are formed constantly, at a rate that is inversely proportional to the number of nuclei. Septation also occurs when the number of nuclei in the apical compartment reaches eight, dividing this compartment in half.

At the moment, there is no advantage to be gained in modeling the kinetics of formation of nuclei in mesoscale models of fungal growth in SSF. Septation could be modeled deterministically, by triggering septum formation at the moment the apical compartment reaches a certain length. This would lead to uniform spacing between septa. Alternatively, a stochastic expression could be used, leading to variable distances between septa. In both cases, the formation of the septa could be instantaneous or it could last for a certain period of time, in the manner of Prosser and Trinci [48]. In the model of Balmant [2], particular interfaces between adjacent tanks could be designated as being septa, with appropriate changes in the diffusivities between the two tanks. Since both the distance behind the tip at which septum formation occurs and the time taken for the septum to form vary from fungus to fungus, any model will need to be calibrated with the use of microscopic observations of the particular fungus being modeled.

4.3.6 Formation of Branches

Branching can be triggered in a model using either a deterministic or a mechanistic approach (Table 5). The deterministic approach would be used in a model that does not describe vesicles. It is based on the principle that the ratio of the total hyphal length to the number of tips within the mycelium, a parameter called the hyphal growth unit (HGU), remains almost constant during development of the mycelium [64]: Whenever the HGU surpasses a chosen value, then branching occurs, either at the tip or at a randomly selected position within the subapical zone. In the mechanistic approach, branching occurs when there is accumulation of vesicles behind a septum (lateral branching) or at a tip (apical branching) [64]. This could be done either by establishing a critical concentration of vesicles above which branching would occur or by expressing the probability of branching as being proportional to the concentration of vesicles.

The mechanistic approach could be incorporated into the model of Balmant [2], since vesicles are already described to determine the extension rate. However, before incorporating branching, it is important to verify which type of branching occurs for each type of hypha for the fungus being modeled. For example, *R. oligosporus* grown on agar seems to form both apical and lateral branches in penetrative hyphae, but only apical branches in aerial hyphae [41, 42]. The choice of direction of growth of the new branch is discussed in Sect. 4.5.

4.3.7 Physiological Differentiation Within the Mycelium

Many mathematical models for fungal growth on solid particles consider only the growth of surface hyphae. However, as explained in Sect. 3.1, growth of microcolonies over the substrate surface is only important in the initial stages of an SSF process. During much of the process, biofilm, aerial, and penetrative hyphae are present simultaneously, with each hyphal type interacting with its environment in a different way. Another important characteristic of a mycelium is physiological differentiation: Active hyphae can become inactive (i.e., losing the ability to extend), vegetative hyphae can give rise to reproductive hyphae, and a particular region of the hypha that is originally producing vesicles can stop producing them as the tip and the accompanying vesicle-producing zone extend away from it. A mesoscale model of SSF should be able to describe these changes.

The importance of describing differentiation of vegetative hyphae into reproductive hyphae was demonstrated by Coradin et al. [9]. These authors simulated density profiles for aerial hyphae of *R. oligosporus*, as a function of height above the surface of a solid substrate, and compared them with experimental results of [42]. Nopharatana et al. [42] obtained an "early profile," before sporangiophores appeared, and a "later profile," after sporangiophores appeared. In the work of Coradin et al. [9], a model including only vegetative hyphae could only describe the early profile; it was only possible to describe the later profile by incorporating reproductive hyphae, with different extension and branching rules than those of the

Table 5 Approaches used by previous models in the description of branching and changes in growth direction

Model	Type of branches	Formation of branches	Angles between branches	Changes in growth direction of unbranched hypha
Ideal approach	Describing the type of branching in the fungus being modeled	Based on vesicle concentration	Angles following a normal distribution	Angles following a normal distribution
Coradin et al. [9]	Apical	In every iteration, at the active tips	Angles depend on the chosen direction of growth, which is determined through probabilities based on the distance from the particle surface	None
Hutchinson et al. [24]	Lateral	In every iteration, with the distance between the active tip and the start of the new branch following a gamma function	Angles following a normal distribution centered at 56° and with a standard deviation of 17°	None
Prosser and Trinci [48]	Apical and lateral	Occurs when concentration of vesicles at the tip or behind a septum surpasses a critical value	90° angles with mother hypha	None
Tunbridge and Jones [65]	Same as Prosser and Trinci [48]	Same as Prosser and Trinci [48]	Angles following a normal distribution centered at 65° and with a standard deviation of 15°	20 % probability of changes of direction occurring, with the new angle following a normal distribution centered at 10° and with a standard deviation of 3°
Yang et al. [69, 70]	Lateral	Occurs a certain time after septation, with the time lapse being described by a normal distribution	Angles following a normal distribution centered at 90° and with a standard deviation of 29.1°	Angles following a normal distribution centered at 0° and with a standard deviation of 10.1°

vegetative hyphae, into the model. This would explain why the pseudohomogeneous model of Nopharatana et al. [40], which considered only vegetative hyphae, gave biomass profiles that were similar to the early profile of Nopharatana et al. [42], but not to the later profile.

The simplest approach to describing different physiological states within a single hypha is that used by Balmant [2]. The model recognizes three separate regions with different physiologies: first, a non-vesicle-producing zone, which only consumes substrate for cellular maintenance and therefore does not contribute to tip extension; second, a vesicle-producing zone, where substrate is consumed for maintenance and vesicle production; and third, a tip, in which vesicles are incorporated, driving hyphal extension. This approach could be further adapted to describe other types of physiological differentiation, for example, the production of hydrolytic enzymes being limited to the tip-tank. In addition, it is possible to have different maintenance coefficients for each type of tank, reflecting their different physiologies.

The trigger used for each differentiation process varies. Inactivation of a hyphal tip should occur when there is either no available internal substrate or no empty space to extend to. For example, in the model of Boswell et al. [6], inactivation of a hyphal segment occurs when the concentration of intracellular substrate is less than the value that is required for cellular maintenance. On the other hand, in the model of Coradin et al. [9], who did not include substrate-related phenomena, inactivation of a tip occurs when it tries to extend, for the second time, into a cube in the lattice that is already occupied by biomass.

A simple way to describe differentiation of vegetative hyphae into reproductive hyphae is to use the age of the mycelium as the trigger, as done by Coradin et al. [9]. However, the formation of reproductive structures can also be triggered by growth conditions, such as a low nutrient concentration or a low water activity in the surroundings. A combination of triggers was used by Georgiou and Shuler [16] in their pseudohomogeneous model for growth of surface and aerial hyphae in a single colony: Reproductive hyphae were only formed from vegetative hyphae that were older than 24 h (representing the time required to accumulate the intracellular carbohydrates used to support the growth of reproductive structures) and only when the concentration of nitrogen was lower than a critical value.

4.4 Phenomena Occurring Within the Matrix of the Substrate Particle

The most important phenomena that occur within the matrix of the solid particle and which, therefore, should be included in a mesoscale model for fungal growth in SSF are as follows: enzyme secretion near the hyphal tip, enzyme diffusion in the matrix of the solid particle, hydrolysis of polymeric carbon and energy sources into soluble hydrolysis products and diffusion of these soluble hydrolysis products, soluble nutrients, and O_2 in the matrix of the solid particle.

Mitchell et al. [33] incorporated several of these phenomena into a pseudohomogeneous model for fungal growth on the surface of a solid medium. Experimentally, formation of penetrative biomass was prevented by a membrane placed directly on the surface of the solid, while nutrients and enzymes could diffuse across the membrane [34].

In the model, processes involving the components within the solid particle (i.e., starch, glucose, and glucoamylase) are described with mass balance equations, forming a system of partial differential equations. The mass balance equation for glucose (S_e) includes diffusion in the medium and formation through the hydrolysis of starch by glucoamylase, following Michaelis–Menten kinetics:

$$\frac{\partial S_e}{\partial t} = \underbrace{D_S^e \frac{\partial^2 S_e}{\partial z^2}}_{\text{diffusion}} + \underbrace{Y_E \frac{k_{cat} E S_p}{K_m + S_p}}_{\text{formation}} \tag{8}$$

where z is the depth in the substrate matrix, D_S^e is the effective diffusivity of glucose in the matrix of the particle, Y_E is the yield of glucose from starch, k_{cat} is the catalytic constant of glucoamylase, S_p is the concentration of starch, and K_m is the saturation constant.

Starch is considered to be immobile; thus, its mass balance equation describes only its hydrolysis by glucoamylase:

$$\frac{\partial S_p}{\partial t} = -\frac{k_{cat} E S_p}{K_m + S_p} \tag{9}$$

The mass balance equation for glucoamylase within the matrix of the particle includes only diffusion:

$$\frac{\partial E}{\partial t} = -D_E \frac{\partial^2 E}{\partial z^2} \tag{10}$$

where D_E is the effective diffusivity of the glucoamylase in the solid particle and E is the glucoamylase concentration.

Secretion of glucoamylase occurs only at the surface of the solid medium, so that the secretion rate is used as a boundary condition:

$$J_E\big|_{z=0} = D_E \frac{\partial E}{\partial z}\bigg|_{z=0} = r_E H(t_E - t) \tag{11}$$

where $J_E\big|_{z=0}$ is the flux of glucoamylase across the surface of the particle, r_E is the rate of secretion of glucoamylase (which has a constant value, while glucoamylase is being secreted), $H(t_E - t)$ is a Heaviside function, which is equal to 1 when $t \le t_E$ and 0 when $t > t_E$, and therefore, t_E is the time when the secretion of glucoamylase ceases. Equation (11) is a simple empirical description of the experimentally observed secretion profiles: In real SSF systems, enzyme production depends on complex induction and repression mechanisms that are, as yet, not sufficiently understood for incorporation into mesoscale mathematical models.

This model was able to describe a key experimental observation, namely the clearing of starch from the region near the surface of the substrate particle due to the action of the glucoamylase. Once this happened, only the glucoamylase that managed to diffuse past the cleared zone could contribute to the release of glucose from the starch within the substrate matrix. As a result, the rate of supply of glucose to the biomass at the surface limited the growth rate for most of the cultivation [33].

This approach to modeling the processes that occur within the matrix of the substrate particle could be incorporated into a discrete model, such as that of Balmant [2]. The extracellular matrix would be divided into well-mixed tanks, of the same size as the tanks used to describe the hypha. Mass balance equations similar to Eqs. (8)–(11) would be written for the extracellular components, except that all equations would be ordinary differential equations in time, since there is no spatial coordinate. It would be necessary to limit enzyme secretion to the tanks representing the hyphal tips, as this is the site at which it occurs [68]. Of course, it would also be necessary to describe the growth of penetrative hyphae, as enzyme secretion from these hyphae reduces the distance over which the enzyme must diffuse in order to reach the polymeric carbon and energy source [50, 66].

4.5 Choice of Hyphal Growth Direction

A mathematical model that proposes to reproduce the morphology of the fungal mycelium in SSF in a realistic manner must allow for changes in the direction of growth of an extending unbranched hypha and also for realistic angles between branches, similar to those observed in micrographs. The direction of growth of hyphae results from a combination of many factors, such as the substrate composition, the fungal species under study, the type of hypha, and tropisms.

In a model, the direction of hyphal extension could be chosen stochastically or deterministically. The stochastic approach was used by Hutchinson et al. [24] and Yang et al. [69, 70], who showed, experimentally, that the angles formed between daughter and mother hyphae followed a normal distribution (Table 5). They then developed lattice-free models in which the angles between branches followed a normal distribution with a mean of 56° and a standard deviation of 17°, for surface hyphae of *Mucor hiemali* [24], and with a mean of 90° (i.e., perpendicular to the direction of the parent hypha) and a standard deviation of 29.1°, for pellet growth of *S. tendae* in submerged liquid fermentation [69].

For an extending unbranched hypha within a pellet, Yang et al. [69] observed that although the hypha does deviate, the overall growth direction changes little. In their model, an extending hypha intermittently changed its growth direction according to a normal distribution centered at 0°, with a standard deviation of 10.1°. Hutchinson et al. [24] did not incorporate changes in growth direction of an extending unbranched hypha in their model.

The advantages of this stochastic approach are its simplicity and its ability to generate patterns that appear similar to real mycelia. The drawback is that these angles must be measured experimentally for each type of hypha. Also, for this approach to be used in a lattice-based model, the measured angles need to be distributed among the limited number of angles that are available in the particular lattice used.

The deterministic approach to controlling branch angles and growth directions of hyphae is based on tropisms. It requires identification of the factors that influence growth direction and the explicit incorporation of these factors in the model. Such factors might include the concentrations of nutrients or inhibitory compounds in the surroundings, but it is difficult to identify all the influences, especially since most studies on tropism are for surface hyphae. As a result, our current knowledge is insufficient for incorporating the deterministic approach into mesoscale SSF models.

4.6 Availability of O_2 and Respirative or Fermentative Metabolism

As the mycelium develops on the particle surface, producing an ever thicker biofilm, a stage will be reached when the supply of O_2 into the interior of the biofilm is lower than the rate of consumption of O_2 by the biomass (Sect. 3.2.3). The lack of O_2 will affect growth of hyphae in the anaerobic regions of the biofilm and of penetrative hyphae within the substrate particle. O_2 diffusion in the extracellular environment and consumption for hyphal growth and maintenance must therefore be incorporated into a mesoscale mathematical model of SSF. As O_2 diffuses freely across the plasma membrane, it is not necessary to include special expressions [i.e., expressions for carrier-mediated transport similar to Eq. (6)] to describe O_2 transport into the hypha; however, a mass balance of O_2 within the hypha should be included. Intracellular phenomena that rely on O_2 can be expressed as functions of the local intracellular O_2 concentration.

The main question is which biological phenomena should be expressed as depending on the concentration of O_2 and how this dependence should be expressed mathematically. The two phenomena that are the most affected by the O_2 concentration are the growth rate and the type of metabolism, which can be "respirative" or "fermentative." The only model to relate the growth rate to O_2 concentration in SSF was that of Rajagopalan and Modak [52], a pseudohomogeneous model that describes the growth of a fungal biofilm around a spherical particle, without hyphal penetration into the particle. The rate of biomass growth (r_X) was expressed as depending on the concentrations of glucose and O_2 in the biofilm layer, according to a double Monod expression:

$$r_X = \frac{\mu_{max} S^f}{K_s + S^f} \frac{C_{O_2}^f}{K_{O_2} + C_{O_2}^f} X \tag{12}$$

where $\mu_{\max}$ is the maximum specific growth rate constant, S^f is the concentration of glucose in the biofilm, K_s is the saturation constant for glucose, $C^f_{O_2}$ is the O_2 concentration in the biofilm, and K_{O_2} is the saturation constant for O_2. In a model that describes hyphal tip extension as depending on the production and consumption of vesicles, the influence of O_2 concentration could be incorporated into the expression for vesicle production. In the model of Balmant [2], this could be done by using the expression of Rajagopalan and Modak [52] within the term for vesicle production in the mass balance of maltose [i.e., within Eq. (1)]. With this modification, the mass balance of maltose would be

$$\begin{aligned}\frac{\mathrm{d}S_i|_j}{\mathrm{d}t} = & \underbrace{\frac{v}{\Delta z}(S_i|_{j-1} - S_i|_j)}_{\text{convection}} + \underbrace{\frac{D_S}{\Delta z^2}(S_i|_{j-1} - 2S_i|_j + S_i|_{j+1})}_{\text{diffusion}} \\ & - \underbrace{m\rho_X}_{\text{maintenance}} - \underbrace{\frac{1}{Y_\phi}\frac{k_p S_i|_j}{K_P + S_i|_j}\frac{C_{O_2}|_j}{K_{O_2} + C_{O_2}|_j}}_{\text{consumption for growth}}\end{aligned} \tag{13}$$

where $C_{O_2}|_j$ is the O_2 concentration in tank j.

As some fungi can switch to fermentative metabolism in regions of the mycelium where O_2 is limiting, the local intracellular O_2 concentration could be used in the model as a trigger for a metabolic shift between respirative and fermentative metabolism. In other words, fermentative metabolism would occur at O_2 concentrations lower than a critical value. In fermentative metabolism, the mass balance equations for nutrient or vesicles would have the same form as in respirative metabolism, but the parameters related to rates and yields would have different values. However, we currently know little about fermentative metabolism in filamentous fungi and this is probably the reason that fermentative metabolism has never been included in a mesoscale model for SSF. In the few models that do include a balance for O_2, such as those of Rajagopalan and Modak [52] and Meeuwse et al. [31], whenever the O_2 concentration reaches zero, biomass growth and any other biological reactions stop completely.

A balance equation for O_2 will include consumption terms. Typically, there will be two consumption terms, one related to the production of biomass or vesicles and the other related to maintenance metabolism. Rajagopalan and Modak [52] considered that O_2 was used only for biomass growth, so that the mass balance for O_2 in the biofilm was

$$\frac{\partial C^f_{O_2}}{\partial t} = \underbrace{\frac{D^f_{O_2}}{r}\frac{\partial}{\partial r}\left(r^2\frac{\partial C^f_{O_2}}{\partial r}\right)}_{\text{diffusion}} - \underbrace{\frac{1}{Y_{X/O_2}}\frac{\mu_{\max}S^f}{K_s + S^f}\frac{C^f_{O_2}}{K_{O_2} + C^f_{O_2}}X}_{\text{consumption for growth}} \tag{14}$$

where $D^f_{O_2}$ is the effective diffusivity of O_2 in the biofilm layer, r is the radial position in the biofilm, and Y_{X/O_2} is the yield of biomass based on O_2 consumption.

A different approach was adopted by Meeuwse et al. [31]. Their pseudohomogeneous model describes the growth of a fungal biofilm over the surface of a substrate particle, without hyphal penetration. In this model, the rate of O_2 consumption (r_{O_2}) is calculated based on the rates of biomass growth, nutrient consumption, and metabolite production:

$$r_{O_2} = 1.1675 r_X + 1.4325 r_{LI} + r_{IP} + 1.5 r_{EP} + 3.75 r_N + r_S \tag{15}$$

where r_{LI} is the rate of production of lipids, r_{IP} is the rate of production of storage carbohydrates, r_{EP} is the rate of production of extracellular products, r_N is the rate of consumption of alanine, and r_S is the rate of consumption of glucose. However, other than a cessation of O_2-consuming processes in the absence of O_2, the rates of growth and production of metabolites were not influenced by the O_2 concentration itself.

There are at least three important questions, the answers to which might influence future model development. Firstly, does the growth rate really depend on the O_2 concentration? Above a certain O_2 concentration, growth is fully respirative, and it may not be useful to express the growth rate as a function of O_2 concentration if K_{O_2} is very low [49]. Of course, if fermentative metabolism does occur in the absence of O_2, then it is still necessary to describe the effect of the type of metabolism on the growth rate. Secondly, how would the shift to anaerobic conditions change the kinetic and yield parameters used for calculating the maintenance coefficient and vesicle production rate? It would be important to determine these parameters for both respirative metabolism and fermentative metabolism, before incorporating fermentative metabolism in the model. Thirdly, do hyphae that are in aerobic regions supply O_2 or metabolites derived from respirative metabolism to the penetrative hyphae? This possibility was raised by Nopharatana et al. [41], who suggested that hyphae penetrating into deep regions of the substrate particle might receive ATP or other precursors from hyphae closer to the surface. Later, te Biesebeke et al. [60] showed that the genomes of *Aspergillus* species contain several genes that code for domains of hemoglobin and that the superexpression of one of these domains in *A. oryzae* resulted in a higher growth rate and greater enzyme production compared to the parent strain. If there is such a transport system for intracellular O_2 or precursors of biosynthesis, then a segment of hypha surrounded by an anaerobic environment might not turn to fermentative metabolism if it receives these precursors. This would affect how the fermentative metabolism is triggered.

5 Concluding Remarks

Many SSF processes involve filamentous fungi and a mathematical model describing what happens at the particle scale could be an important tool for understanding not only how the mycelial mode of growth of these fungi affects bioreactor performance but also how bioreactor operation affects the development of the mycelium. However, there is currently no model that describes, at an appropriate level of detail, all of the relevant mass transfer and biological processes that occur in the spatially heterogeneous bed of solid substrate particles. Notwithstanding, we have shown that many of the individual phenomena that would be included in such a model have been adequately represented in models that have already been developed to describe fungal growth in various systems. Further, we have argued that the Coradin–Balmant approach to representing the system, which involves treating the hyphae as series of well-mixed tanks, has several advantages. First, it allows a realistic representation of the network of hyphae in the mycelium. Second, it can readily be extended, through the incorporation of new expressions, to describe phenomena that one may desire to include in the model. Third, it involves only ordinary differential equations, expressed in terms of time; in other words, the spatial coordinate does not appear in the balance equations, and consequently, there is no need to solve partial differential equations. Fourth, as a consequence of the division of hyphae into tanks, it becomes a simple matter to represent different physiologies along a single hypha. In conclusion, it is our opinion that the current knowledge that is available in the literature can underpin the development of a suitable mesoscale model for describing the growth of filamentous fungi in SSF.

References

1. Auria R, Ortiz I, Villegas E, Revah S (1995) Influence of growth and high mould concentration on the pressure drop in solid state fermentations. Process Biochem 30:751–756
2. Balmant W (2013) Mathematical modeling of the microscopic growth of filamentous fungi on solid surfaces. Ph.D. thesis, Universidade Federal do Paraná
3. Berepiki A, Lichius A, Read ND (2011) Actin organization and dynamics in filamentous fungi. Nat Rev Microbiol 9:876–887
4. Bleichrodt R, Vinck A, Krijgsheld P, van Leeuwen MR, Dijksterhuis J, Wösten HAB (2013) Cytosolic streaming in vegetative mycelium and aerial structures of *Aspergillus niger*. Stud Mycol 74:31–46
5. Boswell GP, Davidson FA (2012) Modelling hyphal networks. Fungal Biol Rev 26:30–38
6. Boswell GP, Jacobs H, Ritz K, Gadd GM, Davidson FA (2007) The development of fungal networks in complex environments. Bull Math Biol 69:605–634
7. Chahal DS, Moo-Young M, Vlach D (1983) Protein production and growth characteristics of *Chaetomium cellulolyticum* during solid state fermentation of corn stover. Mycologia 75:597–603
8. Collinge AJ, Trinci APJ (1974) Hyphal tips of wild-type and spreading colonial mutants of *Neurospora crassa*. Arch Microbiol 99:353–368
9. Coradin JH, Braun A, Viccini G, Luz LFL Jr, Krieger N, Mitchell DA (2011) A three-dimensional discrete lattice-based system for modeling the growth of aerial hyphae of filamentous fungi in solid-state fermentation systems. Biochem Eng J 54:164–171
10. Davidson FA (2007) Mathematical modelling of mycelia: a question of scale. Fungal Biol Rev 21:30–41
11. Du C, Lin SKC, Koutinas A, Wang R, Dorado P, Webb C (2008) A wheat biorefining strategy based on solid-state fermentation for fermentative production of succinic acid. Bioresour Technol 99:8310–8315

12. Edelstein L, Hadar Y, Chet I, Henis Y, Segel LA (1983) A model for fungal colony growth applied to *Sclerotium rolfsii*. J Gen Microbiol 129:1873–1881
13. Edelstein L, Segel LA (1983) Growth and metabolism in mycelial fungi. J Theor Biol 104:187–210
14. Ferrer J, Prats C,·López D (2008) Individual-based modelling: An essential tool for microbiology. J Biol Phys 34: 19–37
15. Fuhr MJ, Schubert M, Schwarze FWMR, Herrmann HJ (2011) Modelling the hyphal growth of the wood-decay fungus *Physisporinus vitreus*. Fungal Biol 115:919–932
16. Georgiou G, Shuler ML (1986) A computer model for the growth and differentiation of a fungal colony on solid substrate. Biotechnol Bioeng 28:405–416
17. Glass NL, Rasmussen C, Roca MG, Read ND (2004) Hyphal homing, fusion and mycelial interconnectedness. Trends Microbiol 12:135–141
18. Gooday GW (1971) An autoradiographic study of hyphal growth of some fungi. J Gen Microbiol 67:125–133
19. Goriely A, Tabor M (2008) Mathematical modeling of hyphal tip growth. Fungal Biol Rev 22:77–83
20. Grimm LH, Kelly S, Krull R, Hempel DC (2005) Morphology and productivity of filamentous fungi. App Microbiol Biotechnol 69:375–384
21. Harris SD (2008) Branching of fungal hyphae: regulation, mechanisms and comparison with other branching systems. Mycologia 100:823–832
22. Heaton LLM, López E, Maini PK, Fricker MD, Jones NS (2010) Growth-induced mass flows in fungal networks. Proc R Soc B 277:3265–3274
23. Hopkins S, Boswell GP (2012) Mycelial response to spatiotemporal nutrient heterogeneity: A velocity-jump mathematical model. Fungal Ecol 5:124–136
24. Hutchinson SA, Sharma P, Clarke KR, MacDonald I (1980) Control of hyphal orientation in colonies of *Mucor hiemalis*. Trans Brit Mycol Soc 75:177–191
25. Ikasari L, Mitchell DA (2000) Two-phase model of the kinetics of growth of *Rhizopus oligosporus* in membrane culture. Biotechnol Bioeng 68:619–627
26. Ito K, Kimizuka A, Okazaki N, Kobayashi S (1989) Mycelial distribution in rice koji. J Ferment Bioeng 68:7–13
27. Jurus AM, Sundberg WJ (1976) Penetration of *Rhizopus oligosporus* into soybeans in tempeh. Appl Environ Microbiol 32:284–287
28. Lew RR (2011) How does a hypha grow? The biophysics of pressurized growth in fungi. Nat Rev Microbiol 9:509–518
29. López-Isunza F, Larralde-Corona CP, Viniegra-González G (1997) Mass transfer and growth kinetics in filamentous fungi. Chem Eng Sci 52:2629–2639
30. Markham P, Collinge AJ (1987) Woronin bodies of filamentous fungi. FEMS Microbiol Lett 46:1–11
31. Meeuwse P, Klok AJ, Haemers S, Tramper J, Rinzema A (2012) Growth and lipid production of *Umbelopsis isabellina* on a solid substrate—mechanistic modeling and validation. Process Biochem 47:1228–1242
32. Mitchell DA, Berovic M, Krieger N (2000) Biochemical engineering aspects of solid state bioprocessing. Adv Biochem Eng/Biotechnol 68:61–138
33. Mitchell DA, Do DD, Greenfield PF, Doelle HW (1991) A semi-mechanistic mathematical model for growth of *Rhizopus oligosporus* in a model solid-state fermentation system. Biotechnol Bioeng 38:353–362
34. Mitchell DA, Doelle HW, Greenfield PF (1989) Suppression of penetrative hyphae of *Rhizopus oligosporus* by membrane filters in a model solid-state fermentation system. Biotechnol Tech 3:45–50
35. Mitchell DA, Krieger N, Berovič M (2006) Solid state fermentation bioreactors: fundamentals of design and operation. Springer, Heidelberg
36. Mitchell DA, von Meien OF, Krieger N (2003) Recent developments in modeling of solid-state fermentation: heat and mass transfer in bioreactors. Biochem Eng J 13:137–147

37. Mitchell DA, von Meien OF, Krieger N, Dalsenter FDH (2004) A review of recent developments in modeling of microbial growth kinetics and intraparticle phenomena in solid-state fermentation. Biochem Eng J 17:15–26
38. Money NP (1995) Turgor pressure and the mechanics of fungal penetration. Can J Botany 73 (S1):96–102
39. Mussatto SI, Ballesteros LF, Martins S, Teixeira JA (2012) Use of agro-industrial wastes in solid-state fermentation processes. In: Show K-Y (ed), Industrial Waste, InTech, Rijeka, Croatia
40. Nopharatana M, Howes T, Mitchell D (1998) Modelling fungal growth on surfaces. Biotechnol Tech 12:313–318
41. Nopharatana M, Mitchell DA, Howes T (2003) Use of confocal microscopy to follow the development of penetrative hyphae during growth of *Rhizopus oligosporus* in an artificial solid-state fermentation system. Biotechnol Bioeng 81:438–447
42. Nopharatana M, Mitchell DA, Howes T (2003) Use of confocal scanning laser microscopy to measure the concentrations of aerial and penetrative hyphae during growth of *Rhizopus oligosporus* on a solid surface. Biotechnol Bioeng 84: 71
43. Oostra J, le Comte EP, van der Heuvel JC, Tramper J, Rinzema A (2001) Intra-particle oxygen diffusion limitation in solid-state fermentation. Biotechnol Bioeng 74:13–24
44. Pandey A, Soccol CR, Mitchell D (2000) New developments in solid-state fermentation: I—bioprocesses and products. Process Biochem 35:1153–1169
45. Papagianni M (2004) Fungal morphology and metabolite production in submerged mycelial processes. Biotechnol Adv 22:189–259
46. Peñalva MA (2010) Endocytosis in filamentous fungi: Cinderella gets her reward. Curr Opin Microbiol 13:684–692
47. Prosser JI (1995) Kinetics of filamentous growth and branching. In: Gow NAR, Gadd GM (ed) The growing fungus, Chapman Hall, London
48. Prosser JI, Trinci APJ (1979) A model for hyphal growth and branching. J Gen Microbiol 111:153–164
49. Rahardjo YSP, Sie S, Weber FJ, Tramper J, Rinzema A (2005) Aerial mycelia of *Aspergillus oryzae* accelerate α-amylase production in a model solid-state fermentation system. Enzyme Microb Technol 36:900–902
50. Rahardjo YSP, Tramper J, Rinzema A (2006) Modeling conversion and transport phenomena in solid-state fermentation: a review and perspectives. Biotechnol Adv 24:161–177
51. Rahardjo YSP, Weber FJ, le Comte EP, Tramper J, Rinzema A (2002) Contribution of aerial hyphae of *Aspergillus oryzae* to respiration in a model solid-state fermentation system. Biotechnol Bioeng 78:539–544
52. Rajagopalan S, Modak JM (1995) Evaluation of relative growth limitation due to depletion of glucose and oxygen during fungal growth on a spherical solid particle. Chem Eng Sci 50:803–811
53. Riquelme M, Bartnicki-García S (2008) Advances in understanding hyphal morphogenesis: ontogeny, phylogeny and cellular localization of chitin synthases. Fungal Biol Rev 22:56–70
54. Riquelme M, Yarden O, Bartnicki-Garcia S, Bowman B, Castro-Longoria E, Free SJ, Fleißner A, Freitag M, Lew RR, Mourino-Pérez R, Plamann M, Rasmussen C, Richthammer C, Roberson RW, Sanchez-Leon E, Seiler S, Watters MK (2011) Architecture and development of the *Neurospora crassa* hypha—a model cell for polarized growth. Fungal Biol 115:446–474
55. Schutyser MAI, de Pagter P, Weber FJ, Briels WJ, Boom RM, Rinzema A (2003) Substrate aggregation due to aerial hyphae during discontinuously mixed solid-state fermentation with *Aspergillus oryzae*: experiments and modeling. Biotechnol Bioeng 83:503–513
56. Smits JP, van Sonsbeek HM, Tramper J, Knol W, Geelhoed W, Peeters M, Rinzema A (1999) Modelling fungal solid-state fermentation: the role of inactivation kinetics. Bioprocess Eng 20:391–404
57. Steinberg G (2007) Hyphal growth: a tale of motors, lipids, and the Spitzenkörper. Eukaryotic Cell 6:351–360

58. Sugden KEP, Evans MR, Poon WCK, Read ND (2007) A model of hyphal tip growth involving microtubule-based transport. Phys Rev E 75:031909
59. Taheri-Talesh N, Horio T, Araujo-Bazán L, Dou X, Espeso EA, Peñalva MA, Osmani SA, Oakley BR (2008) The tip growth apparatus of *Aspergillus nidulans*. Mol Biol Cell 19:1439–1449
60. te Biesebeke R, Boussier A, van Biezen N, Braaksma M, van den Hondel CAMJJ, de Vos WM, Punt PJ (2006) Expression of *Aspergillus* hemoglobin domain activities in *Aspergillus oryzae* grown on solid substrates improves growth rate and enzyme production. Biotechnol J 1:822–827
61. te Biesebeke R, Record E, van Biezen N, Heerikhuisen M, Franken A, Punt PJ, van den Hondel CAMJJ (2005) Branching mutants of *Aspergillus oryzae* with improved amylase and protease production on solid substrates. Appl Genet Mol Biotechnol 69:44–50
62. Tindemans SH, Kern N, Mulder BM (2006) The diffusive vesicle supply center model for tip growth in fungal hyphae. J Theor Biol 238:937–948
63. Trinci APJ (1971) Influence of the width of the peripheral growth zone on the radial growth rate of fungal colonies on solid media. J Gen Microbiol 67:324–344
64. Trinci APJ (1974) A study of the kinetics of hyphal extension and branch initiation of fungal mycelia. J Gen Microbiol 81:225–236
65. Tunbridge A, Jones H (1995) An L-systems approach to the modelling of fungal growth. J Vis Comput Anim 6:91–107
66. Varzakas T (1998) *Rhizopus oligosporus* mycelial penetration and enzyme diffusion in soya bean tempe. Process Biochem 33:741–747
67. Viccini G, Mitchell DA, Boit SD, Gern JC, da Rosa AS, Costa RM, Dalsenter FDH, von Meien OF, Krieger N (2001) Analysis of growth kinetic profiles in solid-state fermentation. Food Technol Biotechnol 39:271–294
68. Wösten HAB, Moukha SM, Sietsma JH, Wessels JGH (1991) Localization of growth and secretion of proteins in *Aspergillus niger*. J Gen Microbiol 137:2017–2023
69. Yang H, King R, Reichl U, Gilles ED (1992) Mathematical model for apical growth, septation, and branching of mycelial microorganisms. Biotechnol Bioeng 39:49–58
70. Yang H, Reichl U, King R, Gilles ED (1992) Measurement and simulation of the morphological development of filamentous microorganisms. Biotechnol Bioeng 39:44–48

Adv Biochem Eng Biotechnol (2015) 149: 223–252
DOI 10.1007/10_2014_300

Published Online: 10 April 2015

Better One-Eyed than Blind—Challenges and Opportunities of Biomass Measurement During Solid-State Fermentation of Basidiomycetes

Susanne Steudler and Thomas Bley

Abstract Filamentous fungi, especially basidiomycetes, produce a wide range of metabolites, many of which have potential biotechnological and industrial applications. Solid-state fermentation (SSF) is very suitable for the cultivation of basidiomycetes since it mimics the natural habitat of these fungi. Some of the major advantages of SSF are the robustness of the process, the use of low-cost residual materials as substrates, and the reduced usage of water. However, monitoring key variables is difficult, which makes process control a challenge. Specifically, it is very difficult to determine the biomass during SSF process involving basidiomycetes. This is problematic, as the biomass is normally a key variable in mass and energy balance equations. Further, the success of fungal SSF processes is often evaluated, in part, based on the growth of the fungus. Direct determination of the dry weight of biomass is impossible and indirect quantification techniques must be used. Over the years, various determination techniques have been developed for the quantification of fungal biomass in SSF processes. The current review gives an overview of various direct and indirect biomass determination methods, discussing their advantages and disadvantages.

Keywords Solid-state fermentation · Monitoring · Biomass determination · Basidiomycetes · Cell components · Microscopy · Metabolic activity

S. Steudler (✉) · T. Bley
Institut für Lebensmittel- und Bioverfahrenstechnik, Technische Universität Dresden, Bergstraße 120, 01069 Dresden, Germany
e-mail: Susanne.steudler@tu-dresden.de

Abbreviations

ATP	Adenosine triphosphate
B_f	Fungal biomass (g/g_{sample})
C	Carbon
CER	CO_2 evolution rate
CO_2	Carbon dioxide
DNA	Deoxyribonucleic acid
e	Density of hypha (g/cm^3)
ELISA	Enzyme-linked immunosorbent assay
IR	Infrared
L	Length of hypha (cm/g_{sample})
Mbp	Mega base pairs
mRNA	Messenger ribonucleic acid
N	Nitrogen
O	Oxygen
PCR	Polymerase chain reaction
PLFA	Phospholipid fatty acids
qPCR	Quantitative polymerase chain reaction
r	Radius of hypha
RNA	Ribonucleic acid
RNP	Ribonucleoprotein
S_c	Solid content
SIR	Substrate-induced respiration
SmF	Submerged fermentation
sRNA	Small ribonucleic acid
SSF	Solid-state fermentation
tRNA	Transfer ribonucleic acid

Contents

1 Introduction

For centuries, filamentous fungi have been used to produce enzymes, organic acids, antibiotics, pigments, and flavorings and have also been consumed as food, in the form of mushrooms and fermented foods [1]. This review focuses on one group of filamentous fungi, the basidiomycetes, which produce a wide range of metabolites, many of which have potential biotechnological and industrial applications, for example, lignocellulose-degrading enzymes, such as laccases, cellulases, and xylanases [1, 2]. Basidiomycetes are highly developed fungi and the second-largest group of fungi, after the ascomycetes [2].

Solid-state fermentation (SSF) is very suitable for the cultivation of basidiomycetes, since it reproduces the natural habitat of these fungi. In this fermentation technique, the microorganism is grown in a bed of solid particles, with the spaces between the particles forming a continuous gas phase. Although some droplets of free water can be present in the inter-particle spaces, the majority of the water in the system is bound within the matrix of the solid particles. This traditional fermentation technique is used at industrial scale in Asia for the production of fermented foods, such as miso or soy sauce [3]. In Western countries, SSF was used for the production of penicillin until 1940, when it was replaced by submerged fermentation (SmF). Over the last few decades, the potential and the possibilities of this fermentation method have been rediscovered. Some of the major advantages of the SSF are the robustness of the process, the use of low-cost residual materials as substrates, and the reduced usage of water [4].

Although SSF has been used for centuries, it still presents significant challenges with respect to process control and monitoring. One of the key variables in any fermentation process is the amount of microbial biomass, which not only appears in mass and energy balance equations but also is used as a criterion for evaluating the success of the process. Direct determination of the biomass of basidiomycetes in SSF by gravimetric techniques is impossible: The fungal hypha penetrate into the matrix of the solid substrate, binding the mycelium so tightly to the solid particles that the two cannot be separated from each other. During growth, the overall mass of dry solids decreases, since the production of new fungal biomass is supported by the consumption of nutrients from the solid substrate. However, due to the complex range of nutrients present in most solid substrates, it is not possible to correlate the decrease in total dry weight directly with fungal growth.

As a result of the impossibility of the direct measurement of fungal biomass in SSF, indirect quantification techniques must be used. Various indirect determination techniques have been developed, which can be classified as follows: (I) optical detection using digital imaging or microscopy; (II) measurement of cell components (which might be specific to fungi or not), such as ergosterol, glucosamine or nucleic acids; and (III) measurement of metabolic activities, such as formation of carbon dioxide, release of metabolic heat, consumption of oxygen, or production of enzymes. Each different types of technique have its own advantages and disadvantages. Some problems that may be faced are masking, sorption, precipitation, inactivation of the compound that is to be measured, the presence of interfering substances, and poor extraction efficiencies [2]. This review evaluates the advantages and disadvantages of various indirect biomass quantification techniques that have been used in SSF, grouped according to the classification above.

2 Optical Determination Methods

The oldest and simplest method to assess the growth of basidiomycetes is direct optical detection. This can be carried out subjectively by direct observation with the naked eye or by digital imaging and analysis using a camera and appropriate software.

Optical determination methods need to take the filamentous mode of growth into account. Basidiomycetes are dikaryotic organisms with vegetative and reproductive growth stages. During the vegetative stage, they produce filamentous mycelia through the apical growth and branching of hyphae. During the reproductive stage, which is triggered by an acute lack of nutrients, they undergo sexual reproduction, forming fruiting bodies. The vegetative mycelium, which is usually of interest for the production of metabolites, consists of branched hypha, normally with a diameter of 2–7 μm, although there can also be microhyphae, from 0.1 to 0.4 μm in diameter or mycelia strands (cords) of up to 60 μm in diameter. The surface area covered by a mycelium can vary from a few mm^2 to several km^2. Depending on the limit of resolution that is desired or the accuracy that is required, the detection can be carried out solely by digital imaging, as is the case for the determination of colony diameters, or with the use of additional magnification, which allows investigations of samples at the hyphal level, as is the case with the various microscopy techniques that are described in detail below.

Optical methods are nondestructive direct techniques for biomass determination, but can only be used if observation is unhindered. In the case of transparent substrates such as agar-based media, the sample can be detected from all sides. In the case of the nontransparent solids that are typically used in SSF systems, only the particle surface can be observed. Here, a good contrast between the background (usually the substrate) and the fungal mycelium is important.

2.1 Determination of the Diameter and Area of Fungal Colonies

The growth of fungal colonies on soft surfaces, such as agar, is often used for estimating growth rates and even for obtaining correlations that can be used later in measuring biomass in real SSF systems. A simple and fast method of optical determination of fungal biomass is the detection of colony diameter and area (see Fig. 1a). An increase in colony diameter correlates with the increase of fungal biomass, as more hyphae are present.

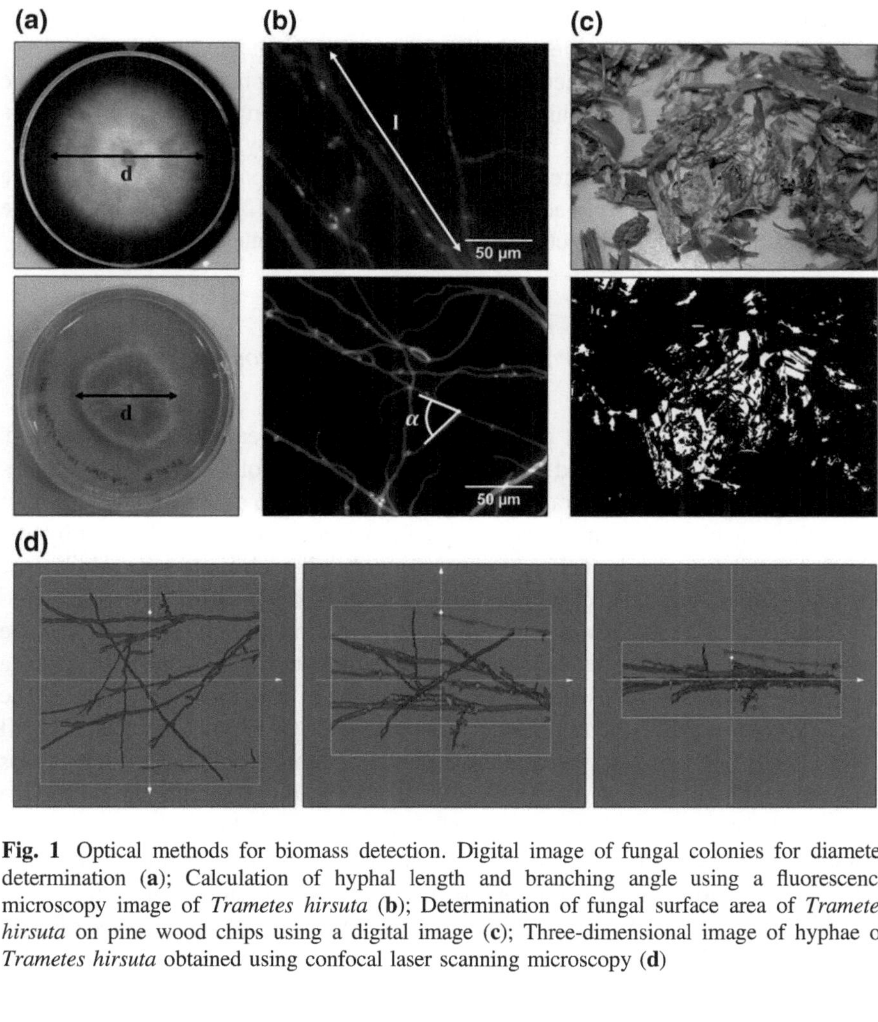

Fig. 1 Optical methods for biomass detection. Digital image of fungal colonies for diameter determination (**a**); Calculation of hyphal length and branching angle using a fluorescence microscopy image of *Trametes hirsuta* (**b**); Determination of fungal surface area of *Trametes hirsuta* on pine wood chips using a digital image (**c**); Three-dimensional image of hyphae of *Trametes hirsuta* obtained using confocal laser scanning microscopy (**d**)

A digital image of the corresponding surface is recorded for the determination of the biomass and converted to grayscale using an image editing software [5]. Subsequently, a fixed threshold value is selected to differentiate the biomass and the background from each other. Usually, the pixels are coded as 1 (white) for biomass and as 0 (black) for background, such that a clear black-and-white image is created for calculation of colony diameter and area. In order to follow colony growth, this can be done at several different times. Li and Wadsö [5] used this simple and non-destructive method to follow successfully the growth of the ascomycete *Penicillium brevicompactum* at various temperatures from 15 to 30 °C. This method can be easily adapted to basidiomycetes. However, Li and Wadsö [5] were not able to differentiate clearly between the growth phase and the stationary phase. The detection of the density of the mycelium within the colony and the height of the colony is not possible using this technique, which may lead to considerable errors in the biomass determination [5]. Also the reflection of incident light at the substrate surface diminishes the quality of the results and requires good image analysis software to minimize interferences. There are also problems in detecting the start of growth due to the poor contrast between the substrate and the young mycelium. Furthermore, it is not possible to differentiate between surface mycelium and hyphae penetrating into the substrate. For this reason, this simple method has been enhanced by various microscopy techniques, for example, stereomicroscopy and epifluorescence microscopy, that enable a more accurate determination of parameters, but require a lot of experience and effort to achieve reproducible results [2]. These techniques are time-consuming, difficult to automate, and susceptible to interference [6].

2.2 Simple Optical Microscopy and Stereomicroscopy

A stereomicroscope is a light microscope with separate beam paths for each eye. A spatial impression is created due to the different angles of the two beams. Stereomicroscopy can help in separating the overlying hyphae in a dense mycelium from each other, enabling determination of key parameters for modeling, such as hyphal length, hyphal diameter, branching points, tip number, branch angle and suchlike (see Fig. 1b), although a sparse mycelium is better suited for estimating key parameters.

Couri et al. [7] and Dutra et al. [8] used stereomicroscopy to determine the biomass of the ascomycete *Aspergillus niger* cultivated on wheat bran, using a procedure similar to that used by Li and Wadsö [5] to determine colony area. The samples were magnified tenfold using a Nikon SMZ 800 [7] or Carl Zeiss STEMI 2000-CS [8] stereomicroscope, and digital images were recorded. Each image was converted into a black-and-white image, as described above, and the fungal surface area was determined (see Fig. 1c). On such heterogeneous substrates, it is important not only have a good illumination of the sample, but also to obtain the right focus plane, in order to obtain a good contrast between background and mycelium [7].

Only surface mycelium is detected using stereomicroscopy. It is not possible to obtain a three-dimensional depiction of the complete mycelium, since hyphae penetrating into opaque substrates are not visible. This makes it difficult to obtain an accurate determination of biomass.

An additional method for determining the fungal surface area is based on the estimation of fungal hyphal length used in the field of soil biology. In this method, the mycelium is stained and analyzed under a microscope with a suitable magnification. Stahl et al. [9] used a magnification of 600X (40X objective, 10X ocular, and 1.5X magnifier) and evaluated 25 randomly selected fields per object slide. The fungal biomass B_f (g/g_{sample}) was then calculated assuming a tubular structure, according to Paul and Clark [10].

$$B_f = \pi \times r^2 \times L \times e \times S_c \quad (1)$$

where r is the radius of the hypha (an average value of 1.5 μm was used), L is the observed total hyphal length in the sample (cm/g_{sample}), e is the density of the hypha (1.3 g/cm^3), and S_c is the solids content (reported to be 0.3) [9]. Observed hyphal length can be very different in different samples. For example, Stahl et al. [9] detected lengths of 146 m per gram of sample, corresponding to 0.06 mg fungal biomass per gram of sample, in an agricultural soil in Iowa, while Sönderström and Bååth [11] detected hyphal lengths of 66.9 km per gram of soil sample, with a resultant biomass of 35.1 mg per gram of sample, in a soil sample of a Swedish coniferous forest.

2.3 Epifluorescence Microscopy

In order to avoid the problem of the lack of contrast between background and mycelium, the hypha can be stained with dyes. One possibility is the use of fluorescent dyes.

Fluorescence microscopy has been used since 1970 [5] and is, like stereomicroscopy, a special form of light microscopy. The sample is stained with a fluorescent dye that binds specifically to fungal biomass. The stained sample is excited with light of a specific wavelength and emits light of a different wavelength. The emitted light is detected using special filters, so, in general, a good contrast is obtained between the bright sample and the dark background. Suitable and frequently used dyes are acridine orange, which has a high affinity for nucleic acids (under excitation with UV-light, acridine orange bound to DNA emits green light, while acridine orange bound to RNA emits orange light) [5, 6], 4′,6-diamidino-2-phenylindole (DAPI), the fluorescence of which correlated well with the fungal biomass according to Madrid and Felice [6], propidium iodide (PI), europium chelate and ethidium bromide, which stain DNA, and also fluorescein isothiocyanate (FITC), which stains proteins [5]. In the particular case of staining fungal hyphae, suitable fluorescent dyes include phenol blue [5], calcofluor-white (also called fluorescence brightener, FB) [9, 12] and Mykoval [13]. It is also possible to use so-called vital fluorescent stains, such as fluorescein diacetate (FDA). This compound does not fluoresce itself, but when cleaved by the fungus, producing fluorescein and acetate, the fluorescein fluoresces and is therefore detectable [2]. In addition, the use of specific fluorescent antibodies (i.e., antibodies coupled with fluorescent dye) is possible [6]. Methods can be combined in order to obtain additional information: For example, if both a general stain and FDA are used, then, in addition to determining morphological parameters, it is possible to distinguish metabolically active and inactive hyphae from one another.

Additives, such as immersion oil, or the substrate should not autofluoresce, which would reduce the quality of the images [6]. In any case, the conversion of fluorescence microscopy data into fungal biomass remains problematic due to difficulties in estimating the hyphal volume [2]. This can be rectified by the use of confocal laser scanning microscopy.

2.4 Confocal Laser Scanning Microscopy

Confocal laser scanning microscopy involves pointwise scanning of the sample by a laser beam and scan mirrors. A three-dimensional image can be generated by assembling the various focus planes (see Fig. 1d). This enables the accurate detection of the three-dimensional hyphal geometry, and thus an accurate determination of the volume occupied by the fungal biomass. Nopharatana et al. [14] used this nondestructive microscopy technique to determine hyphal density within the mycelium of the zygomycete *Rhizopus oligosporus* on potato dextrose agar. Density profiles were obtained both for the growth of the aerial hyphae, as a function of height above the surface, and the growth of penetrative hyphae, as a function of depth below the surface. The conversion of the volume occupied by the biomass into an estimate of dry biomass of the fungus was successfully implemented. Nevertheless, this conversion can be problematic, because different parts of the mycelium can require different conversion factors, due to differences in water content, mycelium age, and hyphal cell wall thickness [14]. Also, it should be noted that this technique can only generate three-dimensional images of penetrative hyphae if the solid substrate is transparent.

2.5 Scanning Electron Microscopy

In order to improve the degree of magnification of the microscopy images, scanning electron microscopy can be used, although this method is expensive and time-consuming. Osma et al. [12] used this technique to quantify the biomass of the basidiomycete *Trametes pubescens* on banana peels and sunflower seed shells. The image analysis software is very important [12], because the software converts the image to gray scale and differentiates the fungal mycelium from the background based on a suitable threshold value. In order to calculate the area and volume, a comparison with structures of known height and width, for example defined tubular structures, was used for the calibration of the height of the hyphae in the sample [12]. Nonetheless, the calculation of the biomass of complete colonies was not possible using this method: Due to limited ability to penetrate into the dense mycelium, the hyphae at the colony bottom were not visible. Furthermore, in order to obtain information that can be used in differentiating mycelium from the background, a position with low density is needed, usually in the peripheral region of the colony [12]. However, this noninvasive technique can be used for studies on agar with individual hypha or for simple detection of fungal contamination [15].

3 Measurement of Fungal Cell Components

It is possible to measure cell components that are found in the fungal biomass, such as ergosterol or glucosamine, but, unfortunately, sometimes also in the solid substrate, such as nucleic acids. However, these cell components can vary significantly, depending on the fungal species, the age of the mycelium, the cultivation temperature, and other growth conditions [16].

Figure 2 gives an overview of the structure of the mycelium of a basidiomycete, indicating the various cell components that can be used for the quantification of the fungal biomass.

The cell wall of the hypha is multilamellar. The inner layer consists mainly of chitin, which contains glucosamine residues, and has a thickness of 10–20 nm. The outer layer consists of a protein layer (thickness 10 nm), a glycoprotein layer (thickness 50 nm) and a 75–100-nm-thick slime layer, which contains a number of carbohydrates, fats, and proteins. The cytoplasm of the hypha contains not only nuclei and other organelles but also glycogen vesicles for storage of carbon, vacuoles for fat storage, as well as amino acids and proteins for nitrogen storage. This cytoplasm is enclosed by the plasma membrane. In order to quantify a particular

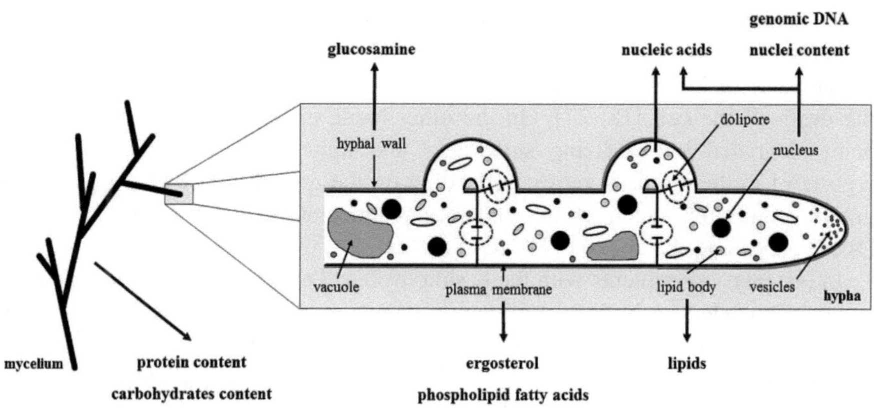

Fig. 2 Schematic depiction of the mycelium structure of a basidiomycete, indicating the various biomass components that can be measured

cell component, it is usually necessary to hydrolyze the sample and extract the target component [17]. The variety of detection methods presented in this section for different cellular components differ with respect to the time required, the cost of the measurement, as well as the interferences that occur [16].

3.1 Ergosterol

One of the most commonly used methods for the quantification of fungal biomass is the determination of the ergosterol content of the sample. Ergosterol is the most abundant sterol in the plasma membrane of hyphae and can represent from 0.7 to 1 % of the dry matter of the fungus [2, 18–21].

Ergosterol is not found in plants, animals, or other organisms and is found only in minimal amounts in some bacteria, protozoa, cyanobacteria, and other microalgae [2, 18–22], so that no interference is expected during biomass determination [18]. The method for determining the ergosterol content is very accurate and reproducible, but it is also very time-consuming and labor-intensive [17, 23, 24]. The analysis is carried out in three steps: (I) hydrolysis and saponification of the sample to liberate the ergosterol from the plasma membrane, (II) extraction of the ergosterol, and (III) quantification of the ergosterol by HPLC [15–18, 20–22, 25], with detection at 282 nm, which is made possible by the shift of UV absorption due to the double bond of ergosterol [18].

There are efforts to reduce the time and cost of this method. In particular, accelerated extraction by microwave treatment gives good results [15, 23, 26]. With this treatment, Zhang et al. [23] obtained extraction efficiencies of over 99 %, while conventional extraction methods have lower efficiencies, ranging from 33 to 83 %. Similarly, Muniroh et al. [26] reduced the saponification and extraction time to 30 s (compared to a normal time of 30 min to 1 h) at 70 °C using microwave treatment during the biomass determination of the basidiomycete *Ganoderma boninense*.

Ergosterol has proven to be a suitable and sensitive indicator for the quantification of biomass of filamentous fungi, especially basidiomycetes [20–22, 25]. It is an indicator of living cells and is reduced relatively quickly (days to weeks) after the death of the cell [18, 27]. On the other hand, ergosterol is somewhat labile, being degraded by oxidizing substances and light [18]. Also, the content of ergosterol in the biomass varies, depending on the growth phase [16, 17, 22], the growth conditions (such as the cultivation temperature and the substrate used) [16, 17, 19], and the species [19, 22].

In our own experiments with the basidiomycete *Trametes hirsuta,* the ergosterol content depended on both the cultivation temperature and the mycelium age (see Fig. 3). In order to determine the extent of the influence of the cultivation temperature and the mycelium age, the biomass determined gravimetrically was compared to the biomass calculated from the ergosterol content using a constant conversion factor. The two biomass values converge with increasing temperature and mycelium age, so that a separate calibration curve must be prepared for each

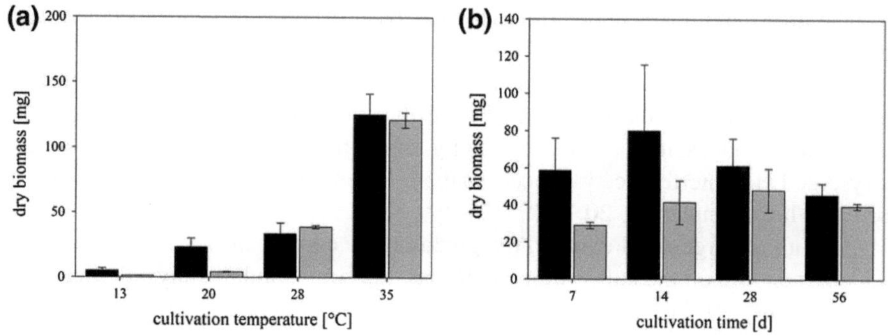

Fig. 3 Comparison of the biomass determined gravimetrically (*black*) and calculated by ergosterol content using a constant conversion factor (*gray*) of *Trametes hirsuta* during fermentation on malt extract agar at different temperatures for 7 days (**a**) and at 28 °C for different cultivation times (**b**) ($n = 3$)

combination of culture temperature and mycelium age. The calibration curves showed a very good correlation between the biomass and ergosterol content (e.g., calibration curve for 7 days and 28 °C: $R^2 = 0.98$).

An influence of the mycelium age on the ergosterol content was also observed for the basidiomycetes *Phanerochaete chrysosporium* and *Schizophyllum commune*; furthermore, the extent of the influence varied with the fungal species [22]. In studies with fungi of other divisions, different results have also been found. The ergosterol content did not vary with the mycelium age for the ascomycete *Phaeocryptopus gaeumannii* [19], but it did for the ascomycete *Beauveria bassiana* [17], various other ascomycetes [22], and the deuteromycete *Coniothyrium minitans* [16]. In the last of these cases, the ergosterol content increased, but then remained constant after the beginning of sporulation, so that the ergosterol content can be used as an indicator of conidia formation. For all these fungi (independent of the division), the ergosterol content of the biomass was affected by the nutritional composition of the medium used.

3.2 Glucosamine and Chitin

The measurement of glucosamine is also commonly used for estimating fungal biomass [7, 8]. Chitin, which contains residues of *N*-acetyl-β-D-glucosamine, is an essential cell wall component of most fungi, representing from 2 to 24 % of the dry weight of the biomass [28–30]. Glucosamine residues also occur in the murein of bacteria and in the chitin of insects and invertebrate soil animals (such as earthworms, nematodes, and snails) [2], but not in plant materials that are commonly used as substrates in SSF processes [21]. Chitin is much more stable than ergosterol and thus is an indicator of both living and dead biomass [30]. In fact, it remains in the cell walls of empty ghost hyphae [16].

The time that it takes to determine the glucosamine content varies, depending on the particular method that it used [20], but the methods tend to be labor-intensive and complex [17, 20, 21]. The analysis is carried out in two steps: (I) the hydrolysis and deacetylation of the chitin to liberate the glucosamine residues and (II) the detection of glucosamine, most commonly using the staining protocol of Ride and Drysdale [31]. Interference will occur if there are any other sources of glucosamine or other hexosamines [7, 20, 30].

As with the ergosterol content, the glucosamine content of the dry biomass may vary, depending on the cultivation conditions, the strain, and the physiological state of the culture [20, 21, 30]. Also, for the same sample of biomass (i.e., with the same underlying glucosamine content), different methods for glucosamine determination will give different results [20, 21, 30]. The glucosamine content of the biomass did not vary with mycelium age for the ascomycetes *B. bassiana* [17] and *A. niger* [29], as well as the zygomycete *Cunninghamella elegans* [32], although it did for the ascomycete *C. minitans* [16]. In the case of *C. elegans* [32], the glucosamine content of the biomass did not vary with cultivation type (SSF or SmF); however, it was influenced by the composition of the growth medium, especially the carbon source used. For the brown rot fungus *Neolentinus lepideus* (a basidiomycete) and the soft rot fungus *Phialophora* sp. (an ascomycete), Nilsson and Bjurman [30] showed that the glucosamine content was affected not only by the carbon and nitrogen source but also by the mycelium age, initial nutrient content, and temperature. Similarly, the glucosamine content of 11 *Ceuteromycotina* species was influenced by the nitrogen content of the substrate [33].

Although Matcham et al. [20] showed that the chitin content of the biomass of the basidiomycete *Agaricus bisporus* was constant, they found it impractical to use glucosamine determinations to estimate biomass in samples consisting of fungi growing on composted wheat straw, due to the high background levels caused by endogenous hexosamines in the compost. Also, in our own investigations with the basidiomycete *T. hirsuta*, a good correlation between biomass and glucosamine content was obtained (R^2 = 0.98); however, this method also proved unsuitable during the analysis of real samples obtained during the growth of this fungus on corn silage, pine wood chips, straw or orange peel, due to the poor reproducibility and sensitivity of the results.

3.3 Nucleic Acids and Genomic DNA

Nucleic acids and genomic DNA can be quantified rapidly, and it is possible to use high-throughput methods. Of course, DNA and RNA are not fungal-specific cell components [34]; they exist in all organic material and thus will be present in the solid substrates that are typically used in SSF processes. However, specific species can be detected using real-time PCR and specific primers.

Basidiomycetes contain the smallest genome of eukaryotes, with a genome size between 20 and 40 Mbp. In fact, the DNA content of the biomass of basidiomycetes is a tenth of that found in bacteria and higher eukaryotes. Consequently, they also have a high ratio of RNA to DNA. Most of the RNA (approximately 80 %) is bound with ribosomes in the cytoplasm or is present in the form of ribonucleoprotein (RNP) or in small quantities in the form of sRNA. The remaining 20 % is also located in the cytoplasm: around 5 % as mRNA and 15 % as tRNA [28].

Determination of the total nucleic acid content is simple and rapid. The nucleic acids are extracted using hot and cold extraction with perchloric acid and then measured photometrically at 260 nm [16]. However, the nucleic acid content of the fungal biomass is not always constant. For example, with the ascomycete *C. minitans*, the content decreased with increasing cultivation time and depended strongly on the medium [16]. In our own investigations using the basidiomycete *T. hirsuta*, the content of nucleic acids in the fungal biomass was constant for biomass of the same mycelium age. As in the studies of Ooijkaas et al. [16], a decrease of the nucleic acid content with increasing mycelium age was observed. In addition, an independence of the cultivation temperature was found.

It is also possible to determine genomic DNA content specifically. Commercial extraction kits are now available for the extraction of genomic DNA, such as the FAST DNA® SPIN Kit for Soil [35] and the "NucleoSpin® Plant II Mini" extraction kit (MACHEREY-NAGEL GmbH & Co. KG, Düren, Germany), which we have used in our experiments. The extracted genomic DNA is measured at 260 nm and quantified. A second measurement is carried out, usually at 280 nm, in order to assess the purity of the DNA by calculation of the quotient "260 nm/280 nm." A ratio of 1.8 would be ideal, indicating pure DNA. A ratio of 2.0 implies the existence of pure RNA and a ratio smaller than 1.8 indicates a contamination with proteins. May et al. [36] used this method to determine the biomass of the oomycete *Lagenidium giganteum* on wheat bran at different temperatures (24–35 °C). A linear relationship was obtained between the concentration of dry biomass of the fungus, which was collected from liquid cultivation, and extracted DNA ($R^2 = 0.91$). This linear relationship was used to quantify the fungal biomass from solid cultures; however, representative sampling was problematic, so the dry biomass estimated by DNA measurement showed a high standard deviation. Further, there was an influence of the mycelium age, as well as interference from the medium. In our studies, we obtained a linear relationship between the dry biomass of the basidiomycete *T. hirsuta*, which was collected from agar culture, and the extracted genomic DNA ($R^2 = 0.93$), but the biomass values estimated by genomic DNA were far above those estimated using reference methods (e.g., ergosterol), probably due to the interference from the organic solid substrate. Also, Blagodatskaya et al. [34] showed that the genomic DNA content of fungal biomass is not constant and depends on the nitrogen content of the substrate.

Another method to quantify fungal biomass is PCR, especially real-time PCR. This technique requires not only appropriate equipment (a thermal cycler), but also a lot of know-how and optimization, as well as a good primer selection [37]. However, after successful establishment of this method, a species-specific quantification of fungal biomass is possible. In order to quantify the biomass of a specific fungus, it is important that the nuclei content of the target organism does not change: Any change of the genomic DNA content with mycelium age or growth conditions will lead to an erroneous determination of the biomass content [38].

Voegele and Schmid [38] successfully used reverse transcription (RT) real-time PCR for rapid, specific, and highly sensitive quantification of the basidiomycete and plant pathogen *Uromyces fabae* within its host plant, using three constitutively overexpressed genes. Also, Tellenbach et al. [39] used locus-based quantitative PCR (qPCR) and obtained a strong linear relationship between the pure dry biomass of the fungus and the qPCR estimates (i.e., DNA amount) of the ascomycetes *Acephala applanata* (R^2 = 0.965), *Phialocephala fortinii* (R^2 = 0.963), and *Phialocephala subalpina* (R^2 = 0.936). For Tellenbach et al. [39], currently no better method of quantification exists, due to the speed and very high specificity.

The high sensitivity of quantitative real-time PCR was also confirmed by Pilgård et al. [40] during a study of the biodegradation of wood by basidiomycetes. As control methods, the ergosterol and glucosamine contents were determined and samples were also evaluated microscopically. In addition to the good correlation of biomass and quantified DNA content by qPCR, this method is much more sensitive than the determination of ergosterol and glucosamine and therefore is suitable for the determination of biomass in the early growth phase.

While the determination of nucleic acids and the genomic DNA using photometric detection is limited due to various interferences, quantitative real-time PCR, after optimization, is an appropriate and sensitive method for quantification of fungal biomass during SSF, provided the content of genomic DNA is independent of the culture conditions.

3.4 Counting of Fungal Nuclei by Flow Cytometry

Flow cytometry is a simple measurement technique that enables a quick quantification of cells, cell components, and particles. For this, the particles pass a laser beam in single file. In addition to pure counting, certain properties of the particles can also be detected, based on fluorescence and light scattering (forward and side scatter).

This method works well with unicellular organisms. Costa-de-Oliveira et al. [41] used it to quantify the glucosamine content of the yeast *Candida* spp. by staining the cells with calcofluor-white. In principle, this approach could be applied to basidiomycetes growing on solid substrates, after sample grinding, but dyes that stain chitin often also stain cellulose, which is present in large amounts in many SSF substrates and would therefore interfere with the measurement. Spores are easily counted by flow cytometry [42]. However, basidiomycetes only form spores in the fruiting bodies, while most SSF processes with basidiomycetes involve only pure mycelium.

In our own work, we have developed a flow cytometry method for quantifying fungal biomass by counting the fungal nuclei [43]. The sample is homogenized and the nuclei are extracted. The nuclei are then stained with a fluorescent dye (SYTOX® Green) and analyzed in the flow cytometer. A linear correlation between

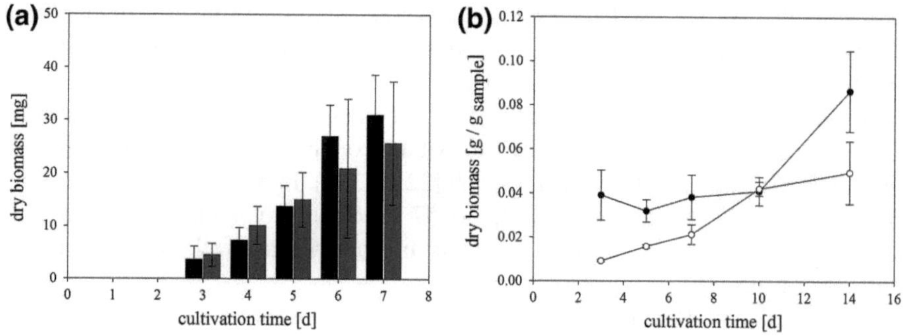

Fig. 4 Comparison of the biomass determined gravimetrically (*black*) and calculated by nuclei counting (*gray*) of *Trametes hirsuta* during fermentation on malt extract agar at 28 °C (**a**) ($n = 10$) and of the biomass determined by measurement of the ergosterol content (*white circles*) and by nuclei counting (*black circles*) during SSF of *Trametes hirsuta* at 28 °C on corn silage (**b**) ($n = 5$)

the pure dry biomass of the basidiomycete *T. hirsuta* and the counted nuclei was obtained ($R^2 = 0.93$). The content of nuclei within the biomass depended on both the culture conditions and the mycelium age. This variation was taken into account by the construction of different calibration curves for different cultivation temperatures and times.

With the use of different calibration curves, good agreement was found between the fungal biomass of *T. hirsuta* determined gravimetrically and biomass estimated by nuclei counting (see Fig. 4a) during a cultivation on malt extract agar at 28 °C for 7 days. The method was also used to monitor SSF of corn silage in a reactor with a working volume of 10 L (developed by the Research Center for Medical Technology and Biotechnology (fzmb GmbH)). Useful and reproducible results were obtained with this novel method. Furthermore, the accuracy of biomass values determined by flow cytometry was similar to that of the values obtained by the reference method (ergosterol content, see Fig. 4b).

The use of flow cytometry to quantify fungal biomass by nuclei counting represents a new, rapid, and simple method of biomass determination during SSF and, when used in combination with other methods, provides additional information about the physiology of the culture.

3.5 Lipids and Phospholipid Fatty Acids

The mycelium of basidiomycetes contains various glycerolipids, which, in turn, can contain both saturated and unsaturated fatty acids. These glycerolipids are located mostly in lipid bodies in the cytoplasm and in the plasma membrane and the membranes of organelles, in the form of acylglycerols and phosphoglycerides [44]. However, the lipid content of the biomass varies greatly, especially due to the fact that the amount of storage lipids varies greatly, depending on the growth phase [22]. Therefore, phospholipid fatty acids (PLFAs) can be used as a biomass indicator. PLFAs are used for rapid and inexpensive determination of microbial diversity in soils [45], but can also be used during SSF. This method is sensitive and highly reproducible, and gives an indication of living biomass [45, 46].

In addition to determine the total PLFA content, it is possible to determine separately the contents of special PLFAs that are contained by certain fungi, such as the PLFA 18:2ω6,9 and the PLFA 18:2ω9 [22, 35, 45]. In basidiomycetes, the main PLFA, 18:2ω6,9, represents from 45 to 57 % of the total PLFA content [22]. This PLFA also occurs in small amounts in bacteria, as well as in plants and animals [22, 45, 46].

Frostegård and Bååth [46] correlated the content of PLFA 18:2ω6,9 in soil samples with the ergosterol content ($R = 0.92$), although influences of the substrate and cultivation conditions were observed [45]. A decrease of PLFA concentration in an SSF sample can imply either cell death or degradation of specific membrane phospholipids due to changes in environmental conditions [45].

It is problematic to use PLFA measurements to estimate fungal biomass since the conversion factor varies greatly. Despite this, Frostegård and Bååth [46] did use this method as a qualitative indicator of fungal biomass in 15 different soil samples, while Klamer and Bååth [22] used PLFA contents during an investigation of the levels of 11 fungi in compost. Klamer and Bååth [22] found a linear correlation between the content of the PLFA 18:2ω6,9 and the content of ergosterol within the samples ($R^2 = 0.782$), although both the ergosterol content and the PLFA 18:2ω6,9 content varied considerably among the tested species. The total PLFA content in fungal biomass grown on potato dextrose agar showed large interspecific variations. It does not depend significantly on the mycelium age, but does depend on the substrate and on the fermentation type (SSF and SmF). A perfect PLFA marker does not exist [45], but the determination of PLFA content (total and specific) is a suitable indicator of fungal biomass, so that this method is also suitable for the quantification of living biomass.

3.6 Protein Content

Proteins are ubiquitous in biomass and the protein content is often used as an indicator of fungal biomass. Protein measurement assays are usually sensitive, simple, and reproducible [29, 47]. In order to determine the insoluble protein in the fungal biomass, it is first necessary to disrupt the sample using, for example, phosphoric acid [47, 48] or sodium hydroxide [16, 29]. The protein that is released can then be determined by a colorimetric method, such as that of Bradford [47] or that of Lowry [29].

Protein determination can be useful when the solid substrate that is used in the process does not contain high protein levels. This method has been used to follow the growth of the ascomycete *A. niger* on various substrates, including sugar cane bagasse and palm kernel cake [29, 47, 48]. The protein content of the biomass of *A. niger* was found to remain constant [28].

Protein determination cannot be used for estimating fungal growth when the solid substrate that is used has a high protein content. This problem occurred in studies involving the ascomycete *C. minitans* [16] and in our studies with the basidiomycete *T. hirsuta*. In both cases, the protein content of early samples was high, but, due to consumption of the protein in the substrate, it decreased to a constant level in later samples, although a linear relationship between the pure fungal biomass and the protein content was found. However, the protein content depended on the substrate and the mycelium age [16].

3.7 Carbohydrate Content

Carbohydrates are located in fungi as sugars, sugar alcohols, polysaccharides, and derivatives. Pure sugars exist only in small amounts in hyphae and are present usually as phosphorylated derivatives [28].

Ooijkaas et al. [16] obtained a total carbohydrate content of up to 9.7 % of the dry biomass of the ascomycetes *C. minitans*. However, the content varied depending on the medium and mycelium age. Also Desgranges et al. [17] observed a variation in the sugar content of the biomass of the ascomycete *B. bassiana* depending on the medium and growth phase.

This method can be used only for special cases, because many substrates that will be used in SSF contain carbohydrates, so the carbohydrate content is inappropriate as an indicator of fungal biomass.

4 Measurement of Metabolic Activity

Normally, measurements of metabolic activity are simple and can be used for continuous monitoring of culture performance. As a result, they are suitable for on-line monitoring of mycelium growth in large-scale bioreactors. Figure 5 indicates the various techniques that will be discussed in this section.

4.1 Respiration and Gas Composition

The measurement of respiratory activity is simple to implement and is quite convenient for use at large scale. Determination of the oxygen (O_2) and carbon dioxide (CO_2) contents in the gas phase enables calculation of O_2 uptake rates and CO_2 evolution rates (CERs). It is also possible to integrate the values to obtain cumulative O_2 consumption and CO_2 formation. The measurement techniques are non-destructive, performable in real time and online [36, 49, 50]. An advantage of measuring both O_2 and CO_2 is that this gives additional information about the

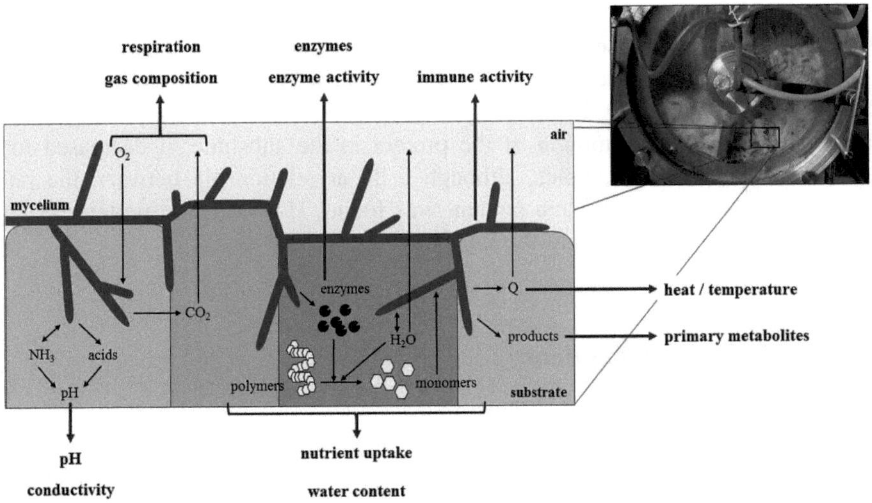

Fig. 5 Schematic depiction of the mycelium structure and metabolic activity of a mycelium, as well as the related determination methods

physiology and the state of the culture [49]. Fungi not only respire actively in the growth phase, but also in the stationary phase. Since it is impossible to differentiate the origin of the respiratory activity, the cultivation should be axenic.

CO_2 evolution was used to follow the growth of the ascomycetes *Penicillium roqueforti* [51] and *B. bassiana* [52] in SSF. A good correlation between the formation of CO_2 and the glucosamine content was obtained. However, although the method is sensitive, it is not suitable for comparing different processes since the relation of CO_2 evolution with growth depends on the medium [52].

May et al. [36] used the CER to follow growth of the oomycete *L. giganteum*. In SmF, the CER increased exponentially, similarly to the fungal biomass, until the stationary phase. Here, the fungal biomass leveled off at a constant value, while the CER peaked and then decreased. However, it is not always the case that CO_2 evolution and O_2 consumption are related to biomass production in a simple manner; therefore, the conversion into an amount of fungal biomass and thus the monitoring of the fungal biomass remain difficult [16, 50, 53]. In fact, Ikasari and Mitchell [50] noted that biomass estimation from O_2 uptake was problematic in fermentations involving the zygomycete *R. oligosporus*, since neither the yield coefficient for O_2 nor the maintenance coefficient for O_2 was constant. However, they did recommend the use of O_2 measurements for online monitoring of overall bioreactor performance, since O_2 consumption is a useful indicator of the metabolic activity of fungal biomass [50, 53].

In our own studies, the CO_2 content and the O_2 content in the gas phase were measured continuously during the fermentation of the basidiomycete *T. hirsuta* on corn silage. A linear relationship between the daily O_2 consumption and CO_2 formation and the fungal dry biomass was found (O_2: $R^2 = 0.68$; CO_2: $R^2 = 0.76$) and was used to convert the respiratory activity into fungal biomass during the SSF

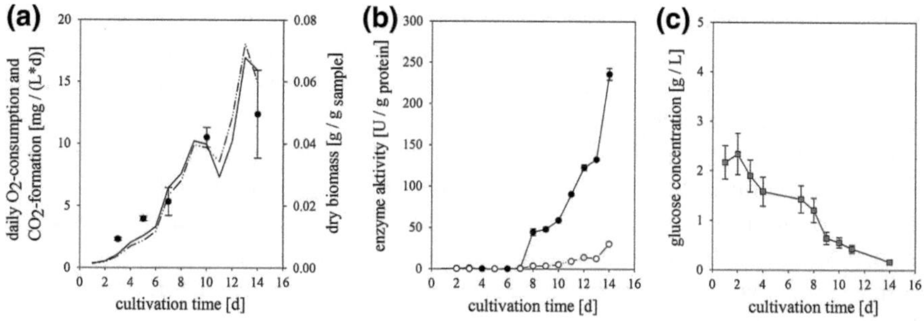

Fig. 6 Comparison of the biomass determined by the measurement of the ergosterol content (*black circles*, $n = 5$) and the corresponding daily O_2 consumption (*dotted line*) and CO_2 formation (*solid line*) (**a**), enzyme activity of laccase (*black circles*) and peroxidase (*white circles*) ($n = 3$) (**b**) and glucose content in the leached liquid phase obtained after extraction of the extracellular enzymes (*gray squares*, $n = 3$) (**c**) during SSF of *Trametes hirsuta* at 28 °C on corn silage

process. The biomass estimated from the daily O_2 consumption and CO_2 formation was quite similar to that estimated through measurement of the ergosterol content of the solids (see Fig. 6a).

According to our studies, the measurement of the respiratory activity is a simple and rapid method of determining the biomass of *T. hirsuta* until the stationary phase. Furthermore, it provides additional information about the physiological state of the culture, for example the type of metabolism (calculating respiratory quotient).

4.2 Enzyme Activity

The use of measurements of enzymatic activity to estimate fungal growth presupposes that the enzymes are fungal-specific, growth-associated, and, usually, extracellular. A few enzymes satisfy these criteria: Most enzymes are not suitable candidates for this approach, since their production is not related to growth in a simple manner; further, production levels can vary with the environmental conditions, such as pH, temperature, and water content [2]. Despite these potential problems, since determining enzyme activity is usually fast and easy, many workers have used this technique to estimate the biomass.

Dutra et al. [8] determined the activity of lipases of the ascomycete *A. niger* and correlated the results with glucosamine contents and microscopic observations. There was a good correlation between the area occupied by the hyphae, the glucosamine content of the solids and the lipase activity ($R > 0.9$). Similarly, for the growth of *A. niger* on palm kernel cake, the activity of β-mannanase correlated well with the biomass based on measuring the protein concentration ($R^2 = 0.941$) [29]. For the ascomycetes *Stachybotrys chartarum* and *Aspergillus versicolor* growing on gypsum board, the activity of β-*N*-acetylhexosaminidase correlated well with the biomass, which was estimated by measuring the ergosterol content, with R^2 values of 0.935 and 0.968, respectively [24]. On the other hand, the activity of extracellular proteases of *Aspergillus tamarii* was unsuitable as a biomass indicator due to the strong influence of the substrate and physical factors, such as temperature, aeration, and hyphal density [54].

Basidiomycetes produce interesting extracellular enzymes such as laccases, peroxidases, cellulases, and xylanases. Matcham et al. [20] used the laccase activity of the basidiomycete *A. bisporus* to estimate the biomass during SSF on cereal grains. Although the laccase activity increased in direct proportion to the fungal biomass, it was problematic to determine the conversion factor, because the absolute enzyme activities were highly dependent on the substrate used. The determination of the enzyme activity was recommended as a qualitative indicator of biomass. In our studies of the SSF of the basidiomycete *T. hirsuta*, the maximum activities of laccases and peroxidases also varied with the substrate and cultivation conditions, but the activity profiles were proportional to the biomass level as determined using the ergosterol method (see Fig. 6b). On the other hand, the activities of xylanases and cellulases were unsuitable as indicators of biomass.

The usefulness of enzyme activity as a biomass indicator depends on the selected enzyme and the fermentation conditions. Enzyme activities can only be used as indicators under the same conditions as the correlation was originally determined.

4.3 Nutrient Uptake

The uptake of nutrients or the current nutrient concentration has been used in SmF as an indicator of growth and also has the potential to be used in SSF. Often the carbon source is used, such as glucose or other sugars.

Desgranges et al. [17] used the uptake of sucrose as an indicator of growth during the fermentation of the ascomycete *B. bassiana*. The amount of sucrose consumed correlated well with the content of ergosterol and glucosamine of the fungal biomass ($R > 0.81$) and was independent of the C/N ratio of the substrate. In contrast, the glucose uptake was not suitable for biomass estimation during the studies of Larroche and Gros [51] with the ascomycete *P. roqueforti*. They determined the glucose content of wheat starch after hydrolysis with amyloglucosidase using the dinitrosalicylate method. The glucose was taken up into both the mycelium and spores, so that only a portion of the glucose consumption was used for mycelium growth. Further, in studies undertaken with *A. niger*, Favela-Torres et al. [48] found that the amount of glucose taken up depended on the initial concentration in the medium and on the fermentation type (SmF or SSF). In SSF, higher initial glucose concentrations could be tolerated, and higher glucose uptake rates were obtained. In our own investigations with the basidiomycete *T. hirsuta*, extracellular enzymes were leached from the solid substrate daily with water. The glucose concentration of the leachate was also determined, in order to obtain an indication of growth. The amount of glucose extracted decreased as the fermentation progressed (see Fig. 6c). It also depended on the substrate that was used. For example, over the first days of the SSF process, the amount of glucose extracted was around 2 g/L when corn silage was used, but around 0.3 g/L when pine wood chips were used.

It is usually fast and simple to determine the concentrations of sugars or other nutrients, enabling the use of nutrient uptake as a biomass indicator. However, the method must be carefully calibrated for each specific application.

4.4 Production of Antigens

Immunological methods for the detection of fungal antigens can be used to obtain an indication of growth. The enzyme-linked immunosorbent assay (ELISA) is often used. In this method, an antibody against a fungal antigen is added to a sample in a microtiter plate. A second antibody is then added; this antibody, which has an enzyme linked to it, reacts with the first antibody. Finally, a substrate for the enzyme is added, such that a colorogenic reaction occurs. The intensity of the color produced depends on the original amount of fungal antigen. The development of this method requires a lot of know-how, but, once it is optimized, it is very sensitive and quite specific.

Anand and Rati [55] used ELISA to quantify the biomass of the ascomycete *Aspergillus ochraceus* on coffee bean extract, chili extract, and poultry feed. The sensitivity was 0.2 mg biomass per gram substrate. The results were influenced by the particular substrate used and by the moisture content, such that a new calibration curve had to be established for each substrate. The background signal of the uninoculated substrate was much lower than the lower detection limit, so it did not cause any problems [55]. In a similar manner, Dubey et al. [56] used ELISA to monitor growth of the ascomycete *A. niger*. The suitable choice of a fungal cell wall antigen, which was not present in the substrate, meant that the biomass could be determined quickly in a substrate-independent manner. Furthermore, the fungus was detectable at all morphological stages.

Several factors make the use of ELISA to determine biomass interesting: not only is it rapid, reliable, and sensitive, but also it is inexpensive and requires only small amounts of substrate [56]. However, it does require a suitable antigen to be available, and the development of the assay requires a lot of know-how and optimization.

4.5 Primary Metabolites (ATP)

The quantification of fungal biomass based on primary metabolites, especially ATP, is often used in soil analysis. ATP is an indicator of living biomass. After sample drying, ATP is extracted and quantified by bioluminescence using the luciferin–luciferase system [57]. The method is sensitive, but reports about the contents of ATP in actively growing biomass and live but nongrowing biomass are inconsistent [57]. Also, the amount of ATP extracted is highly dependent on the extracting agent. An internal standard is advisable, since a certain proportion of ATP is hydrolyzed during the extraction or adsorbed on the substrate matrix [57]. A further factor to consider is that the substrate can also contain ATP, which will interfere with the measurement of the ATP from microbial biomass [57, 58]. West et al. [58] used the determination of the ATP content for quantification of the microbial biomass in soil samples, but this method was only suitable for biomass estimation at equal cultivation conditions (e.g., same substrate, temperature, and cultivation time) due to the variable ATP content in the biomass.

The usefulness of ATP as a biomass indicator depends on many factors, including the physiological state of the culture. The determination is simple and fast; nevertheless, this method is unsuitable for the determination of fungal biomass during SSF due to the many influencing factors and fluctuations.

4.6 Changes of Dielectric Permittivity

A simple and rapid method for general biomass quantification is the measurement of the dielectric conductivity. This measurement is continuous, noninvasive and can be performed in situ and on-line [6, 59]. Living biomass is polarized by low radio frequencies with the degree of polarization depending on the cell structure in a phenomenon called beta-dispersion. The degree of beta-dispersion is linearly proportional to the volume of the biomass. Since the biomass volume changes during microbial growth, the dielectric conductivity will also change. It can be measured and subsequently correlated with the amount of biomass in the sample or bioreactor [6, 59].

Davey et al. [59] analyzed the change of the conductivity, capacitance, and pH during SSF of the zygomycete *R. oligosporus* on soy beans, lupine, and quinoa. The conductivity of the fungal biomass increased linearly with the total hyphal length, which was measured microscopically, during the growth phase and did not change during the death phase. The capacitance was also related to growth: It increased during the growth phase and dropped during cell death.

The measurement of the dielectric conductivity is simple and fast, which means that it has good potential to be used online for biomass quantification during SSF. However, this method requires a liquid film at the particle surface in order to ensure a proper contact with the electrode. Consequently, this method can only be used conditionally, due to the poor contact of the electrode with many of the porous substrates and substrate beds that are used in classical SSF.

4.7 Metabolic Heat Production

The release of waste metabolic heat from respiration and maintenance metabolism will affect the ambient temperature of the bed of solids within which the fungus is growing, and consequent temperature changes can be measured online in a continuous and nondestructive manner. However, although it is theoretically possible to quantify biomass from measured temperature changes, such "calorimetry" is rarely used for biomass estimation in bioreactors [5] since the temperature is also affected by many other phenomena that occur in the bed (i.e., conduction, convection, and evaporation) [5, 54].

Li and Wadsö [5] used heat production, measured by isothermal calorimetry, as an indicator of the activity of the ascomycete *Penicillium brevicumpactum*. During the growth phase, the total heat produced correlated well with the colony area, which was recorded digitally [5]. According to Li and Wadsö [5], the method is suitable for all substrates and organisms and is usually very sensitive. Dhandapani et al. [54] used heat measurement to observe the growth of the ascomycete *A. tamarii* and described this method as a robust and suitable reference technique for the monitoring of fungal activity during SSF.

The measurement of heat production can be done rapid and continuously in a calorimeter. However, the quantification of the production of waste metabolic heat by calorimetry is more suitable for academic studies than for online monitoring of a bioreactor.

5 Other Determination Methods

There are other techniques that can be used to quantify fungal biomass in the laboratory or at large scale. They will briefly mention in this section.

5.1 Plate Count Technique

The plate count technique is a routine procedure in industry and research laboratories for determination of colony-forming units [2, 55]. However, for fungi the number of colony-forming units varies greatly depending on medium composition and cultivation conditions, which makes it difficult to compare the results with other studies. This method is only suitable for cultivable fungi, does not detect dead fungi, and is time-consuming [2, 55]. Only the level of sporulation is represented, instead of the current biomass. Therefore, this method is unsuitable for the determination of the biomass of basidiomycetes, as well as other filamentous fungi [12, 55].

5.2 Agar Film Technique

The agar film technique is an optical identification method that was developed in 1948 by Jones and Mollison [60]. The sample is mixed with agar after several washing steps, heated, and poured into a counting chamber (hemocytometer).

The piece of agar with uniform volume is released after solidification, dried, and analyzed after staining by microscopy. Subsequently, the biomass concentration is calculated using a conversion factor. Since 1948, the method has been further developed and optimized, conversion factors improved, and handling simplified, so that important biomass structures are not destroyed [61]. The conversion factor is influenced by the hyphal length, density, magnification, and substrate [62]. Usually, this method is used in parallel with other determination methods, because it is very complicated and requires expertise [63].

5.3 Light Reflectance and IR-Spectroscopy

Fungal biomass can be estimated using light reflectance and infrared spectroscopy (IR-spectroscopy).

Murthy et al. [64] used the light reflection method to measure the color change of the medium during production of *koji* with *A. niger*. The substrate (rice) took on a darker color during the fermentation. The principle of trichromatic generalization (i.e., that each color consists of 3 primary colors) was used to detect the color change. The reflection, absorption, and scattering of the light that is incident on the sample vary, depending on the color. Visible light (360–800 nm) was used as the incident light and the total color difference in the reflected light between the sample and the white standard (barium sulfate powder) was calculated. The total color difference correlated well with biomass in an experiment in which pure biomass was mixed with dried and autoclaved wheat bran in different proportions ($R = 0.98$), until a biomass concentration of 20 % in the sample.

IR-spectroscopy uses infrared light (800 nm–1 mm) and can be used to quantify known substances or for structural analysis. The IR-radiation is absorbed by certain bonds, which lead to a vibration of the bond. Different types of bonds give vibration peaks at different wavelengths in the IR spectrum. The interpretation and assignment of the individual spectra and absorption bands requires a lot of experience, but this method is an interesting technique for the determination of fungal biomass, because it is noninvasive and can also be used to quantify several components simultaneously.

Desgranges et al. [52] used IR-spectroscopy for the direct determination of the cell components, ergosterol and glucosamine, of the ascomycete *B. bassiana* and the medium components, sucrose and nitrogen. The values determined by IR-spectroscopy correlated well with contents determined in chemical assays. In the case of glucosamine, the nutrient medium interfered partially. The differentiation of the growth stages was possible. However, the accuracy of the measurement decreased with increasing biomass content in the sample being analyzed.

Zornoza et al. [65] used near-infrared (NIR) spectroscopy in an attempt to quantify fungal biomass, by detecting fungal-specific PLFAs in leaf litter. The values of the PLFAs predicted by NIR reflectance correlated well with the PLFA values that were determined by a standard chemical assay. However, the interpretation of the spectra that was obtained during the measurement of the leaf litter, and thus, the estimation of the fungal biomass content was not directly possible due to interferences and superposition of spectra from various sample components.

In contrast to Zornoza et al. [65] and Desgranges et al. [52], Brandl [66] used IR-spectroscopy for qualitative determination of the ascomycete *Neotyphodium lolii* in perennial ryegrass and not for quantification. The detection limits were too high, so that small amounts of biomass remained undetected.

The use of IR-spectroscopy requires a lot of experience for proper interpretation of the spectra, and it is also time-consuming to calibrate the method properly. Additionally, the method does not differentiate between live and dead biomass. Nevertheless, it is an interesting alternative for the quantification of biomass during SSF since it is robust and nondestructive and the measurement itself is rapid. Also, sample preparation is easy and does not require the use of additional chemicals, thereby avoiding the generation of hazardous waste [52]. IR-spectroscopy is already used routinely in chemistry and can be used with relatively small samples. It might even be possible to use this method online for transparent reactors. On the other hand, the determination of the color change of the substrate as a biomass indicator is suitable only for special cases.

5.4 Substrate-Induced Respiration (SIR)

Biomass quantification using substrate-induced respiration (SIR) is a commonly used standardized method (ISO 14240-1) from soil biology and can be carried out with fresh samples. In this method, active biomass is estimated based on the determination of the CO_2 that is formed after the addition of a readily available carbon source, such as glucose. A conversion factor, obtained through calibration, is used to convert the CO_2 formed into biomass [57]. After optimization, this method is very sensitive, highly reproducible, automatable, and performable with large sample throughput [2, 67].

There are two general approaches. In the Heinemeyer approach, the CO_2 formation is measured continuously throughout the lag phase of the microbial growth using infrared gas analysis (measurement of the maximum initial respiratory response). In the Isermeyer approach, the CO_2 is absorbed in a sodium hydroxide solution. After 4 h of incubation, the carbonate is precipitated and the unused sodium hydroxide is determined by titration. The SIR method is commonly used and inexpensive, but only detects active fungi that are capable of metabolizing the added glucose. It is not suitable for samples with a high content of young cells due to the higher CO_2 formation compared to older cells, which leads to an overestimation of the fungal biomass [57]. Furthermore, this method is highly influenced by the humidity of the sample and the pH value [68]. When nonsterile samples are used, the percentage of fungal biomass can be quantified by the addition of streptomycin [67, 68]. However, the inhibitor concentration must be optimized for the particular conditions since it greatly affects the results.

6 Conclusions

Various biomass determination methods are available for the quantification of fungal biomass in SSF processes. Each method has its advantages and drawbacks. A perfect method does not yet exist, but some methods are more suitable than others.

Optical methods enable a deeper understanding of morphology and, with the use of appropriate staining techniques, a deeper understanding of the physiological state of the fungus. It is also possible to distinguish active and inactive hypha and quantify them separately. However, quantification of the biomass by imaging techniques is difficult during SSF, because usually only surface mycelium can be detected and the methods are very time-consuming.

The determination of cell components is commonly used to quantify fungal biomass. Fungal-specific components can be used in nonsterile processes, as well as in mixed cultures. New techniques, such as real-time PCR, enable a species-specific determination of the biomass. Most of these methods are very complicated and time-consuming, and require invasive sampling. However, the use of alternative measurement techniques, such as IR-spectroscopy, enables the nondestructive and online detection of certain components of the fungus and even of the substrate. Most methods involving measurement of cell components require careful calibration, and it is crucial to check whether the conversion factors depend on mycelium age and the cultivation conditions.

Measurement of the metabolic activity of the fungal biomass is usually suitable for continuous and online monitoring and is therefore appropriate for industrial bioreactors. However, it is usually difficult to convert the measured metabolic activities into reliable estimates of the biomass, so these methods should be used as qualitative indicators of the biomass and its physiological state. Also, species-specific differentiation of the biomass is usually not possible, so these methods are less suitable for nonsterile processes.

In most cases, it would be advantageous to use a combination of different methods. This has two advantages. Firstly, it can increase the reliability of the results. Secondly, it can provide additional information about the state of culture.

In our own work, we tested a variety of different methods to monitor growth of the basidiomycete *T. hirsuta* in cultivation systems that included a reactor of up to 10-L working volume. We obtained good results for the measurement of ergosterol content, the counting of cell nuclei, the measurement of respiratory activity, and the determination of the activity of lignolytic enzymes. The determination of the ergosterol content is expensive, but the results are accurate and reproducible. The specially developed method of counting the fungal cell nuclei by flow cytometry is influenced by many factors, but this can be resolved by the use of different calibration curves. The results correlated well with the biomass calculated by ergosterol measurement and were also reproducible. Furthermore, this method is characterized by rapidity, straightforward handling and enables additional information about the physiology of the culture. The measurement of respiratory activity was very suitable for continuous and online biomass estimation until the stationary phase.

The activities of lignolytic enzymes, such as laccases and peroxidases, were useful as qualitative indicators of biomass and gave additional information about the state of the culture. The enzyme activity assays were rapid and easy to perform.

To conclude, the most suitable methods for quantification of fungal biomass in SSF depend on the particular requirements for accuracy, cost, and throughput: We have given you some advice, now you must make your choice!

References

1. Krishna C (2005) Solid-state fermentation systems-an overview. Crit Rev Biotechnol 25(1–2):1–30
2. Ottow JCG (2011) Mikrobiologie von Böden. Springer, Berlin, 508 p
3. Böhmer U, Frömmel S, Bley T, Müller M, Frankenfeld K, Miethe P (2011) Solid-state fermentation of lignocellulotic materials for the production of enzymes by the white-rot fungus *Trametes hirsuta* in a modular bioreactor. Eng Life Sci 11(4):395–401
4. Mitchell DA, Krieger N, Berovic M (2006) Solid-state fermentation bioreactors: fundamentals of design and operation. Springer Science & Business Media, Berlin, 481 p
5. Li Y, Wadsö L (2011) Simultaneous measurements of colony size and heat production rate of a mould *Penicillium brevicompactum* growing on agar. J Therm Anal Calorim 104(1):105–111
6. Madrid RE, Felice CJ (2005) Microbial biomass estimation. Crit Rev Biotechnol 25(3):97–112
7. Couri S, Mercês EP, Neves BCV, Senna LF (2006) Digital image processing as a tool to monitor biomass growth in *Aspergillus niger* 3T5B8 solid-state fermentation: preliminary results. J Microsc 224(3):290–297
8. Dutra J, da C Terzi S, Bevilaqua J, Damaso M, Couri S, Langone M, Senna LF (2008) Lipase production in solid-state fermentation monitoring biomass growth of *Aspergillus niger* using digital image processing. Appl Biochem Biotechnol 147(1):63–75
9. Stahl PD, Parkin TB, Christensen M (1999) Fungal presence in paired cultivated and uncultivated soils in central Iowa. USA Biol Fertil Soils 29(1):92–97
10. Paul EA, Clark FE (1989) Soil microbiology and biochemistry. Academic Press, Waltham, 298 p
11. Söderström B, Bååth E (1979) Fungal biomass and fungal immobilization of plants nutrients in swedish coniferous forest soils. Rev Ecol Biol Sol 16(4):477–489
12. Osma JF, Toca-Herrera JL, Rodríguez-Couto S (2011) Environmental, scanning electron and optical microscope image analysis software for determining volume and occupied area of solid-state fermentation fungal cultures. Biotechnol J 6(1):45–55
13. Haus J (2014) Optische Mikroskopie: Funktionsweise und Kontrastierverfahren. Wiley, New York, 240 p
14. Nopharatana M, Mitchell DA, Howes T (2003) Use of confocal scanning laser microscopy to measure the concentrations of aerial and penetrative hyphae during growth of *Rhizopus oligosporus* on a solid surface. Biotechnol Bioeng 84(1):71–77
15. Horbik D, Łowińska-Kluge A, Górski Z, Stanisz E, Zgoła-Grześkowiak A (2013) Microwave-assisted extraction combined with HPLC-MS/MS for diagnosis of fungal contamination in building materials. J Braz Chem Soc 24(9):1478–1486
16. Ooijkaas LP, Tramper J, Buitelaar RM (1998) Biomass estimation of *Coniothyrium minitans* in solid-state fermentation. Enzyme Microb Technol 22(6):480–486
17. Desgranges C, Vergoignan C, Georges M, Durand A (1991) Biomass estimation in solid state fermentation I. Manual biochemical methods. Appl Microbiol Biotechnol 35(2):200–205
18. Gessner, Newell SY (2002) Manual of environmental microbiology, 2nd edn. ASM Press, Washington DC
19. Manter DK, Kelsey RG, Stone JK (2001) Quantification of *Phaeocryptopus gaeumannii* colonization in Douglas-fir needles by ergosterol analysis. For Pathol 31(4):229–240

20. Matcham SE, Jordan BR, Wood DA (1985) Estimation of fungal biomass in a solid substrate by three independent methods. Appl Microbiol Biotechnol 21(1):108–112
21. Zelles L, Hund K, Stepper K (1987) Methoden zur relativen Quantifizierung der pilzlichen Biomasse im Boden. Z Für Pflanzenernähr Bodenkd 150(4):249–252
22. Klamer M, Bååth E (2004) Estimation of conversion factors for fungal biomass determination in compost using ergosterol and PLFA 18:2ω6,9. Soil Biol Biochem 36(1):57–65
23. Zhang H, Wolf-Hall C, Hall C (2008) Modified microwave-assisted extraction of ergosterol for measuring fungal biomass in grain cultures. J Agric Food Chem 56(23):11077–11080
24. Reeslev M, Miller M, Nielsen KF (2003) Quantifying mold biomass on gypsum board: comparison of ergosterol and beta-N-acetylhexosaminidase as mold biomass parameters. Appl Environ Microbiol 69(7):3996–3998
25. Snajdr J, Valásková V, Merhautová V, Cajthaml T, Baldrian P (2008) Activity and spatial distribution of lignocellulose-degrading enzymes during forest soil colonization by saprotrophic basidiomycetes. Enzyme Microb Technol 43(2):186–192
26. Muniroh MS, Sariah M, Abidin MAZ, Lima N, Paterson RRM (2014) Rapid detection of *Ganoderrna*-infected oil palms by microwave ergosterol extraction with HPLC and TLC. J Microbiol Methods 100:143–147
27. Deng Z-L, Yuan J-P, Zhang Y, Xu X-M, Wu C-F, Peng J, Wang J-H (2013) Fatty acid composition in ergosteryl esters and triglycerides from the fungus *Ganoderma lucidum*. J Am Oil Chem Soc 90(10):1495–1502
28. Schmidt O (2006) Wood and tree fungi: biology, damage, protection, and use. Springer Science & Business Media, Berlin, 336 p
29. Abd-Aziz S, Hung GS, Hassan MA, Abdul Karim MI, Samat N (2008) Indirect method for quantification of cell biomass during solid-state fermentation of palm kernel cake based on protein content. Asian J Sci Res 1(4):385–393
30. Nilsson K, Bjurman J (1998) Chitin as an indicator of the biomass of two wood-decay fungi in relation to temperature, incubation time, and media composition. Can J Microbiol 44(6):575–581
31. Ride J, Drysdale R (1972) Rapid method for chemical estimation of filamentous fungi in plant-tissue. Physiol Plant Pathol 2(1):7–15
32. Scotti CT, Vergoignan C, Feron G, Durand A (2001) Glucosamine measurement as indirect method for biomass estimation of *Cunninghamella elegans* grown in solid state cultivation conditions. Biochem Eng J 7(1):1–5
33. Roche N, Venague A, Desgranges C, Durand A (1993) Use of chitin measurement to estimate fungal biomass in solid state fermentation. Biotechnol Adv 11(3):677–683
34. Blagodatskaya E, Blagodatsky S, Anderson T-H, Kuzyakov Y (2014) Microbial growth and carbon use efficiency in the rhizosphere and root-free soil. PLoS ONE 9(4):e93282
35. Jirout J, Šimek M, Elhottová D (2011) Inputs of nitrogen and organic matter govern the composition of fungal communities in soil disturbed by overwintering cattle. Soil Biol Biochem 43(3):647–656
36. May BA, VanderGheynst JS, Rumsey T (2006) The kinetics of *Lagenidium giganteum* growth in liquid and solid cultures. J Appl Microbiol 101(4):807–814
37. Knoll S, Mulfinger S, Niessen L, Vogel RF (2002) Rapid preparation of *Fusarium* DNA from cereals for diagnostic PCR using sonification and an extraction kit. Plant Pathol 51(6):728–734
38. Voegele RT, Schmid A (2011) RT real-time PCR-based quantification of *Uromyces fabae* in planta. FEMS Microbiol Lett 322(2):131–137
39. Tellenbach C, Grünig CR, Sieber TN (2010) Suitability of quantitative real-time PCR to estimate the biomass of fungal root endophytes. Appl Environ Microbiol 76(17):5764–5772
40. Pilgård A, Alfredsen G, Björdal CG, Fossdal CG, Børja I (2011) qPCR as a tool to study basidiomycete colonization in wooden field stakes. Holzforschung 65(6):889–895
41. Costa-de-Oliveira S, Silva AP, Miranda IM, Salvador A, Azevedo MM, Munro CA et al (2013) Determination of chitin content in fungal cell wall: an alternative flow cytometric method. Cytometry A 83(3):324–328

42. Throndset W, Kim S, Bower B, Lantz S, Kelemen B, Pepsin M et al (2010) Flow cytometric sorting of the filamentous fungus Trichoderma reesei for improved strains. Enzyme Microb Technol 47(7):335–341
43. Steudler S, Böhmer U, Weber J, Bley T (2014) Biomass measurement by flow cytometry during solid-state fermentation of basidiomycetes. Cytometry A. doi:10.1002/cyto.a.22592
44. Griffin DH (1996) Fungal physiology. Wiley, New York, 476 p
45. Frostegård A, Tunlid A, Bååth E (2011) Use and misuse of PLFA measurements in soils. Soil Biol Biochem 43(8):1621–1625
46. Frostegård A, Bååth E (1996) The use of phospholipid fatty acid analysis to estimate bacterial and fungal biomass in soil. Biol Fertil Soils 22(1):59–65
47. Córdova-López J, Gutiérrez-Rojas M, Huerta S, Saucedo-Castañeda G, Favela-Torres E (1996) Biomass estimation of *Aspergillus niger* growing on real and model supports in solid state fermentation. Biotechnol Tech 10(1):1–6
48. Favela-Torres E, Cordova-López J, García-Rivero M, Gutiérrez-Rojas M (1998) Kinetics of growth of *Aspergillus niger* during submerged, agar surface and solid state fermentations. Process Biochem 33(2):103–107
49. Saucedo-Castañeda G, Trejo-Hernández MR, Lonsane BK, Navarro JM, Roussos S, Dufour D et al (1994) On-line automated monitoring and control systems for CO_2 and O_2 in aerobic and anaerobic solid-state fermentations. Process Biochem 29(1):13–24
50. Ikasari L, Mitchell DA (1998) Oxygen uptake kinetics during solid state fermentation with *Rhizopus oligosporus*. Biotechnol Tech 12(2):171–175
51. Larroche C, Gros JB (1992) Characterization of the growth and sporulation behavior of *Penicillium roquefortii* in solid substrate fermentation by material and bioenergetic balances. Biotechnol Bioeng 39(8):815–827
52. Desgranges C, Georges M, Vergoignan C, Durand A (1991) Biomass estimation in solid state fermentation II. On-line measurements. Appl Microbiol Biotechnol 35(2):206–209
53. Sakurai Y, Misawa S, Shiota H (1985) Growth and respiratory activity of *Aspergillus oryzae* grown on solid state medium. Agric Biol Chem 49(3):745–750
54. Dhandapani B, Mahadevan S, Dhilipkumar SS, Rajkumar S, Mandal AB (2012) Impact of aeration and agitation on metabolic heat and protease secretion of *Aspergillus tamarii* in a real-time biological reaction calorimeter. Appl Microbiol Biotechnol 94(6):1533–1542
55. Anand S, Rati ER (2006) An enzyme-linked immunosorbent assay for monitoring of *Aspergillus ochraceus* growth in coffee powder, chilli powder and poultry feed. Lett Appl Microbiol 42(1):59–65
56. Dubey AK, Suresh C, Kumar SU, Karanth NG (1998) An enzyme-linked immunosorbent assay for the estimation of fungal biomass during solid-state fermentation. Appl Microbiol Biotechnol 50(3):299–302
57. Kaufmann K, Rossier N, Oberholzer H (2010) Niederhäusern A v. Vergleich dreier Methoden zur Langzeitbeobachtung der biologischen Bodenaktivität. Publ Bodenkd Ges Schweiz 30:75–80
58. West AW, Ross DJ, Cowling JC (1986) Changes in microbial C, N, P and ATP contents, numbers and respiration on storage of soil. Soil Biol Biochem 18(2):141–148
59. Davey CL, Peñaloza W, Kell DB, Hedger JN (1991) Real-time monitoring of the accretion of *Rhizopus oligosporus* biomass during the solid-substrate tempe fermentation. World J Microbiol Biotechnol 7(2):248–259
60. Jones PCT, Mollison JE, Quenouille MH (1948) A technique for the quantitative estimation of soil micro-organisms. J Gen Microbiol 2(1):54–69
61. Hill DR (1996) Thin agar film for enhanced fungal growth and microscopic viewing in a new sealable fungal culture case. J Clin Microbiol 34(9):2140–2142
62. Kasai K, Horikoshi T (1997) Estimation of fungal biomass in the decaying cones of Pinus densiflora. Mycoscience 38(3):313–322
63. Lodge DJ, Ingham ER (1991) A comparison of agar film techniques for estimating fungal biovolumes in litter and soil. Agric Ecosyst Environ 34(1):131–144
64. Ramana Murthy MV, Thakur MS, Karanth NG (1993) Monitoring of biomass in solid state fermentation using light reflectance. Biosens Bioelectron 8(1):59–63

65. Zornoza R, Guerrero C, Mataix-Solera J, Scow KM, Arcenegui V, Mataix-Beneyto J (2008) Near infrared spectroscopy for determination of various physical, chemical and biochemical properties in Mediterranean soils. Soil Biol Biochem 40(7):1923–1930
66. Brandl H (2013) Detection of fungal infection in *Lolium perenne* by fourier transform infrared spectroscopy. J Plant Ecol 6(4):265–269
67. Beare MH, Neely CL, Coleman DC, Hargrove WL (1990) A substrate-induced respiration (SIR) method for measurement of fungal and bacterial biomass on plant residues. Soil Biol Biochem 22(5):585–594
68. Lin Q, Brookes PC (1999) An evaluation of the substrate-induced respiration method. Soil Biol Biochem 31(14):1969–1983

Adv Biochem Eng Biotechnol (2015) 149: 253–274
DOI 10.1007/10_2015_305

Published Online: 5 February 2015

Ramified Challenges: Monitoring and Modeling of Hairy Root Growth in Bioprocesses—A Review

Felix Lenk and Thomas Bley

Abstract The review presents a comprehensive overview on available solutions for the monitoring and modeling of various aspects of hairy root growth processes. Several online and offline measurement principles get explained exemplary and are being compared. It was found that no direct online measurement principle for hairy root biomass in submerged and solid-state culturing environment is available. However, certain indirect methods involving one or more measurement values have been developed for biomonitoring of hairy roots especially in bioreactors. In the field of modeling of hairy root growth processes, four independent architectures (continuous models, metabolic flux analysis, agent-based models, and artificial neural networks) are described and compared including literature references. The discussion is structured into microscopic model approaches, addressing only certain aspects of growth, and macroscopic model approaches, describing the hairy root network as a whole. An agent-based macroscopic model based on phenomenological data acquired with systematic imaging of hairy roots on culture dishes together with a 3D visualization of simulation results is presented in detail.

Keywords Hairy roots · Biomonitoring · Growth modeling · Artificial neural networks · Agent-based models

Abbreviations

ABM	Agent-based model
ANN	Artificial neural network
CFD	Computational fluid dynamics
DART	Direct analysis in real time
EFSA	European Food Safety Authority
FDA	Federal Drug Administration
GC-MS	Gas chromatography mass spectrometry
HP-LC	High-performance liquid chromatography

F. Lenk (✉) · T. Bley
Institute of Food Technology and Bioprocess Engineering, Technische Universität Dresden, Bergstrasse 120, 01069 Dresden, Germany
e-mail: felix.lenk@tu-dresden.de

HRC	Hairy root culture
L-DOPA	L-3,4-Dihydroxyphenylalanin
MFA	Metabolic flux analysis
MS	Murashige and Skoog (medium)
ODD	Overview—design concept—details
PCR	Polymerase chain reaction
Ri	Root-inducing
STR	Stirred tank reactor
TCA	Tricarboxylic acid cycle
UML	Unified modeling language

Contents

1 Introduction

Plants offer a wide range of secondary metabolites which are not primarily necessary for growth and cell division but rather act as, e.g., protective agents. For centuries these active agents are used as food additives, in phyto-therapeutics and in the cosmetics industry [1].

Beside the extraction of phyto-products from conventionally cultivated full plants the in vitro cultivation of plant tissue cultures such as hairy roots cultures (HRC) is considered a promising alternative and is nowadays a main research focus. Advantages with the use of in vitro cultures are the independence from seasonal and metrological influences and the traceability and control of the whole production process necessary for regulatory approval by the FDA or EFSA for many substances. Additionally higher concentrations in the bulk material and a higher space-time yield can be achieved together with a simplified down-streaming process [2, 3].

The naturally occurring hairy root disease resulting in tissue-like growth of root networks of infected plant cells has been adapted by the directed infection of untreated plant cells with *Agrobacterium rhizogenes*. With the transfer of plasmids from the agrobacterium into the plant cells the uncontrolled proliferation of distinct root hairs, the so-called hairy roots, gets induced [4]. Interestingly to know that the name *hairy* does not refer to the finely structured network itself but rather to the tiny hairs on the tips of the hairy root network assumingly used for nutrient detection. The advantages of hairy roots consist especially in the specifically simple separation of biomass from the nutrient medium after the cultivation is finished and that no phyto-hormones are needed to stabilize the cultures during the cultivation [1].

Prominent examples using plant in vitro tissue cultures on an industrial scale are the production of medically active Ginsenosides using hairy roots of *Panax ginseng* and shikonin with *Lithospermum erythrorhizon* by the Nitto Denko Corporation, Osaka, Japan [5, 6]. Other promising applications of the active agent production with hairy roots in pilot scale have been compiled by Curtis et al. [7]. For the production of α-Tocopherol (Vitamin E) from hazel, rosmarinic acid (anti-inflammatory) from lavender and basil as well the betalains (pigments) from red beet (*Beta vulgaris*) auspicious results have been obtained [8–12].

Consequently, beside the necessity to provide tailored bioreactor systems for the cultivation of hairy root cultures [1, 13, 14], the topics of monitoring of growth and metabolic processes as well as the mathematical modeling of all growth and production related processes need to be investigated and addressed. The lack of solutions for monitoring and modeling of hairy root growth currently hinders the implementation of production processes with hairy root plant in vitro cultures on an industrial scale despite that a broad variety of different types of hairy root cultures is already available from a research perspective [11, 15]. In order to outline the progress and the current state of the art in terms of hairy root growth monitoring and modeling, this review has been compiled.

2 Monitoring of Hairy Root Growth

For an effective development and optimization of hairy root culturing conditions, robust and reliable solutions for monitoring of growth processes and the production processes of the target metabolites are essential. Due to the nature of inhomogeneous distribution of biomass and the root-like structure, conventional methods for the biomonitoring of hairy roots are often insufficient (see Fig. 1 and [16]).

Generally, the development process of a new hairy root type can be divided into two elementary stages. After the infection and co-cultivation with *Agrobacterium rhizogenes* in Erlenmeyer shaking flasks, each proliferating new hairy root line gets transferred onto a culture dish with solid MS-medium. In the following steps, vital and productive lines should be identified and first optimizations of medium composition are carried out. For the second stage, only a few productive and stable lines get transferred from the solid medium cultivation environment into the liquid medium. Comparable to the first-stage medium composition and cultivation condition, optimizations take place before the scale-up into a higher cultivation volume need to be performed.

Fig. 1 Hairy roots of Devil's claw (Harpagophytum procumbens) in a bubble column bioreactor [16]

2.1 Monitoring and Screening on Culture Dishes (Solid Medium) and in Erlenmeyer Flasks (Liquid Medium)

Starting point for culture line development is a successfully transformed and stable in vitro tissue culture of a certain plant species. In order to narrow down the selection of available line, all of them are screened for highest productivity of the target metabolites and for their general growth behavior. This multi-stage process takes places firstly in culture dishes on solid nutrient medium and secondly on Erlenmeyer flasks with a working volume of up to 250 mL [1]. Generally, the growth monitoring of hairy roots on solid medium can be divided into invasive, off-line, and non-invasive, on- and at-line methods. In particular, with cultivation cycles of 14–21 days and effective dry weights of under 1 g per sample line, non-invasive methods are superior to invasive methods. However, invasive methods are more common and only a couple of non-invasive methods exist.

A fundamental invasive approach to biomass measurement and metabolic profiling is the harvesting of hairy roots after a distinct cultivation period, disrupting them for a homogeneous release of cell components followed by a freeze drying or other storage method. Consequently, measurements with PCR for verification purposes or HP-LC, GC-MS, and spectrophotometry for metabolite analysis or flow cytometry to determine, e.g., the ploidy level are carried out. Typically, detection targets need to be stained or extracted with appropriate solvents beforehand. In order to determine culture development over time, parallel experiments with, e.g., daily harvesting need to be set up from one initial sample. Considering the need for replicate determinations under the same cultivation conditions (minimally 3–6 fold to derive statistically significant results), this approach results in an extensive number of experiments while exhibiting the disadvantage that the precise course of growth of one single hairy root line cannot be investigated.

As an initial step mainly to verify the insertion of the Ri-plasmid but also for further characterization and monitoring of plasmid stability, PCR methods are used, e.g., by Zhang et al. [17] to investigate the biosynthetic pathway in *Hyoscyamus niger* hairy root cultures. The genetic stability of *Salvia miltiorrhiza* hairy roots has been studied with a PCR method by Yuan et al. [18]. High-performance liquid chromatography (HP-LC) is commonly used for metabolite profiling with hairy root cultures such as presented by Zhi et al. [19] for the monitoring of diterpenoid production in hairy root cultures of *S. miltiorrhiza* based on a protocol initially derived by Kamada et al. [20]. The detection of target metabolites such as the very popular flavonoids an HP-LC method for their determination in *Scutellaria baicalensis Georgi* hairy roots has been developed by Kovács et al. [21]. Sugar concentrations were determined by a HP-LC method presented by Lanoue et al. [22] to compare the growth properties and alkaloid production of two *Datura* hairy root lines.

The application of GC-MS for target metabolite screening is also a common method such as presented by Bais et al. [23] using solid-phase microextraction sample preparation to detect volatile compounds in hairy root cultures of *Cichorium intybus* L. For the optimization of the culture medium composition and to improve the production of hyoscyamine, hairy roots of *Datura stramonium* L. were analyzed by Amdoun et al. [24] using GC-MS.

A spectrophotometrical method for the detection of silymarin in *Silybum marianum* hairy root cultures has been developed by Rahimi et al. [25]. Weber et al. investigated diploid and tetraploid hairy root cultures of *Datura stramonium, H. niger,* and *B. vulgaris* with a novel flow cytometry method [26, 27] based on the findings of Neumann et al. [28]. A very effective method of screening plant in vitro cultures is presented by Geipel et al. [29], however, not for hairy root cultures but for suspension cultures of *Helianthus annuus* producing α-tocopherol using the RAMOS® for continuous monitoring of oxygen and carbon dioxide transfer rates.

Non-invasive monitoring methods with only a small delay from sampling to result evaluation are a highly relevant topic also in the field of biomonitoring with hairy root cultures. Inspired by monitoring solutions composed from an imaging unit and an image post-processing method in the field of plant root investigation such as the WinRhizo [30, 31] or the GiARoots [32] solution, Lenk et al. [33] proposed the manual imaging method PetriCam (see Fig. 2) together with an

Fig. 2 Image taken with PetriCam of a hairy root network of *Beta vulgaris* in a culture dish [33]

automated image analysis algorithm to investigate the growth process and secondary metabolite accumulation of *B. vulgaris* hairy roots on culture dishes with solid medium. Unique advantage of the method is due to the non-invasive imaging through the cover of the sealed culture dish, the continued monitoring of the same hairy root line over the whole cultivation period. Currently, images consist of planar 2D representations orthogonally taken from the top of the culture dish which do not provide any information on the distribution of the hairy root culture in the z-axis. Therefore, a unit for stereo imaging which offers a complete new range of post-processing solutions needs to developed.

Once systematic imaging of the culture dishes is finished such as described by Lenk et al. [34] which also offers the opportunity of automatic imaging of culture dishes for high sample throughput [35, 36], images need to be post-processed in order to extract numerical information such as, e.g., segment length values, branching point distribution, or information on local secondary metabolite accumulation. For such applications, numerous open-source and commercial solutions such as presented by Clark et al. [37], Leitner et al. [38], or Lobet et al. [39] are available. To assess the herbicidal toxicity of photoautotrophic hairy roots of *Ipomoea aquatica,* Ninomiya et al. [40] developed a non-destructive measurement method of the local chlorophyll content using color image analysis.

2.2 Monitoring on Bioreactor Scale (Liquid Medium)

Once a specific hairy root line is cultivated in a bioreactor system, the main focus lies in a continuous monitoring of growth [1, 41]. Analysis on target metabolite contents is typically carried out after the cultivation for comparison to the screening stage. For the continuous monitoring of biomass growth in the cultivation vessel, two methods of measurement need to be divided: indirect and direct measurement.

While direct measurement is the desired method due to the inhomogeneous distribution of the hairy root network, a direct quantification of biomass accumulation is often impossible. In fact in order to precisely measure the biomass concentration directly, the whole cultivation vessel would needed to be scanned by, e.g., a tomographic device. However, several indirect methods for biomass measurement and estimation exist but their accuracy depends on the complexity of the underlying mathematical model. Typically, the trend of one or several global process variables such as the carbon source concentration (e.g., glucose), the CO_2/O_2 transfer rates, the electrical conductivity, dissolved oxygen, the culture broth level, or the pH gets monitored continuously and the observed gradients are inserted into mathematical models in order to determine the actual biomass concentration.

Already in the late 1980s, Taya et al. [42, 43] developed a comprehensive model based on conductivity measurements to determine the biomass concentration of hairy root cultures of horseradish and carrot in stirred tank and airlift loop bioreactors. The effects of oxygen, carbon dioxide, and ethylene on growth and bioactive compound production with *P. ginseng* hairy roots have been directly measured and studied by Jeong et al. [44, 45] in various cultivation environments. As the level of the culture broth changes during the cultivation period as nutrients are consumed especially during high-density cultivations, Jung et al. [46] proposed to use the change in biomass volume as a new parameter for biomass estimation as a deviation of about 20 % from the actual biomass concentration was observed during experiments with *Catharanthus roseus* hairy roots applying a model solely considering the electrical conductivity.

Huang et al. [47] have measured the electrical conductivity, pH, and dissolved oxygen tension in the culture broth to elucidate the kinetic response of essential constituents along with the growth of the hairy roots of the L-DOPA producing strain *Stizolobium hassjoo*.

Madhusudanan et al. [48] developed an at-line direct analysis in real-time (DART) mass spectrometric technique to determine two nitrogen-containing compounds in *Rauvolfia serpentina* hairy roots. An offline flow cytometry method has been developed by Homova et al. [49] for Devil's claw hairy roots in a stirred tank bioreactor to investigate the relationship between the different cell division stages and the production of phenylethanoid glycosides.

A comprehensive working biomass estimation system based on the refractive index or the electrical conductivity of the liquid medium along with liquid medium osmolality has been established by Ramakrishnan et al. [50]. The derived model gives an accurate online estimation of dry weight, fresh weight, and liquid volume of hairy root cultures and was validated with experiments in 125-mL shaking flasks and 2-L bioreactor cultivations.

3 Hairy Root Growth and Biosynthesis Modeling

While there are several mathematical models to describe the growth of suspended plant single-cell cultures [51, 52], only very few global, macroscopic descriptions for the growth of hairy root plant tissue cultures exist due to their complexity. In the field of modeling yeast growth behavior, several models to describe mycelial growth, which can be compared to hairy roots, have been proposed [53–55]. Available models for hairy roots are typically limited to certain parts and are referred to as microscopic models. Modeling efforts follow the aim to understand processes scientifically and in order to optimize culturing conditions and the composition of the nutrient medium. However, practically until now, cultures are only investigated experimentally but in a very systematic way resulting in a huge number of experiments. The following two sections outline an overview on available microscopic and macroscopic models which have been applied to hairy root growth and other metabolic processes.

3.1 Microscopic Models

Efforts in modeling, simulating, and consequently understanding the metabolism of hairy root cultures have increased significantly over the last years. In particular, methods of the MFA have been applied to hairy root cultures of *C. roseus*. Sriram et al. [56] determined the biomass composition of *C. roseus* hairy roots in order to perform a MFA to improve the alkaloid production rate. The resulting flux quantification in the central carbon metabolism of *C. roseus* hairy roots is presented in [57] and introduced the application of "bondomers" for flux evaluation in plants. Bondomers are a computationally efficient and intuitive alternative to the commonly used isotopomer concept. A metabolic model including a MFA was also developed for *C. roseus* hairy roots by Leduc et al. [58] which included a primary metabolic network consisting of only 31 metabolic fluxes and 20 independent pathways. Experiments for model calibration were performed on culture dishes and comparisons showed that the model is efficient for growth rate estimation. In hairy root cultures of *H. niger,* Zhang et al. [17] showed the engineering of the tropane biosynthetic pathway to enhance the scopolamine yield. A biogenetic organizational approach for the quantification of metabolic fluxes in the plant secondary metabolism is presented by Morgan et al. [59]. The main idea was to group metabolites of similar biosynthetic origin. The proposed model was then again validated toward metabolic pathways of hairy roots of *C. roseus*. A comparable approach for the plant primary metabolism is outlined by Cloutier et al. [60] who defined a dynamic flux model which includes nutrient uptakes (phosphate, sugars, and nitrogen sources), the glycolysis and pentose phosphate pathways, the TCA cycle, amino acid biosynthesis, the respiratory chain, and biosynthesis of cell building blocks (see Fig. 3).

Lan et al. [61] address in his work an approach to engineer the salidroside biosynthetic pathway in hairy root cultures of *Rhodiola crenulata* based on the metabolic characterization of tyrosine decarboxylase which plays an important role in thesalidroside biosynthesis. It was found that the overexpression of only the RcTYDC gene resulted in an increased production of the end product salidroside. All presented MFA approaches had the aim to improve the overall target metabolite production and showed that MFA is a powerful tool for hairy root growth and biosynthesis characterization and even small changes to metabolic pathways can lead to significantly higher production yields.

A structured model for the growth of *C. roseus* and *Daucus carota* hairy roots has been derived by Cloutier et al. [62]. The model describes the growth rates taken the intracellular nutrient concentrations of, e.g., sugars, nitrogen, and phosphate into account. Michaelis–Menten kinetics describes the nutrient uptake, and for model calibrations, data from batch cultures were used. A distinct cellular growth model for root tip elongation has been developed by Chavarría-Krauser et al. [63], and its simulation results have been compared to measurements of *Arabidopsis thaliana*, *Nicotiana tabacum,* and *Pisum sativum* roots. Han et al. [64] proposed a population balance model that provides a better understanding of the increase in

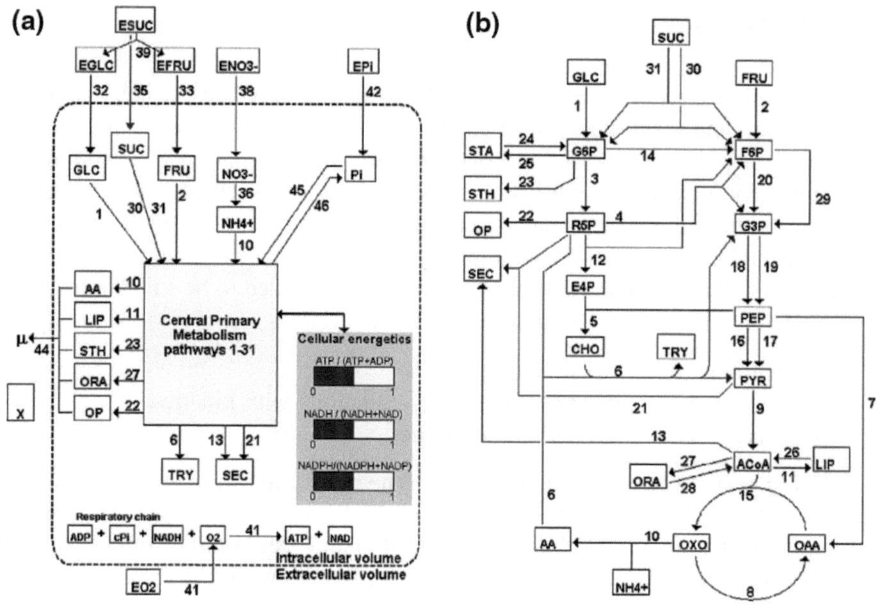

Fig. 3 Schematic view of the dynamic flux model for *Catharanthus roseus* by Cloutier et al. [60]

biomass of hairy root cultures. The model was calibrated with six different hairy root types, and experimental results could be reproduced by model simulations. Another investigation by Asplund et al. [65] focused on the intrinsic oxygen kinetics in *Hyoscyamus muticus*, *Solanum tuberosum*, and *Brassica juncea* hairy roots. Palavalli et al. [66] derived an unstructured model for the growth and oxygen transfer in *Azadirachta indica* hairy roots. The model enables simulations taking the effect of oxygen transfer limitations into account and is able to predict the beginning of an oxygen transfer limitation in the center of the growing hairy root matrix which is a major bottleneck during high-density cultivations. This indicator can then be used to increase the aeration of the medium or to end the cultivation cycle.

A model for tanshinone synthesis and phase distribution in *S. miltiorrhiza* hairy root cultures is presented by Yan et al. [67] including a parameter sensitivity analysis showing that the tanshinone accumulation in the resin heavily depends on the transport rate of tanshinones from the hairy roots to the medium. The specific symbiosis of *Daucus carota* hairy roots and the arbuscular mycorrhizal fungus *Glomus intraradices* has been investigated and transformed into a structured growth model by Jolicoeur et al. [68] revealing through the model that the fungus competes for soluble root sugars with the hairy roots itself. Comparison experiments with cultures on cultures dishes and in Erlenmeyer flasks showed confirmed the model simulation findings.

3.2 Macroscopic Models

While distinct aspects of the growth and biosynthesis processes in hairy root cultures are an important field of research, also macroscopic models need to be developed in order to describe the overall hairy root morphology formation and their interaction with the cultivation environment. This chapter addresses the importance to find macroscopic descriptions for hairy root growth in various environments in order to improve biomass yield, culturing conditions, and culturing environment. Three major architectures have been identified to be suitable for hairy root modeling and will be discussed:

- Artificial neural networks (ANN),
- Continuous models based on differential equations with kinetics,
- Agent-based modeling (ABM) approaches.

ANN appears to be a feasible method for the modeling of the growth as well as biomass estimation and have been applied in the field of plant in vitro cultures for the first time already in 1992 by Albiol et al. [69] and seem to fit also for modeling purposes for hairy root networks. An extensive overview on the applications and the potentials of ANN in plant tissue culture has been compiled by Prasad et al. [70]. Osama et al. [71] studied the growth of *Artimisia annua* hairy roots in a nutrient mist bioreactor and developed an artificial neural network based on the inoculum size, mist on/off time, initial packing density, media volume, initial sucrose concentration in media, and time of culture to describe and to predict the final biomass. By training the ANN with three different algorithms, it was found that the Levenberg–Marquardt algorithm fits best with a correlation coefficient of above 96 %. A slightly different approach with an ANN is presented by Prakash et al. [72] who consider the inoculum density, pH, and volume of growth medium per culture vessel and sucrose content of the growth medium in 50-mL shaking flasks to estimate biomass concentration over the cultivation period for *Glycyrrhiza* hairy root cultures. It was found that the regression neural network fits better than the back propagation neural network for the prediction of optimal culture conditions for maximum hairy root biomass yield.

Many approaches to model the macroscopic and morphological development of hairy root networks are derived from mycelia or roots itself as they altogether comprise of filamentous growth processes. Already in the late 1960s, Lindenmayer developed a mathematical model—the so-called L-systems—to describe filamentous growth processes and processes related to the formation of filaments [73, 74]. Fundamental idea was that cell changes their state during the mitose and forms apical and band-like structures. Based on that mathematical composition of agents to form a specific geometry, it was for the first time possible to reproduce plant-like structures [75, 76] (see also Fig. 4).

Comparable to other systems in nature (e.g., blood vessels or filamentous fungi), HRC is part of a dense network which can be characterized through growth and branching. Important for modeling and simulation according to Dupuy et al. [77] is

Fig. 4 Examples for plant-like structures resolved with L-systems [76]

distinct knowledge on the growth processes itself and on metabolic and transport processes. Bastian et al. [78] demonstrated that the growth of dense root networks and their morphology can be described as a response to external chemical stimuli.

Nevertheless, there are certain approaches to model hairy root networks as a whole system. One so-called multi-scale approach is presented by Bastian et al. [79] where on the macroscopic scale, a continuous model is in place that handles growth and nutrient transport. The water transport is considered on the microscopic scale. The permeability of the root segments is computed by a Galerkin scheme which effectively links the microscopic with the macroscopic scale. The derived model simulations were compared with experimental results of flask cultivations of hairy roots of *Ophiorrhiza mungos*.

Another macroscopic approach is presented by Mairet et al. [80] to describe the growth of hairy roots of *Datura innoxia* in a bioreactor. The derived kinetic model enables to improve the feeding strategy and has been validated during bioreactor cultivations.

A typical approach toward macroscopic modeling of the mentioned problems is the description with balances and kinetics and applying the states on homogeneous grids for deriving the differential equations and the consequent numerical solution as outlined above with various references [77]. However, as experimental data material for modeling is typically taken in discrete intervals, a matching model architecture needs to be found. Final step of the modeling and simulation process should be the aim for an optical comparison of experimental results and simulation output as well as to derive a macroscopically description of the whole hairy root network. Therefore, a completely different approach needs to be used: ABM approaches seem to fit for these challenges.

Hjortso et al. [81] firstly discussed basic approaches toward the modeling of hairy root cultures and outlines that basically from a kinetic standpoint, only two types of cells exist in a hairy root network:

1. Cells in the apical meristem which perform the tip elongation through continuous cell division.
2. The remaining cells that are non-dividing. However, these cells elongate and differentiate into root tissue.

Due to the ability of the meristematic cells to form new branches also outside, the initial tip to form new tips allows the formation of lateral branches. Models of the kinetics of branching point formation are therefore an important aspect with macroscopic modeling of hairy root network. This issue gets also addressed by Kim et al. [82] who proposed a set of verbal rules describing the branching kinetics in detail and then converts these verbal rules into a mathematical model (see Fig. 5). In the timed simulation, the developing hairy root network also gets correlated with an age distribution. An implementation of the branching model is compared with experimental results of hairy roots of *T. erecta*.

The application of classical grid- or state-based models is limited due to the high complexity and the inconstancy of the carriers of the properties. That means that the carrier of the property—in this case, the hairy root network itself—changes its appearance constantly from one step to another when branching and elongating. Due to this fact, the possible occurring constellations would need a practically infinite high resolution of the grid together with a very complex analysis of the step calculations to accurately reflect the structure and morphology of a hairy root network. The steps could be forecasted within defined borders, but it would not be possible to derive a descriptive model for the whole problem.

Consequently, another approach should be investigated: The carrier of the properties, the so-called agents, should only describe the organism in parts and are designed to change their appearance. An object-oriented implementation of such an

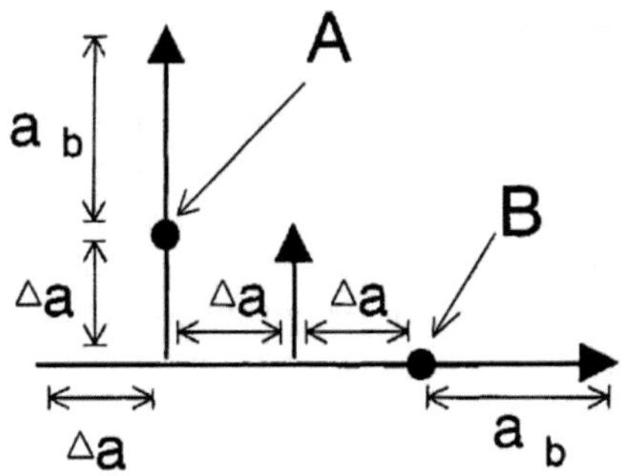

Fig. 5 Schematic drawing of a hairy root network. Old tips are denoted by *arrows*, and newborn tips are denoted by *filled circles*. The newborn tip can form either as the first tip on a branch, the event marked *A*, or as a "sibling" tip on a parent branch that has branched previously, the event marked *B* [82]

agent-based approach is a viable solution toward such a problem. The fundamental equivalents are described in classes—in the presented case, the hairy root network itself and the underlying vessel (e.g., culture dish or bioreactor) as a reservoir for nutrients. These classes are identical in terms of their structure as only their instances differ in the defined properties and values assigned. This approach is a consequent advancement of the work presented in [74, 82] and includes finding from the field of the behavior of microbial communities presented by Lardon et al. [83] to be incorporated in the agent-based model for hairy root networks. Furthermore, it allows the efficient reproduction of a real hairy root structure and a discrete evolution from TIME-STEP T to TIME-STEP $T + 1$. The refinement for a verifiable iterative modeling needs to be variable in terms of time and the structure of the objects.

The agent-based character of the model showed several advantages during the development process. Typically, the number and the cross-linking are not that variable as in the described hairy root model. At first the description seems to be complex, as the object-oriented approach uses unified modeling language (UML) diagrams to express a static class structure. However, as several parts of a hairy root network can be interpreted as agents, the overview-design concept-details ODD-approach proposed by Grimm et al. [84, 85] can be used. This approach basically consists of three blocks: OVERVIEW, DESIGN CONCEPT, and DETAILS.

3.2.1 Overview

In the case of modeling and simulation of complex root networks, three independent growth processes can be observed [38, 81]:

1. Formation of new root tips and new branching points,
2. Tip elongation,
3. Secondary thickening of existing root parts.

To macroscopically model the growth processes of hairy root networks, a timely discrete, recursive description system was used by Lenk et al. [34] that only depends on the previous TIME-STEP (one-step procedure). Figure 6 shows this process flow for TIME-STEP T toward TIME-STEP $T + 1$ horizontally. Vertically the agent-based approach gets expressed as a single-way state machine which describes the phases of an agent in states. Another information exchange and interaction between two agents are implemented using the interaction of resources through the environment.

3.2.2 Design Concept

Starting point of the modeling process was sets of images of hairy root cultures placed on solid nutrient medium in standard culture dishes. The growth characteristics and the evolution of the morphology should be described mathematically.

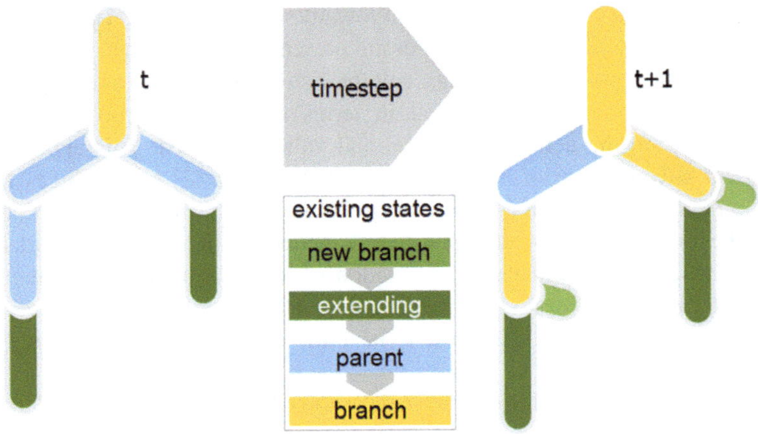

Fig. 6 Fundamental model scheme showing the time-step horizontally and the state machine vertically [34]

Beginning with these experimental images treated with a tailored image analysis algorithm [33], every initial counterpart of a hairy root segment, defined as the connection between two branching points or the connection between a branching point and the root tip, can be mapped to the class rootElement (see Fig. 7). To derive a model architecture and a closely connected representation in a data structure, the hairy root network was fragmented into small parts, where the smallest part is representing cell conglomerates (see Fig. 7d, e). Figure 7a shows a small hairy root sample, while in Fig. 7b, a typical branching point situation is outlined and magnified in Fig. 7c.

The instances of rootElement are equipped with several properties such as length, width as physical parameters as well as oxygen-, pigment- and nutrient-concentrations. With calculated values such as the surface area, volume, and diffusion pressures of the agents, properties can be retrieved easily.

Many cross-linked rootElements compose of the hairy root network. There are numerous methods to calculate meta-data similar to the results from the image analysis solution. An information exchange can be achieved by the use of transient

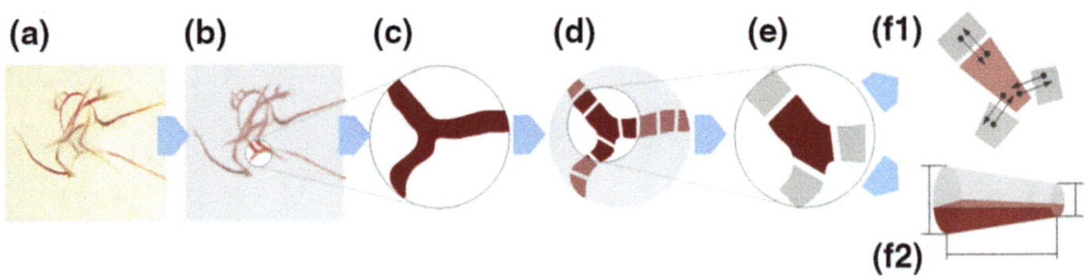

Fig. 7 Derivation of the agent-based architecture from the whole hairy root network (**a**) to the single agent (**e**) with its properties (**f1** and **f2**) [34]

processes, e.g., for the nutrient concentrations as long as there is a connection between the two agents. There is also information exchanged regarding the geometry as every agent gets displayed as adjoining bi-frustums which diameters need to be matched between the instances (see Fig. 7f).

The arrangement of the instances or agents, respectively, corresponds to a cross-linked, incomplete binary tree into both directions. For the implementation of a branch and the aligned formation of branching points, two levels of ROOTELEMENTS are defined: BRANCH1 is the progressive main strand, while BRANCH2 refers to a potential offshoot. The group of nutrient concentrations is managed as cumulative capacity and is used as the consumption reservoir for biomass formation. Additionally, oxygen is necessary which is supplied unlimited above the solid nutrient medium in the virtual culture dish but is available only very limited in the nutrient medium. The transport processes between those phases are part of the implementation. Calling different methods, results in gathering of relevant information on every instance or even the whole network, get assessed. The TIME-STEP-method pushes the simulation one step further.

3.2.3 Details

Due to the nature of the used iterative simulation approach, it is necessary to start the simulation under defined starting conditions. Originating from an initial agent with defined properties concerning, e.g., length that is placed on the virtual nutrient medium, the simulation can be started at TIME-STEP T_0. Lenk et al. [34] in line with Palavalli et al. [66] described also that the supply of the hairy roots with nutrients and oxygen is crucial for the growth processes. Therefore, the class representing the environment contains a scalar 3D field as a carrier for nutrients. Using linear interpolation on this 3D grid agents absorb nutrients and oxygen. The complexity within the agents due to the high affinity of parameters is simplified by a single-way state machine to meet the idea of an agent-based simulation. The complexity should be achieved through the interaction of simple entities. A comprehensive overview on the transition of the agents from one state to another can be found in [34].

Important part of every modeling and simulation task is the visualization of the simulation results to get a visual impression of the calculation results. Very illustrative visualizations of hairy root morphology simulations are also shown by Bastian et al. [79]. The advantages an object-oriented program allows the design of a very efficient visualization engine. In the presented ABM for hairy roots of *B. vulgaris* [34], only one visual counterpart for each class needs to be constructed as its properties parameterized in the instances. Therefore, it is possible to display the properties of each agent individually and rapidly. Another important advantage of implementing the visualization together with the agents is that it results in a faster redevelopment and allows for a rapid parameterization. Simple programming errors do not need to be analyzed using debug tools keeping in mind that the recursive structure of the agents produces an excessive number of instances in a very short computation time (see Fig. 8a). It is obvious that the simulation approach

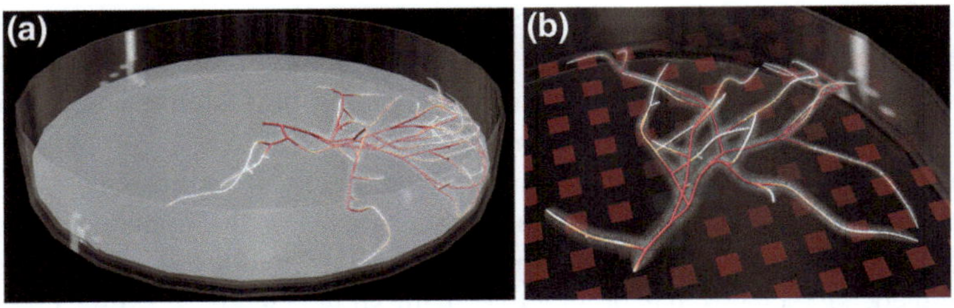

Fig. 8 Examples for the 3D visualization of Beta vulgaris hairy roots (**a**) and the nutrient matrix (**b**) [34]

implemented in Java 1.7 is able to generate millions of single agents within seconds on a commercial personal computer. Consequently, a high-performance strategy to visualize this amount of data to the user is necessary. Considered by Lenk et al. [34], the most efficient solution was the hardware-based OpenGL interface. The agents itself get displayed as frustums between the respective agent and its parental agents. The color gradients were derived directly from the concentration of secondary metabolites, in the case of hairy roots of *B. vulgaris,* the red pigments from the group of betalains, and also have been parameterized through the orientation of the light source. Basically, such a free implementation with only a couple of fixed interfaces principally allows a very efficient simulation and display of the computed data. The compiled framework is open source and runs universally of a device with Java™ support. For the user, there is no need to care, e.g., the movement in space or the pacing as the linear programming of OpenGL gets translated into the object orientation. With this solution, the visualization of about 105 agents with more than 6×10^6 surfaces (polygons) could be achieved with a frame rate of 30 fps on a commercial decided graphics adapter.

The developed visualization engine allows completely free movement throughout the described space. Additionally, the engine is capable of displaying the underlying nutrient matrix (see Fig. 8b). The local nutrient concentration distribution (e.g., for the carbon source) within the nutrient matrix is documented in tiles that change their color from red (high concentration available) to dark blue (depleted). These functions allow the directed, model-based investigation of growth limitation effects through an even locally resolved shortening of necessary resources. Consequently, in silico experiments of a certain experimental setup of an entire cultivation can be carried out before doing in vitro implementations [1, 34].

4 Conclusions

It becomes clear that innovative strategies for biomonitoring and modeling of growth and biosynthesis processes of hairy root cultures are an emerging and challenging field of research and development. Driven by the demand for productive, robust and stable hairy root cultures for the production of active agents for the food, cosmetics, and pharmaceutical industry, the development of a direct online measuring method for the biomass concentration of hairy root cultures in liquid medium still does not exist. However, certain indirect measurement principles are well established and thoroughly investigated. The necessary underlying mathematical models for calculation of the desired target measuring values (e.g., biomass) are in the main focus and have been refined over the course of the last 25 years. A promising direction of development is the increasing interlinking of different measurement value and incorporation in mathematical models with high complexity. The investigation of hairy roots and their properties with well-established methods such as PCR, HP-LC, GC-MS, or flow cytometry is very common but bears the disadvantage of being typically an offline or at least at-line measuring principle which does not provide the user with real-time data necessary, e.g., a dynamic process control.

The description of detailed aspects related to the growth and biosynthesis of hairy root cultures is challenging topic with several exciting research findings published which can be rooted back into the 1960s when early description models for the formation of root networks emerged. From an architectural standpoint, descriptive continuous models based on differential equations, metabolic flux analysis, models based on ANN, and agent-based models are applied to derive actions to improve the culture medium composition, the applied culturing conditions as well as the target metabolite concentrations. Efforts to describe the growth of hairy root cultures completely are right now limited to phenomenological descriptions of the macroscopic morphology that can be observed experimentally.

Main goal for future research should not only be an integration of the available detailed models, e.g., for nutrient uptake or root tip elongation into a full-scale model but also the transition of the simulated cultivation environment from the culture dish to, e.g., shaking flasks or STRs incorporating models for turbulent mixing and aeration (CFD). Ideally, this creates a framework which is available to culture development and optimization for different stages in the upscaling process. Consequently, this approach would certainly speed up culture line developments significantly, would reduce the need for systematic experiments, and would enable the sustainable use of hairy root plant in vitro cultures for a broader variety of applications.

References

1. Steingroewer J, Bley T, Georgiev V, Ivanov I, Lenk F, Marchev A, Pavlov A (2013) Bioprocessing of differentiated plant in vitro systems. Eng Life Sci 13(1):26–38
2. Hanson JR (2003) Natural products: the secondary metabolites. Royal Society of Chemistry, Cambridge
3. Srivastava S, Srivastava AK (2007) Hairy root culture for mass-production of high-value secondary metabolites. Crit Rev Biotechnol 27(1):29–43

4. Chilton M-D, Tepfer DA, Petit A, David C, Casse-Delbart F, Tempé J (1982) *Agrobacterium rhizogenes* inserts T-DNA into the genomes of the host plant root cells. Nature 295 (5848):432–434
5. Lenk F (2012) Innovative Zusatzstoffe aus dem Bioreaktor—hairy roots als Produktionssysteme: Maßgeschneiderte Lösungen vom Screening bis zur Produktionsumgebung. Food-Lab 12(5–6):26–28
6. Misawa M (1994) Plant tissue culture: an alternative for production of useful metabolites. FAO, Rome
7. Curtis WR (2010) Hairy roots, bioreactor growth. In: Encyclopedia of industrial biotechnology. Wiley, Hoboken
8. Giri A, Narasu ML (2000) Transgenic hairy roots—recent trends and applications. Biotechnol Adv 18(1):1–22
9. Matkowski A (2008) Plant in vitro culture for the production of antioxidants—a review. Biotechnol Adv 26(6):548–560
10. Yesil-Celiktas O, Gurel A, Vardar-Sukan F (2010) Large scale cultivation of plant cell and tissue culture in bioreactors. Transword Res Netw 1(1):1–54
11. Guillon S, Tremouillauxguiller J, Pati P, Rideau M, Gantet P (2006) Hairy root research: recent scenario and exciting prospects. Curr Opin Plant Biol 9(3):341–346
12. Eibl D, Eibl R, Poertner R, Catapano G, Czermak P (2009) Cell and tissue reaction engineering. Springer, Berlin
13. Choi Y, Kim Y, Paek K (2006) Types and designs of bioreactors for hairy root culture. In: Gupta SD, Ibaraki Y (eds) Plant tissue culture engineering, vol 2. Springer, Dordrecht, pp 161–172
14. Eibl R, Eibl D (2007) Design of bioreactors suitable for plant cell and tissue cultures. Phytochem Rev 7(3):593–598
15. Doran PM (ed) (1997) Hairy roots: culture and applications. Harwood Academic Publishers, Amsterdam
16. Ludwig-Müller J, Georgiev M, Bley T (2008) Metabolite and hormonal status of hairy root cultures of Devil's claw (*Harpagophytum procumbens*) in flasks and in a bubble column bioreactor. Process Biochem 43(1):15–23
17. Zhang L, Ding R, Chai Y, Bonfill M, Moyano E, Oksman-Caldentey K-M, Xu T, Pi Y, Wang Z, Zhang H, Kai G, Liao Z, Sun X, Tang K (2004) Engineering tropane biosynthetic pathway in *Hyoscyamus niger* hairy root cultures. Proc Natl Acad Sci 101(17):6786–6791
18. Yuan Y, Liu Y, Lu D, Huang L, Liang R, Yang Z, Chen S (2009) Genetic stability, active constituent, and pharmacoactivity of *Salvia miltiorrhiza* hairy roots and wild plant. Z Naturforsch C J Biosci 64(7–8):557–563
19. Zhi BH, Alfermann AW (1993) Diterpenoid production in hairy root cultures of *Salvia miltiorrhiza*. Phytochemistry 32(3):699–703
20. Kamada H, Okamura N, Satake M, Harada H, Shimomura K (1986) Alkaloid production by hairy root cultures in *Atropa belladonna*. Plant Cell Rep 5(4):239–242
21. Kovács G, Kuzovkina IN, Szoke É, Kursinszki L (2004) HPLC determination of flavonoids in hairy-root cultures of *Scutellaria baicalensis Georgi*. Chromatographia 60(1):81–85
22. Lanoue A, Boitel-Conti M, Dechaux C, Laberche J, Christen P, Sangwan-Norreel B (2004) Comparison of growth properties, alkaloid production and water uptake of two selected *Datura* hairy root lines. Acta Biol Cravoviensia Ser Bot 46:185–192
23. Bais HP, Dattatreya B, Ravishankar G (2003) Production of volatile compounds by hairy root cultures of *Cichorium intybus* L. under the influence of fungal elicitors and their analysis using solid-phase micro extraction gas chromatography-mass spectrometry. J Sci Food Agric 83 (8):769–774
24. Amdoun R, Khelifi L, Khelifi-Slaoui M, Amroune S, Asch M, Assaf-Ducrocq C, Gontier E (2010) Optimization of the culture medium composition to improve the production of hyoscyamine in elicited *Datura stramonium* L. Hairy roots using the response surface methodology (RSM). Int J Mol Sci 11(11):4726–4740

25. Rahimi S, Hasanloo T, Najafi F, Khavari-Nejad RA (2011) Enhancement of silymarin accumulation using precursor feeding in *Silybum marianum* hairy root cultures. J Plant Biol Omics 4(1):34–39
26. Weber J, Georgiev V, Pavlov A, Bley T (2008) Flow cytometric investigations of diploid and tetraploid plants and in vitro cultures of Datura stramonium and Hyoscyamus niger. Cytom Part J Int Soc Anal Cytol 73(10):931–939
27. Weber J, Georgiev V, Haas C, Bley T, Pavlov A (2010) Ploidy levels in *Beta vulgaris* (red beet) plant organs and in vitro systems. Eng Life Sci 10(2):139–147
28. Neumann P, Lysák M, Doležel J, Macas J (1998) Isolation of chromosomes from *Pisum sativum* L. hairy root cultures and their analysis by flow cytometry. Plant Sci 137(2):205–215
29. Geipel K, Socher ML, Haas C, Bley T, Steingroewer J (2013) Growth kinetics of a *Helianthus annuus* and a *Salvia fruticosa* suspension cell line: shake flask cultivations with online monitoring system. Eng Life Sci 13(6):593–602
30. Himmelbauer M, Loiskandl W, Kastanek F (2004) Estimating length, average diameter and surface area of roots using two different Image analyses systems. Plant Soil 260(1/2):111–120
31. Zobel RW (2008) Hardware and software efficacy in assessment of fine root diameter distributions. Comput Electron Agric 60(2):178–189
32. Galkovskyi T, Mileyko Y, Bucksch A, Moore B, Symonova O, Price CA, Topp CN, Iyer-Pascuzzi AS, Zurek PR, Fang S, Harer J, Benfey PN, Weitz JS (2012) GiA roots: software for the high throughput analysis of plant root system architecture. BMC Plant Biol 12(1):116
33. Lenk F, Vogel M, Bley T, Steingroewer J (2012) Automatic image recognition to determine morphological development and secondary metabolite accumulation in hairy root networks. Eng Life Sci 12(6):588–594
34. Lenk F, Sürmann A, Oberthür P, Schneider M, Steingroewer J, Bley T (2014) Modeling hairy root tissue growth in in vitro environments using an agent-based, structured growth model. Bioprocess Biosyst Eng 37(6):1173–1184
35. Lenk F, Vogel M, Lauterbach T, Stange R, Bley T, Boschke E (2014) Innovative lab-automation solution for culture dish handling and imaging using the petrijet technology platform. Chem Ing Tech 86(9):1375–1375
36. Lenk F (2014) Automatic handling of culture dishes. Food-Lab 14(1):7–9
37. Clark RT, Famoso AN, Zhao K, Shaff JE, Craft EJ, Bustamante CD, Mccouch SR, Aneshansley DJ, Kochian LV (2013) High-throughput two-dimensional root system phenotyping platform facilitates genetic analysis of root growth and development: root phenotyping platform. Plant Cell Environ 36(2):454–466
38. Leitner D, Schnepf A (2012) Image analysis of 2-dimensional root system architecture. In: Proceedings of algoritmy 2012, Vysoké Tatry—Podbanské, vol 1, pp 113–119
39. Lobet G, Pagès L, Draye X (2011) A novel image-analysis toolbox enabling quantitative analysis of root system architecture. Plant Physiol 157(1):29–39
40. Ninomiya K, Oogami Y, Kino-Oka M, Taya M (2003) Assessment of herbicidal toxicity based on non-destructive measurement of local chlorophyll content in photoautotrophic hairy roots. J Biosci Bioeng 95(3):264–270
41. Eibl R, Brandli J, Eibl D (2012) Plant cell bioreactors. In: Doelle HW, Rokem S, Berovic M (eds) Encyclopedia in life support systems, vol 6. EOLSS Publishers, Oxford, pp 33–51
42. Taya M, Hegglin M, Prenosil JE, Bourne JR (1989) On-line monitoring of cell growth in plant tissue cultures by conductometry. Enzyme Microb Technol 11(3):170–176
43. Taya M, Yoyama A, Kondo O, Kobayashi T, Matsui C (1989) Growth characteristics of plant hairy roots and their cultures in bioreactors. J Chem Eng Jpn 22(1):84–89
44. Jeong C-S, Chakrabarty D, Hahn E-J, Lee H-L, Paek K-Y (2006) Effects of oxygen, carbon dioxide and ethylene on growth and bioactive compound production in bioreactor culture of ginseng adventitious roots. Biochem Eng J 27(3):252–263
45. Jeong G-T, Park D-H, Hwang B, Woo J-C (2003) Comparison of growth characteristics of *Panax ginseng* hairy roots in various bioreactors. Appl Biochem Biotechnol 107(1–3): 493–504

46. Jung K-H, Kwak S-S, Liu JR (1998) Procedure for biomass estimation considering the change in biomass volume during high density culture of hairy roots. J Ferment Bioeng 85(4):454–457
47. Huang S-Y, Hung C-H, Chou S-N (2004) Innovative strategies for operation of mist trickling reactors for enhanced hairy root proliferation and secondary metabolite productivity. Enzyme Microb Technol 35(1):22–32
48. Madhusudanan KP, Banerjee S, Khanuja SPS, Chattopadhyay SK (2008) Analysis of hairy root culture of *Rauvolfia serpentina* using direct analysis in real time mass spectrometric technique. Biomed Chromatogr BMC 22(6):596–600
49. Homova V, Weber J, Schulze J, Alipieva K, Bley T, Georgiev M (2010) Devil's claw hairy root culture in flasks and in a 3-L bioreactor: bioactive metabolite accumulation and flow cytometry. Z Naturforsch C J Biosci 65(7–8):472–478
50. Ramakrishnan D, Luyk D, Curtis WR (1999) Monitoring biomass in root culture systems. Biotechnol Bioeng 62(6):711–721
51. Bley T (2010) From single cells to microbial population dynamics: modelling in biotechnology based on measurements of individual cells. In: Müller S, Bley T (eds) High resolution microbial single cell analytics, vol. 124 Springer, Berlin, pp 211–227
52. Maschke RW, Geipel K, Bley T (2015) Modeling of plant in vitro cultures—overview and estimation of biotechnological processes: modeling in plant biotechnology. Biotechnol Bioeng 112(1):1–12
53. Walther T, Reinsch H, Grosse A, Ostermann K, Deutsch A, Bley T (2004) Mathematical modeling of regulatory mechanisms in yeast colony development. J Theor Biol 229(3):327–338
54. Walther T, Reinsch H, Weber P, Ostermann K, Deutsch A, Bley T (2011) Applying dimorphic yeasts as model organisms to study mycelial growth: part 1. Experimental investigation of the spatio-temporal development of filamentous yeast colonies. Bioprocess Biosyst Eng 34(1):13–20
55. Walther T, Reinsch H, Ostermann K, Deutsch A, Bley T (2011) Applying dimorphic yeasts as model organisms to study mycelial growth: part 2. Use of mathematical simulations to identify different construction principles in yeast colonies. Bioprocess Biosyst Eng 34(1):21–31
56. Sriram G, González-Rivera O, Shanks JV (2006) Determination of biomass composition of *Catharanthus roseus* hairy roots for metabolic flux analysis. Biotechnol Prog 22(6): 1659–1663
57. Sriram G, Fulton DB, Shanks JV (2007) Flux quantification in central carbon metabolism of *Catharanthus roseus* hairy roots by ^{13}C labeling and comprehensive bondomer balancing. Phytochemistry 68(16–18):2243–2257
58. Leduc M, Tikhomiroff C, Cloutier M, Perrier M, Jolicoeur M (2006) Development of a kinetic metabolic model: application to *Catharanthus roseus* hairy root. Bioprocess Biosyst Eng 28 (5):295–313
59. Morgan J (2002) Quantification of metabolic flux in plant secondary metabolism by a biogenetic organizational approach. Metab Eng 4(3):257–262
60. Cloutier M, Perrier M, Jolicoeur M (2007) Dynamic flux cartography of hairy roots primary metabolism. Phytochemistry 68(16–18):2393–2404
61. Lan X, Chang K, Zeng L, Liu X, Qiu F, Zheng W, Quan H, Liao Z, Chen M, Huang W, Liu W, Wang Q (2013) Engineering salidroside biosynthetic pathway in hairy root cultures of *Rhodiola crenulata* based on metabolic characterization of tyrosine decarboxylase. PLoS ONE 8(10):e75459
62. Cloutier M, Bouchard-Marchand É, Perrier M, Jolicoeur M (2008) A predictive nutritional model for plant cells and hairy roots. Biotechnol Bioeng 99(1):189–200
63. Chavarría-Krauser A, Schurr U (2004) A cellular growth model for root tips. J Theor Biol 230 (1):21–32
64. Han B, Linden JC, Gujarathi NP, Wickramasinghe SR (2004) Population balance approach to modeling hairy root growth. Biotechnol Prog 20(3):872–879
65. Asplund PT, Curtis WR (2001) Intrinsic oxygen use kinetics of transformed plant root culture. Biotechnol Prog 17(3):481–489

66. Palavalli RR, Srivastava S, Srivastava AK (2012) Development of a mathematical model for growth and oxygen transfer in in vitro plant hairy Root cultivations. Appl Biochem Biotechnol 167:1831–1844
67. Yan Q, Wu J, Liu R (2011) Modeling of tanshinone synthesis and phase distribution under the combined effect of elicitation and in situ adsorption in *Salvia miltiorrhiza* hairy root cultures. Biotechnol Lett 33(4):813–819
68. Jolicoeur M, Bouchard-Marchand E, Bécard G, Perrier M (2002) Regulation of mycorrhizal symbiosis: development of a structured nutritional dual model. Ecol Model 158(1–2):121–142
69. Albiol J, Campmajo C, Casas C, Poch M (1995) Biomass estimation in plant cell cultures: a neural network approach. Biotechnol Prog 11(1):88–92
70. Prasad VSS, Gupta SD (2006) Applications and potentials of artificial neural networks in plant tissue culture. In: Gupta SD, Ibaraki Y (eds) Plant tissue culture engineering, vol 6. Springer, Dordrecht, pp 47–67
71. Osama K, Somvanshi P, Pandey A (2013) Modelling of nutrient mist reactor for hairy root growth using artificial neural network. Eur J Sci Res 97(4):516–526
72. Prakash O, Mehrotra S, Krishna A, Mishra BN (2010) A neural network approach for the prediction of in vitro culture parameters for maximum biomass yields in hairy root cultures. J Theor Biol 265(4):579–585
73. Lindenmayer A (1968) Mathematical models for cellular interactions in development. I. Filaments with one-sided inputs. J Theor Biol 18(3):280–299
74. Lindenmayer A (1968) Mathematical models for cellular interactions in development. II. Simple and branching filaments with two-sided inputs. J Theor Biol 18(3):300–315
75. Prusinkiewicz P, Lindenmayer A (1990) The algorithmic beauty of plants. Springer, New York
76. Böttcher A (2014) Grundlagen der Lindenmayer-Systeme; Einfache pflanzliche Strukturen. Available Online: http://olli.informatik.uni-oldenburg.de/lily/LP/flow1/page8.html. Accessed 14 Aug 2014
77. Dupuy L, Gregory PJ, Bengough AG (2010) Root growth models: towards a new generation of continuous approaches. J Exp Bot 61(8):2131–2143
78. Bastian P, Chavarría-Krauser A, Engwer C, Jäger W, Marnach S, Ptashnyk M (2008) Modelling in vitro growth of dense root networks. J Theor Biol 254(1):99–109
79. Bastian P, Bauer J, Chavarría-Krauser A, Engwer C, Jäger W, Marnach S, Ptashnyk M, Wetterauer B (2008) Modeling and simulation of hairy root growth. In Krebs H-J, Jäger W (eds) Mathematics—key technology for the future. Springer, Berlin, pp 101–115
80. Mairet F, Villon P, Boitel-Conti M, Shakourzadeh K (2010) Modeling and optimization of hairy root growth in fed-batch process. Biotechnol Prog 26(3):847–856
81. Hjortso M (1997) Mathematical modelling of hairy root growth. In: Hairy roots: culture and applications Amsterdam. Harwood Academic Publishers, The Netherlands
82. Kim S, Hopper E, Hjortso M (1995) Hairy root growth models: effect of different branching patterns. Biotechnol Prog 11(2):178–186
83. Lardon LA, Merkey BV, Martins S, Dötsch A, Picioreanu C, Kreft J-U, Smets BF (2011) iDynoMiCS: next-generation individual-based modelling of biofilms. Environ Microbiol 13 (9):2416–2434
84. Grimm V (2005) Pattern-oriented modeling of agent-based complex systems: lessons from ecology. Science 310(5750):987–991
85. Grimm V, Berger U, Bastiansen F, Eliassen S, Ginot V, Giske J, Goss-Custard J, Grand T, Heinz SK, Huse G, Huth A, Jepsen JU, Jørgensen C, Mooij WM, Müller B, Pe'er G, Piou C, Railsback SF, Robbins AM, Robbins MM, Rossmanith E, Rüger N, Strand E, Souissi S, Stillman RA, Vabø R, Visser U, DeAngelis DL (2006) A standard protocol for describing individual-based and agent-based models. Ecol Model 198(1–2):115–126

Adv Biochem Eng Biotechnol (2015) 149: 275–324
DOI 10.1007/10_2014_298

Published Online: 13 January 2015

Using Hairy Roots for Production of Valuable Plant Secondary Metabolites

Li Tian

Abstract Plants synthesize a wide variety of natural products, which are traditionally termed secondary metabolites and, more recently, coined specialized metabolites. While these chemical compounds are employed by plants for interactions with their environment, humans have long since explored and exploited plant secondary metabolites for medicinal and practical uses. Due to the tissue-specific and low-abundance accumulation of these metabolites, alternative means of production in systems other than intact plants are sought after. To this end, hairy root culture presents an excellent platform for producing valuable secondary metabolites. This chapter will focus on several major groups of secondary metabolites that are manufactured by hairy roots established from different plant species. Additionally, the methods for preservations of hairy roots will also be reviewed.

Keywords Hairy roots · Secondary metabolites · Preservation · Terpenoid · Phenolic · Alkaloid

Contents

L. Tian (✉)
Department of Plant Sciences, University of California, Davis, CA 95616, USA
e-mail: ltian@ucdavis.edu

1 Introduction

Roots play many important roles in plants—they anchor plants to the ground, take up minerals and water from the soil, store nutrients for perennial plants, and produce a diverse array of chemicals for symbiotic interactions or defensive with other plants or microbes in the rhizosphere. These plant-produced chemicals have traditionally been referred to as secondary metabolites and more recently tagged as specialized metabolites. Many secondary metabolites not only protect plants from pathogens, insects, and environmental stresses, but also are invaluable for animal and human health. However, plant cultivation is often time-consuming and metabolite extraction from plant roots is destructive to plant growth. To this end, hairy roots induced from different plant tissues generally grow fast, are genetically stable, and often, but not always, simulate the biochemical profiles of plant roots, which makes hairy roots an attractive system for producing valuable secondary metabolites.

Agrobacterium rhizogenes can infect wounded plants and play a fundamental role in hairy root induction. As tallied in 1991, 463 plant species of 109 families had already been transformed by *A. rhizogenes* [129]. To date, hairy roots established from at least 155 plant species of 41 families, by *A. rhizogenes* strains of diverse host ranges and virulence levels, reportedly produce secondary metabolites (Table 1; Fig. 1). However, a systemic comparative analysis of *A. rhizogenes* strains on effective induction of hairy root growth and secondary metabolite yield has not been performed. It should be noted that, though most of the hairy root cultures resulted from *A. rhizogenes* transformation, high-quality hairy roots were also obtained for *Atropa belladonna*, *Catharanthus roseus*, *Kalanchoe diagremontiana*, and *Nicotiana tabacum* plants infected by *A. tumefaciens* harboring *rolABC* genes [18, 60, 158].

There have been many excellent reviews in the literature on a wide variety of hairy root applications, such as metabolite or therapeutic protein production, biotransformation of core skeletons of secondary metabolites into novel compounds, gene discovery and metabolic pathway characterization, and phytoremediation [13, 20, 39, 45, 47, 54, 52, 53, 124, 150] (and many other reviews). This chapter focuses specifically on production of different classes of secondary metabolites in hairy roots, their bioactivities, and preservation of hairy roots. Modeling and monitoring of hairy root production of valuable compounds will be reviewed and discussed in Chap. "Ramified challenges: Monitoring and Modeling of Hairy Root Growth in Bioprocesses – A review" of this book.

Table 1 Secondary metabolites produced in hairy root cultures

Species	Common names	Agrobacterium strains	Major secondary metabolites produced	Biological activities	References
Amaranthaceae					
Beta vulgaris L.	Red beet	A4	Betalains	Antioxidant, antimicrobial, antifungal, anticancer, anti-inflammatory, neuroprotective, radioprotective	[169]
Apiaceae (*Umbelliferae*)					
Ammi majus L.	Bishops weed	A4, LBA 9402	Coumarins, furanocoumarins	Anti-inflammatory, treatment of leucoderma, psoriasis, and vitiligo	[88]
Ammi visnaga Lam.	Toothpick weed	A4, ATCC 15834, R1601	Furanochromone (visnagin)	Vasodilation, diuretic	[89]
Anethum graveolens L.	Dill	LBA 9402	Essential oils (apiole, falcarinol, dill apiole)	Antiflatulent, antiamenorrhea	[140]
Angelica gigas Nakai	Korean angelica	R1000	Pyranocoumarins (decursin, decursinol angelate)	Neuroprotective, anticancer, antiandrogen-receptor signaling	[194]
Bupleurum falcatum L.	Thorow-wax	A4	Triterpene saponins (saikosaponins)	Anti-inflammatory, antipyretic, antitussive	[2]
Centella asiatica (L.) Urban	Asiatic pennywort	R1000	Triterpenoids (asiaticoside)	Anti-inflammatory, antipyretic	[75]
Daucus carota L.	Carrot	LBA 9402	Phenolic acid (*p*-hydroxybenzoic acid)	Chemical intermediates of synthetic drugs	[156]

(continued)

Table 1 (continued)

Species	Common names	Agrobacterium strains	Major secondary metabolites produced	Biological activities	References
Levisticum officinale W.D.J. Koch	Lovage	A4	Essential oils (falcarinol, (Z)-ligustilide, (Z)-3-butylidenephthalide, *trans*-β-farnesene, β-phellandrene, n-octanal, γ-elemene, n-heptanal)	Spasmolytic, diuretic, carminative	[141]
Perezia cuernavacana B.L.Rob. and Greenm.		AR12	Sesquiterpene quinones (perezone)	Natural pigment, laxative, antimicrobial	[5]
Pimpinella anisum L.	Anise	A4	Essential oils (*trans*-epoxypseudoisoeugenyl 2-methylbutyrate, geijerene, pregeijerene, zingiberene, β-bisabolene)	Flavoring and scent agents	[142]
Apocynaceae					
Catharanthus roseous (L.) G. Don	Madagascar periwinkle	ATCC 15834	Indole alkaloids	Anticancer	[149]
Catharanthus trichophyllus (Baker) Pichon		ATCC 15834	Indole alkaloids	Anticancer	[30]
Rauvolfia micrantha Hook. f.	Small-flowered snakeroot	ATCC 15834	Indole alkaloids (ajmalicine, ajmaline)	Treatment of nervous system disorders	[160]
Rauvolfia serpentina (L.) Benth. ex Kurz	Indian snakeroot	ATCC 15834	Indole alkaloids (ajmaline, serpentine)	Treatment of nervous system disorders	[14]
Vinca minor L.	Lesser periwinkle	DC-AR2 (derived from MAFF301724)	Indole alkaloids (vincamine)	Cerebral vasodilator	[167]

(continued)

Table 1 (continued)

Species	Common names	Agrobacterium strains	Major secondary metabolites produced	Biological activities	References
Araliaceae					
Panax ginseng C.A. Meyer	Asian ginseng	A4	Triterpene saponins (ginsenosides)	Immunomodulatory, adaptogenic, antiaging	[197]
Panax quinquefolium L.	American ginseng	A4, LBA 9402, ATCC 11325, ATCC 15834	Triterpene saponins (ginsenosides)	Immunomodulatory, adaptogenic, antiaging	[105]
Panax hybrid (*P. ginseng* X *P. quinqifolium*)		ATCC 15834	Triterpene saponins (ginsenosides)	Immunomodulatory, adaptogenic, antiaging	[188]
Asclepiadaceae					
Gymnema sylvestre R.Br.	Gur-mar	KCTC 2703	Triterpene saponins (gymnemic acid)	Antidiabetic, antimicrobial, diuretic, stomachic, anti-hypercholesterolemic, hepatoprotective, antisaccharine	[116]
Asteraceae (Compositae)					
Ambrosia artemisiifolia L.	Common ragweed	ATCC 15834, ATCC 11325, TR105	Polyacetylenes, thiophenes	Cytotoxic, antiviral, biocidal	[40]
Artemisia absinthium L.	Wormwood	AR1855, LBA 9402	Essential oils	Antihelminthic, insecticidal, stomachic	[118]
Artemisia annua L.	Sweet wormwood	ATCC 15834	Sesquiterpenes (artemisinin)	Antimalarial	[189]
Bidens sps. [B. sulphurea (Cav.) Sch. Bip, B. pilosa L., B. ferulaefolia (Jacq.) DC]		ATCC 15834, ATCC 11325, TR105	Polyacetylenes, thiophenes	Cytotoxic, antiviral, biocidal	[40]

(continued)

Table 1 (continued)

Species	Common names	Agrobacterium strains	Major secondary metabolites produced	Biological activities	References
Carthamus tinctorius L. (var. Biggs, var. N-10)	Safflower	ATCC 15834, ATCC 11325, TR105	Polyacetylenes, thiophenes	Cytotoxic, antiviral, biocidal	[40]
Chaenactis douglasii (Hook.) Hook. and Arn.	Douglas' dustymaiden	TR7	Polyines (thiarubrines)	Antifungal	[29]
Cichorium intybus L.	Common chicory	LBA 9402	Sesquiterpene lactones	Anti-inflammatory, hepatoprotective	[102]
Coreopsis tinctoria Nutt.	Golden tickseed	ATCC 15834	Phenolics	Antidiabetic	Thron et al. [172]
Echinacea sps. [*E. angustifolia* DC; *E. pallida* (Nutt.) Nutt.; *E. purpurea* (L.) Moench]	Purple coneflower	A4	Alkamides	Anti-inflammatory, immune-stimulatory	[133]
Lactuca virosa L.	Wild lettuce	LBA 9402	Sesquiterpene lactones	Analgesic, antitussive, sedative	[82]
Leontopodium alpinum Cass.	Edelweiss	LBA 9402	Anthocyanins, hydroxycinnamic acid esters, essential oils	Chemoprotective against toxins	[61]
Rudbeckia hirta L. var. *plucherrima*	Black-eyed susan	ATCC 15834, ATCC 11325, TR105	Polyacetylenes, thiophenes	Cytotoxic, antiviral, biocidal	[40]
Saussurea involucrate Karel. and Kir.	Snow lotus	R1601, R1000, LBA 9402	Syringin	Anti-inflammatory, anti-hypersensitive	[42]

(continued)

Table 1 (continued)

Species	Common names	Agrobacterium strains	Major secondary metabolites produced	Biological activities	References
Saussurea medusa Maxim.	Snow lotus	R1601	Flavones (jaceosidin)	Antitumor, anti-inflammatory	[202]
Serratula tinctoria L.	Saw-wort	A4	Triterpenoids (phytoecdysteroids)	Potential use in chemotherapy	[31]
Silybum marianum (L.) Gaertn.	Milk thistle	ATCC 15834	Flavolignans (isosilybin, silychristin, silydianin)	Antihepatotoxic	[3]
Solidago altissima L.	Late goldenrod	A4	Polyacetylenes (*cis*-dehydromatricaria ester)	Cytotoxic, antiviral, biocidal	[65]
Solidago nemoralis Aiton	Gray goldenrod	R1000	Complex alkaloid metabolites	Targeting the insect and human nicotinic receptor for acetylcholine	[55]
Tanacetum parthenium (L.) Sch. Bip. [syn. *Chrysanthemum parthenium* (L.) Bernh.]	Feverfew	LBA 9402	Coumarins (isofraxidin, isofraxidin drimenyl ether)	Antitumor	[81]
Tagetes erecta L.	Mexican marigold	ATCC 15834, ATCC 11325, TR105	Polyacetylenes, thiophenes	Cytotoxic, antiviral, biocidal	[40]
Tagetes patula L.	French marigold	ATCC 43057	Thiophenes (α-terthienyl)	Cytotoxic, antiviral, biocidal	[92]
Bignoniaceae					
Paulownia tomentosa Steud.	Empress tree	LBA 9402	Phenolic glycosides (verbascoside)	Antiviral, antibacterial, antiproliferative, analgesic, anti-hypertensive	[193]

(continued)

Table 1 (continued)

Species	Common names	Agrobacterium strains	Major secondary metabolites produced	Biological activities	References
Bixaceae					
Bixa orellana L.	Achiote	Not described	Bixin, tocopherol, ellagic acid, etc.	Antimalarial	[200]
Boraginaceae					
Echium acanthocarpum Svent.	Gomeranneidonkieli	LBA 1334	Polyunsaturated fatty acids	Nutrients, membrane structural components	[25]
Lithospermum erythrorhizon Sieb. et Zucc.	Purple gromwell	ATCC 15834	Naphthoquinones (shikonin)	Anti-inflammatory, antitumor, antiviral	[152]
Brassicaceae					
Arabis caucasica L.	Mountain rock cress	LBA 9402	Glucosinolates (gluconasturtiin, glucoiberverin)	Anticancer, antifungal, antibacterial, antinematode, anti-insect	[190]
Armoracia rusticana P. Gaert., B. Meyer et Scherb.	Horseradish	Not specified	Diterpenoids (fucicoccin)	Unknown	[8]
Barbarea verna (Mill.) Asch.	Land cress	LBA 9402	Glucosinolates (gluconasturtiin, glucoiberverin)	Anticancer, antifungal, antibacterial, antinematode, anti-insect	[190]
Crambe abyssinica R.E. Fr.	Abyssinian mustard	A4, ATCC 15834	Lipids, fatty acids	Nutrients	[56]
Nasturtium officinale (L.) R.Br.	Watercress	LBA 9402	Glucosinolates (gluconasturtiin, glucotropaeolin)	Anticancer, antifungal, antibacterial, antinematode, anti-insect	[190]

(continued)

Table 1 (continued)

Species	Common names	Agrobacterium strains	Major secondary metabolites produced	Biological activities	References
Tropaeolum majus L.	Garden nasturtium	LBA 9402	Glucosinolates (glucotropaeolin)	Anticancer, antifungal, antibacterial, antinematode, anti-insect	[191]
Campanulaceae					
Campanula glomerata L. var. *dahurica* Fish.	Bellflower	MAFF 03-01724	Polyacetylenes (lobetyol, lobetyolin, lobetyolinin) Phenolics (anthocyanins, rutin, coniferin)	Cytotoxic, antiviral, biocidal	[166]
Campanula medium L.	Canterbury-bells	A13	Polyacetylenes	Cytotoxic, antiviral, biocidal	[163]
Lobelia cardinalis L.	Cardinal flower	ATCC 15834, MAFF 03-01724	Polyacetylene glucosides (lobetyolin, lobetyolinin)	Cytotoxic, antiviral, biocidal	[195]
Lobelia chinensis Lour.	Chinese lobelia	ATCC 15834	Polyacetylenes	Cytotoxic, antiviral, biocidal	[164]
Lobelia inflata L.	Indian tobacco	ATCC 15834	Piperidine alkaloids (lobeline)	Respiratory stimulant	[196]
Platycodon grandiflorum (Jacq.) A.DC.	Broad bell	ATCC 15834	Polyacetylenes	Cytotoxic, antiviral, biocidal	[71]
Trachelium caeruleum L.	Blue throatwort	ATCC 15834	Polyacetylenes (lobetyol, lobetyolin, and lobetyolinin)	Cytotoxic, antiviral, biocidal	[114]
Caprifoliaceae					
Centranthus ruber (L.) DC.	Red valerian	R1601	Iridoids (valepotriate)	Sedative, spasmolytic	[50]

(continued)

Table 1 (continued)

Species	Common names	Agrobacterium strains	Major secondary metabolites produced	Biological activities	References
Cannabaceae					
Cannabis sativa L.	Hemp	Not specified	Choline, atropine	Parasympatolytic, anticholinergic, spasmolytic, antiemetic, precursor of membrane lipids and signaling molecules	[184]
Convolvulaceae					
Calystegia sepium (L.) R.Br.	Hedge bindweed	A4, 8196	Tropane alkaloids	Anticholinergic	[72]
Cornaceae (*Nyssaceae*)					
Camptotheca acuminate Decne.	Happy tree	ATCC 15834, R1000	Quinoline alkaloids (camptothecin)	Antitumor, anticancer, antiviral	[99]
Cucurbitaceae					
Gynostemma pentaphyllum (Thunb.) Makino	Southern ginseng; Jiaogulan	ATCC 15834	Triterpene saponins (gypenosides)	Antitumor, cholesterol lowering, immunopotentiating, antioxidant, hypoglycemic, antidiabetic	[26]
Trichosanthes kirilowii Maxim var *japonicum*	Chinese cucumber	R1601	Triterpenoids (bryonolic acid, chondrillasterol)	Growth inhibition of B-16 melanoma cells	[165]
Euphorbiaceae					
Phyllanthus amarus Schum. and Thonn.	Carry me seed	ATCC 15834	Crude hairy extracts were used for bioactivity assays and were not analyzed for chemical constituents	Anticancer, antiviral	[1, 17]

(continued)

Table 1 (continued)

Species	Common names	Agrobacterium strains	Major secondary metabolites produced	Biological activities	References
Fabaceae (Leguminosae)					
Abrus precatorius L.	Indian licorice	MTCC 532, MTCCC 2364, NCIM 5140	Triterpene saponins (glycyrrhizin)	Emetic, diuretic, anthelmintic, alexeteric	[73]
Arachis hypogaea L.	Peanut	ATCC 15834, R1000, EHA105	Resveratrol, pterostilbene, and derivatives	Cardioprotective, cancer chemopreventive, antiatherosclerosis, antiaging	[108]
Astragalus mongholicus Bunge	Huangqi	LBA 9402, ATCC 15834, R1601, TR105	Triterpene saponins	Immunostimulant	[66]
Cassia sps. (*C. obtusifolia* L.; C. *occidentalis* L.; C. *torosa* Cav.)	Chinese senna	ATCC 15834, A4	Anthraquinones, polypeptide pigments	Antioxidant, laxative, treatment of fungal skin diseases	[87]
Clitoria ternatea L.	Butterfly pea	A4T, 8196	Triterpenoids (taraxerol)	Laxative, diuretic, anthelmintic, anti-inflammatory	[161]
Glycyrrhiza glabra L.	Licorice	ATCC 15834	Flavonoids, isoprenylated flavonoids	Antioxidant, antimicrobial	[6]
Glycyrrhiza uralensis Fisch.	Chinese licorice	A4	Flavonoids	Antimutagenic, antiulcer, antitumor, antimicrobial protective action against hepatotoxicity	[201]
Lotus corniculatus L.	Bird's-foot trefoil	LBA 9402	Condensed tannins (i.e., proanthocyanidins)	Anthelmintic	[23]

(continued)

Table 1 (continued)

Species	Common names	Agrobacterium strains	Major secondary metabolites produced	Biological activities	References
Lupinus mutabilis Sweet	Andean lupin	R1601	Isoflavones (genistein, 2′-hydroxygenistein, wighteone)	Antioxidant, antiproliferative	[9]
Psoralea corylifolia Linn.	Indian bread root	LBA 9402	Isoflavones (daidzein, genistein)	Anticancer, anti-inflammatory, antiallergic, antioxidant, antiangiogenic, inhibition of DNA polymerase, topoisomerase II activities	[153]
Psoralea sps. (*P. canescens* Michx.; *P. cinerea* Lindl.; *P. lachnostachys* F. Muell.; *P. leucantha* F. Muell.; *P. plumosa* F. Muell; P. *pustulata* F. Muell.; *P. obtusifolia* D.C.)		LBA 9402	Phenolics (daidzein, genistein, coumestrol)	Anticancer, anti-inflammatory, antiallergic, antioxidant, antiangiogenic, inhibition of DNA polymerase, topoisomerase II activities	[19]
Pueraria candollei Wall. ex Benth.	Thai kudzu	ATCC 15834	Isoflavonoids (puerarin, daidzin, genistin, daidzein, genistein)	Anticancer, anti-inflammatory, antiallergic, antioxidant, antiangiogenic, inhibition of DNA polymerase, topoisomerase II activities	[177]
Pueraria lobata (Willd.) Ohwi	Kudzu	KCTC 2744	Isoflavones (puerarin, daidzin)	Hypothermic, spasmolytic, hypotensive, antiarrhythmic	[76]
Pueraria phaseoloides (Roxb.) Benth	Tropical kudzu	ATCC 15834	Isoflavones (puerarin)	Hypothermic, spasmolytic, hypotensive, antiarrhythmic	[151]

(continued)

Table 1 (continued)

Species	Common names	Agrobacterium strains	Major secondary metabolites produced	Biological activities	References
Swainsona galegifolia (Andrews) R.Br.	Darling pea	LBA 9402	Indolizidine alkaloid (swainsonine)	α-mannosidase inhibitor; antimetastatic, antiproliferative, immunomodulatory	[113]
Trigonella foenum-graecum L.	Fenugreek	A4	Steroidal sapogenin (diosgenin)	Precursor for synthesis of steroidal hormones	[110]
Gentianaceae					
Gentiana macrophylla Pall.	Large leaf gentian	R1000	Secoiridoid glycosides (gentiopicroside)	Anti-inflammatory, analgesic, antirheumatic, antipyretic, diuretic, hypoglycemic	[173]
Gentiana punctate L.	Spotted gentian	A4M70GUS	Secoiridoid glycosides (gentiopicrin)	Anti-inflammatory, analgesic, antirheumatic, antipyretic, diuretic, hypoglycemic	[109]
Gentiana scabra Bunge	Scabrous gentian	ATCC 15834	Iridoids and secoiridoids (loganic acid, gentiopicrosides, sertiamarin)	Anti-inflammatory, analgesic, antirheumatic, antipyretic, diuretic, hypoglycemic	[63]
Swertia japonica (Roem. and Schult.) Makino	Semburi	ATCC 15834	Xanthones (bellidifolin, methylbellidifolin, swertianolin, 8-*O*-primeverosylbellidifolin), iridoidal glycosides (amarogentin, amaroswerin)	Bitter stomachic, treatment of hepatitis and digestive disorders	[68]

(continued)

Table 1 (continued)

Species	Common names	Agrobacterium strains	Major secondary metabolites produced	Biological activities	References
Geraniaceae					
Geranium thunbergii Siebold ex Lindl. and Paxt.	Cranesbill	A4	Hydrolyzable tannins	Antiviral, antimicrobial, antitumor	[67]
Ginkgoaceae					
Gingko biloba L.	Ginkgo	A4	Diterpenes (ginkgolide)	Treatment of aging-related disorders	[7]
Hypericaceae					
Hypericum perforatum L.	St. John's wort	A4	Flavonoids, xanthones	Antioxidant, antimicrobial, cytotoxic, hepatoprotective	[176]
Lamiaceae (Labiatae)					
Agastache rugosa Kuntze.	Korean mint	R1000	Phenolics (rosmarinic acid)	Antitumor, antimicrobial, anti-inflammatory, antiviral, antimutagenic	[95]
Ajuga reptans L. var. *atropurpurea*	Bugle	MAFF 03-01724	Phytoecdysteroids (20-hydroxyecdysone, norcyasterone B, cyasterone, isocyasterone)	Potential use in chemotherapy	[106]
Coleus sps. (*C. blumei* Benth.; *C. forskohlii* Briq.)		MAFF 03-01724	Diterpenoids (forskolin)	Positive inotropic, positive chronotropic, hypotensive, inhibition of thrombocyte aggregation, reduction of intraocular pressure	[143]
Gmelina arborea Roxb. (formerly Verbenaceae)	Beechwood	ATCC 15834	Phenylpropanoid glycoside (verbascoside)	Anti-inflammatory, wound healing, inhibition of platelet aggregation	[33]

(continued)

Table 1 (continued)

Species	Common names	Agrobacterium strains	Major secondary metabolites produced	Biological activities	References
Ocimum basilicum L.	Sweet basil	ATCC 15834	Phenolics (rosmarinic acid)	Antioxidant	[10]
Salvia austriaca Jacq.	Austrian sage	A4	Diterpenoids (royleanone, 15-deoxyfuerstione, taxodione)	Antioxidant, anti-inflammatory	[90]
Salvia broussonetii Benth	Stiff Canary Island sage	ATCC 15834	Diterpenoids (brussonol, iguestol, 7-oxodehydroabietane, 11-hydroxy-12-methoxyabietatriene, taxodione, inuroyleanol, ferruginol, deoxocarnosol 12-methyl ether, cryptojaponol, pisiferal, sugiol, isomanool, 14-deoxycoleon U, 6R-hydroxydemethylcryptojaponol, demethylsalvicanol, demethylcryptojaponol)	Antioxidant, anti-inflammatory	[41]
Salvia cinnabarina M.Martens and Galeotti	Cinnabar sage	ATCC 15834	Diterpenoids	Antimutagenic	[148]
Salvia involucrata Cav.	Roseleaf sage	R1601	Flavonoids (apigenin)	Anticancer	[96]
Salvia miltiorrhiza Bunge	Red sage or Chinese sage	LBA 9402, ATCC 15834, TR 105, R 1601, A 4 1027	Diterpenoid naphthoquinones (tanshinone), phenolic acids	Antioxidant, promotion of blood circulation	[204]
Salvia officinalis L.	Garden sage or Common sage	ATCC 15834, A4	Phenolics (rosmarinic acid)	Neuroprotective, antioxidant, anti-inflammatory, immunomodulatory, photoprotection, melanogenic	[48]

(continued)

Table 1 (continued)

Species	Common names	Agrobacterium strains	Major secondary metabolites produced	Biological activities	References
Salvia sclarea L.	Clary sage	LBA 9402	Diterpenes	Antioxidant, antibacterial, anti-inflammatory	[91]
Scutellaria baicalensis Georgi.	Baikal skullcap	ATCC 15834	Flavonoids, phenylethanoids (baicalin, wogonin)	Antiviral, antitumor	[205]
Linaceae					
Linum flavum L.	Golden flax	LBA 9402, TR105	Lignans	Anticancer	[97]
Lythraceae					
Lawsonia inermis L.	Henna	NCIB 8196	Naphthoquinone (lawsone)	Coloring agent, UV sunscreen	[11]
Punica granatum L. (formerly Punicaceae)	Pomegranate	MSU440	Hydrolyzable tannins, other phenolics	Anticancer	[123]
Malvaceae					
Gossypium sps. (*G. barbadense* L.; *G. hirsutum* L.)	Cotton	ATCC 15834	Disesquiterpene (gossypol)	Anticancer, antiviral, antiparasitic, antifungal, antimicrobial	[175]
Meliaceae					
Azadirachta indica A. Juss.	Neem	LBA 920	Triterpenes (azadirachtin)	Biopesticide	[159]
Papaveraceae					
Papaver bracteatum Lindl.	Persian poppy	ATCC 15834	Benzylisoquinoline alkaloids	Antitumor, antimicrobial, sedative	[134]
Papaver somniferum L. var. *album*	Opium poppy	ATCC 15834, LBA 9402	Benzylisoquinoline alkaloids	Antitumor, antimicrobial, sedative	[38]

(continued)

Table 1 (continued)

Species	Common names	Agrobacterium strains	Major secondary metabolites produced	Biological activities	References
Pedaliaceae					
Harpagophytum Procumbens (Burch.) DC. ex Meisn.	Devil's claw	ATCC 15834, A4	Iridoid glycosides	Treatment of degenerative rheumatoid arthritis, osteoarthritis, tendonitis, kidney inflammation and heart disease	[51]
Sesamum indicum L.	Sesame	ATCC 15834	Naphthoquinone (2-isopropenylnaphthazarin-2,3-epoxide), anthraquinones [2(4-methyl-1,3pentadienyl) anthraquinone and 2(4-methyl-3pentenyl) anthraquinone]	Antimicrobial	[121]
Plantaginaceae					
Bacopa monnieri (L.) Pennell	Indian pennywort	A4, LBA9402	Triterpene saponins	Anti-inflammatory, analgesic, antipyretic, anticancer	[101]
Digitalis lanata Ehrh.	Woolly foxglove	A4, ATCC 15834, TR105	Anthraquinones, flavonoids	Positive inotropic	[130]
Digitalis purpurea L.	Foxglove	ATCC 15834	Cardiac glycosides	Treatment for heart failure	[137]
Picrorhiza kurroa Royle ex Benth.	Kutki	LBA 9402, A4	Iridoid glycosides (kutkoside, picroside)	Hepatoprotective, anticholestatic, antiulcerogenic, antiasthematic, antidiabetic, anti-inflammatory, immunoregulatory	[182]

(continued)

Table 1 (continued)

Species	Common names	Agrobacterium strains	Major secondary metabolites produced	Biological activities	References
Plumbaginaceae					
Plumbago indica L.	Scarlet leadwort	ATCC 15834	Naphthoquinones (plumbagin)	Antimicrobial, abortifacient, vesicant diuretic	[44]
Plumbago zeylanica L.	Ceylon leadwort	A4	Naphthoquinones (plumbagin)	Antimicrobial, abortifacient, vesicant diuretic	[181]
Polygonaceae					
Fagopyrum esculentum Moench.	Buckwheat	ATCC 15834	Flavonoids (rutin)	Antioxidant, anti-inflammatory, anticancer, antithrombotic, cytoprotective, vasoprotective	[79]
Fagopyrum tataricum (L.) Gaertn	Tartary buckwheat	Ri1601	Flavonoids (Rutin, quercetin)	Antioxidant, hypocholesterolemic, antidiabetic, antimicrobial, antitumor	[203]
Polygonum multiflorum Thunb.	Chinese knotweed	KCTC2703	Hydroxybenzoic acids, hydroxycinnamic acids, flavonols	Antitumor, anti-HIV, antibacterial, hemostatic, spasmolytic, analgesics, antioxidant	[171]
Polygonum tinctorium Lour.	Chinese indigo	A4	Indigo	Treatment of leukemia	[199]

(continued)

Table 1 (continued)

Species	Common names	Agrobacterium strains	Major secondary metabolites produced	Biological activities	References
Ranunculaceae					
Aconitum heterophyllum Wall.	Atis	LBA 9402, LBA 9360, A4	Diterpene alkaloids (heteratisine, atisine, hetidine)	Bitter tonic; treatment of diarrhea, dysentery, cough, dyspepsia and chronic enteritis	[46]
Rosaceae					
Fragaria x *ananassa* cv. Reikou	Strawberry	ATCC 15834	Polyphenols (proanthocyanidins, flavonoids, hydrolyzable tannin)	Antioxidant, anticancer	[112]
Rubiaceae					
Cinchona ledgeriana Moens ex Trimen	Quinine bark	LBA 9402	Quinoline and indole alkaloids	Antiarrhythmic	[57]
Ophiorrhiza pumila L.	Dwarf Ophiorrhiza	ATCC 15834	Quinoline alkaloids (camptothecin)	Antitumor, antineoplastic	[136]
Rubia akane Nakai	Asian madder	R1000	Anthraquinones (alizarin, purpurin)	Natural dyes; anticancer, antimalarial, antimicrobial, antifungal, antioxidant	[126]
Rubia peregrina L.	Wild madder	LBA 9402	Anthraquinones (alizarin)	Natural dyes; anticancer, antimalarial, antimicrobial, antifungal, antioxidant	[98]
Rubia tinctorum L.	Common madder	ATCC 15834	Anthraquinones	Natural dyes; anticancer, antimalarial, antimicrobial, antifungal, antioxidant	[144]

(continued)

Table 1 (continued)

Species	Common names	Agrobacterium strains	Major secondary metabolites produced	Biological activities	References
Rutaceae					
Fagara zanthoxyloides Lam. [syn. Zanthoxylum zanthoxyloides (Lam.) Zepernick and Timle]	Senegal prickly-ash	LBA 9402	Benzophenanthridine alkaloids	Antileukemic, inhibitor of DNA topoisomerases I and II, and protein kinase C, sedative and hypothermic	[37]
Ruta graveolens L.	Common rue	LBA 9402, A4	Coumarins, furanocoumarins, alkaloids	Antimicrobial, treatment of skin diseases	[155]
Solanaceae					
Anisodus luridus Link and Otto	Himalayan scopolia	C58C1	Tropane alkaloids	Anticholinergic	[132]
Atropa belladonna L.	Deadly nightshade	ATCC 15834	Tropane alkaloids	Anticholinergic	[18]
Atropa caucasica Kreyer	Caucasian belladonna	A4	Tropane alkaloids	Anticholinergic	[86]
Brugmansia candida Pers.	Angel's trumpet	LBA 9402	Tropane alkaloids (scopolamine, hyoscyamine) polyamines (cadaverine, putrescine, spermidine, spermine)	Anticholinergic	[22]
Datura candida (Pers.) Safford	Hybrid variety between two cultivars of the same species	ATCC 15834	Tropane alkaloids	Anticholinergic	[28]
Datura candida x Datura aurea	Hybrid variety	Not described	Tropane alkaloids (hyoscyamine, scopolamine)	Anticholinergic	[120]
Datura stramonium L.	Jimson weed	TR105	Tropane alkaloids (hyoscyamine)	Anticholinergic	[162]

(continued)

Table 1 (continued)

Species	Common names	Agrobacterium strains	Major secondary metabolites produced	Biological activities	References
Datura sps. (*D. chlorantha* Hook; *D. fastuosa* L. var. *violacea*; *D. ferox* L.; *D. innoxia* Mill.; *D. metel* L.; *D. meteloides* DC. ex Dun.; *D. quercifolia* H.B.K.; *D. rosei* Saff.; *D. sanguinea* Ruiz. et Pav.; *D. stramonium* L.; *D. stramonium* var. *inermis* Timm.; *D. stramonium* var. *stramonium* Dun.; *D. stramonium* var. *tatula* Torr. *Duboisia*-hybrid)		A4	Tropane alkaloids	Anticholinergic	[86]
Duboisia leichhardtii F. Muell.	Corkwood	ATCC 15834, A4	Tropane alkaloids (scopolamine, hyoscyamine)	Anticholinergic	[104]
Duboisia myoporoides R.Br.	Corkwood	HRI	Tropane alkaloids	Anticholinergic	[32]
Hyoscyamus albus L.	White henbane	MAFF 03-01724	Tropane alkaloids (7β-hydroxyhyoscyamine, littorine, scopolamine, 6β-hydroxyhyoscyamine, hyoscyamine)	Anticholinergic	[146]

(continued)

Table 1 (continued)

Species	Common names	Agrobacterium strains	Major secondary metabolites produced	Biological activities	References
Hyoscyamus niger L.	Black henbane	ATCC 15834	Tropane alkaloids	Anticholinergic	[69]
Hyoscyamus sps. (*H. albus* L.; *H.aureus* L.; *H. bohemicus* F.W. Schmidt; *H. muticus* L.; H. niger L.; *H. niger* var. *pallidus* Rchb.)		A4	Tropane alkaloids	Anticholinergic	[86]
Nicotiana sps. (*N. tabacum* L.; N. *rustica* L.; *N. occidentalis* subsp. *hesperis* (N.T.Burb.) P.Horton; *N. Africana* Merxm; *N. umbratica* N.T. Burb; *N. velutina* H.-M. Wheeler; *N. cavicola* N.T. Burb.)		LBA 9402	Pyridine alkaloids (nornicotine, nicotine, anatabine, anabasine)	Insecticidal	[127]
Scopolia sps. (*S. carniolica* Jacq.; *S. straminifolia* Shrestha)		A4	Tropane alkaloids	Anticholinergic	[86]
Scopolia japonica Maxim.	Japanese belladonna	ATCC 15834	Tropane alkaloids (scopolamine, hyoscyamine)	Anticholinergic	[103]
Solanum aculeatissimum Jacq.	Indian nightshade	ATCC 15834	Steroidal saponins (aculeatiside A, aculeatiside B)	Hemolytic	[64]

(continued)

Table 1 (continued)

Species	Common names	Agrobacterium strains	Major secondary metabolites produced	Biological activities	References
Solanum aviculare Forst.	Poroporo	A4, ATCC 11325, ATCC 15834, ATCC 43057	Steroidal glycoalkaloids (solasodine)	Precursor of steroidal drugs and hormones	[84]
Solanum chrysotrichum Schltdl.	Devil's fig	Not described	Triterpene saponins	Antifungal	[24]
Solanum laciniatum Aiton	Kangaroo apple	ATCC 15834	Steroidal glycoalkaloids (solasodine)	Precursor of steroidal drugs and hormones	[122]
Withania somnifera (L.) Dunal	Indian ginseng	R1601	Steroidal lactones (withanolide A)	Anticancer	[115]
Taxaceae					
Taxus x media var. Hicksii Rehd.	Hicks yew	ATCC 15834, TR 105, LBA 9402	Diterpenoid taxanes (paclitaxel, 10-deacetybaccatin III)	Anticancer	[43]
Valerianaceae					
Valeriana officinalis L. var. sambucifolia Mikan	Valerian	R1601	Iridoids (valepotriates)	Sedative, spasmolytic	[49]
Valeriana wallichii DC	Indian valerian	A4, LBA 9402	Iridoids (valepotriates)	Sedative, spasmolytic	[12]

(continued)

Table 1 (continued)

Species	Common names	Agrobacterium strains	Major secondary metabolites produced	Biological activities	References
Verbenaceae					
Lippia dulcis Trevir. [*Phyla dulcis* (Trevir.) Moldenke]	Sweet Aztec herb	A4	Sesquiterpene (hernandulcin)	Natural sweetener	[147]
Vinceae					
Amsonia elliptica (Thunb. ex Murray) Roem. and Schult.	Japanese bluestar	A4	Indole alkaloids	Anticancer	[145]

When different varieties of the same species were reported by multiple groups or with different stimulation methods, only a representative report is listed in the references. Common names of the plant species are indicated when available

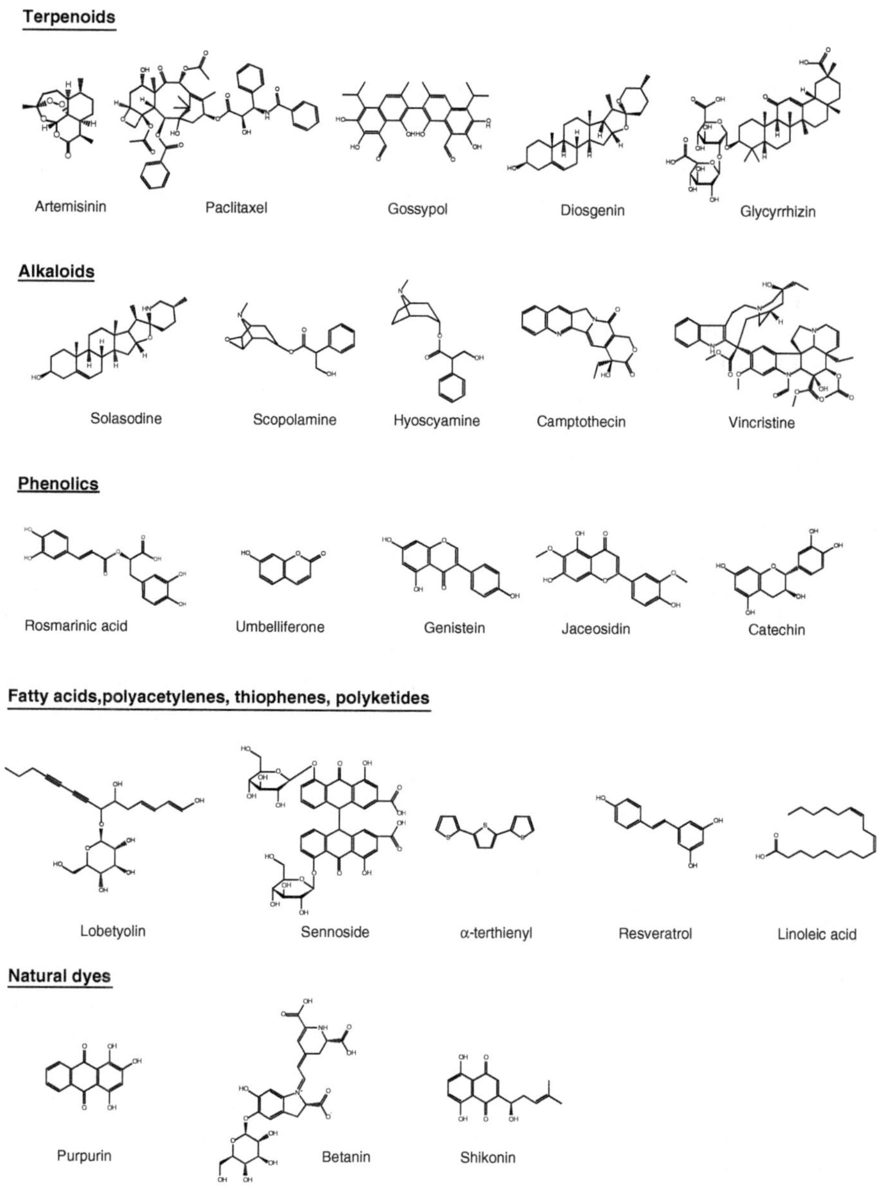

Fig. 1 Chemical structures of selected plant secondary metabolites produced in hairy root cultures

2 Production of Valuable Secondary Metabolites

2.1 Terpenoid Production in Hairy Roots

Terpenoids (also known as isoprenoids) are synthesized from the dimethylallyl diphosphate (DMAPP) and isopentenyl diphosphate (IPP) precursors, and constitute the largest group of secondary metabolites in plants. Based on the number of C_5 isoprene units in their structures, terpenoids can be divided into hemiterpenes (C_5), monoterpenes (C_{10}), sesquiterpenes (C_{15}), diterpenes (C_{20}), sesterterpenes (C_{25}), triterpenes (C_{30}), sesquarterpenes (C_{35}), and tetraterpenes (C_{40}). Polymers of isoprenes are also found in plants, such as natural rubber. In addition, terpenoids can conjugate with molecular structures derived from other biochemical pathways to form additional groups of secondary metabolites; for instance, isoprene units are transferred to phenolic compounds to obtain prenylated phenolics.

Iridoids are monoterpenes that arise from geraniol through cyclization reactions; cleavage of iridoids leads to the formation of secoiridoids. Iridoids and secoiridoids often accumulate in plants in the glycosylated forms. The root extracts of the Gentianaceae family plants have been used for treating many diseases and disorders due to the bioactivities (e.g., hypoglycemic, anti-inflammatory) of secoiridoid glycosides and the active principles of the root tissue [168]. Explants of several Gentiana species, including *Gentiana macrophylla*, *Gentiana punctate*, *Gentiana scabra*, were transformed with *A. rhizogenes,* and in general, the hairy roots grew at a faster pace than the non-transformed control roots [63, 109, 173]. The increase in hairy root biomass also correlated with an average of 2.5-fold higher accumulation of the secoiridoid gentiopicroside in hairy roots than the non-transformed roots [63, 173]. A similar elevated production of iridoids glycosides was also observed for a *Picrorhiza kurroa* hairy root clone, which manufactured 4-fold higher kutkoside and picroside in liquid culture than roots of 3-year-old plants [182]. Besides glycosylation, (seco)iridoids can also conjugate with alkaloids to form (seco)iridoid alkaloids, which are classified as terpene indole alkaloids (TIAs). The TIA valepotriates are valued for their sedative and spasmolytic activities and are mainly found in roots of plants that produce these compounds [36, 183]. It was shown that *A. rhizogenes* transformed *Centranthus ruber*, *Valeriana officinalis*, and *V. wallichii* hairy roots grew rapidly and produced significantly more concentrated valepotriates than the non-transformed roots [12, 49, 50]. In addition, valepotriates were retained in the hairy roots, unlike the cell cultures of these plants that released them into the growth medium. Overall, the above-mentioned hairy root clones provide an alternative means for isolating the valuable TIAs other than the more destructive plant root extraction method, which facilitates the preservation of these endangered medicinal plant species.

Plant essential oils are often used for fragrances, flavoring, and toning. Bioactivities have also been found for various classes of essential oils [34]. While the parent *Anethum graveolens* (dill) plant contains monoterpenes as the major essential oil components in roots and fruits, its hairy roots produced essential oils that had the signature of phenylpropanoid derivatives [140]. Additional examples can be drawn from *Pimpinella anisum* (anise), *Artemisia absinthium,* and *Levisticum officinale* (lovage) where the hairy roots exhibited different essential oil profiles than that of the parent plants [118, 141, 142]. Although not suitable for synthesizing the characteristic essential oils from the parent plants, these hairy roots could still be employed, in the future, for producing the hairy root-type essential oils.

The tropical disease malaria is caused by parasitic infections. According to the World Health Organization (WHO), there were over 200 million cases of malarial illness reported worldwide and more than 600,000 people died from malaria in 2012 [192]. Artemisinin, a sesquiterpene derivative, is the only Food and Drug Administration (FDA)-approved antimalarial drug on the market, and an artemisinin-based combination therapy (ACT) is currently being used for treating malaria. Artemisinin has traditionally been extracted from leaves and flowers of the annual herb *Artemisia annua* where low concentrations of artemisinin are found. Total chemical synthesis of artemisinin, though possible, poses high costs to patients. Complementary to the whole-plant extraction, semi-chemical synthesis, and synthetic biology approaches, hairy root cultures of *A. annua* were established to provide year-round production of this valuable antimalarial metabolite [70, 131, 189]. Additionally, gas-phase and liquid bioreactors have been designed, implemented, and improved for large-scale, hairy root-based production of artemisinin. A 1-L mist (gas phase) bioreactor was modified with a flexible-wall growth chamber, which allowed a dense inoculum bed at the early stage of culture growth and resulted in more proliferative hairy root growth than shake flasks [157]. When the growing *A. annua* hairy roots were separated from the impeller of a 3-L stir tank liquid bioreactor, damage of the growing roots was avoided. A combination of this improved stir tank bioreactor design and application of the elicitor methyl jasmonate (MeJA) led to an enhanced yield of 10.33 mg L^{-1} artemisinin [128]. However, despite these technological advancements, to date, commercial production of artemisinin from hairy root cultures has not been realized.

Although the ACT has been effective in combating malaria, artemisinin-resistant plasmodium parasites were reported recently [35] and there is an immediate need for the discovery of new antimalarial drugs. The seed extracts of *Bixa orellana* are commonly used in food coloring and cosmetics. Ethnopharmacy records showed that several Bixa species were also among a few indigenous plants adopted for treating malaria by Amazonians [135]. In contrast to a previous report where in vitro antiplasmodial activity assays were negative for crude Bixa extracts [135], a recent study indicated that crude extracts and purified compounds (e.g., stigmasterol) from Bixa hairy roots exhibited EC_{50} values on a micromolar scale against two plasmodial strains [200]. The discrepancy in antiplasmodial activities reported in these two studies could be due to different metabolite profiles, chemical extraction methods, assay procedures, and parasite strains that were used. Further studies are necessary to verify the observations made by Zhai et al. [200] and identify the antiplasmodial compounds from the hairy root extracts. As documented by Kaur et al. [74], there are continuous efforts in the search for additional antimalarial drugs from natural sources; many of the lead compounds have already been identified from hairy roots induced from various plant species (Table 1), and hairy roots will find great utility in the development of new antimalarial drugs.

The dimeric sesquiterpene compound gossypol is uniquely localized in the specialized glandular tissues of cotton plants. Gossypol could inhibit the growth of several types of cancer cells, while antiviral, antiamoebic, and antiprotozoan effects were also ascribed to this compound [186]. Hairy roots induced from *Gossypium hirsutum* and *Gossypium barbadense* were highly productive, with an average yield of 15 mg g^{-1} and a maximum yield of over 40 mg g^{-1} gossypol on a dry culture mass basis [175]. Gossypol-related compounds that contain additional functional groups were also present in these hairy root clones. Between 60 and 95 % of gossypol and derivatives were retained in the hairy roots, which allows for targeted extraction of these valuable compounds from hairy root tissues.

The renowned diterpene anticancer drug paclitaxel was directly extracted from the bark of pacific yew trees (*Taxus brevifolia*). The lengthy growth cycle of the trees, the destructive extraction procedure, and the low yield (0.01 %) of paclitaxel prompted the use of a cell culture system for paclitaxel production (0.6 % yield). Complementary to the cell culture-based approach, hairy roots of *T. brevifolia* were induced and could synthesize 69–210 μg g^{-1} paclitaxel (i.e., up to 0.02 % yield) upon elicitation by MeJA [43]. Though the hairy roots are twice as productive as the tree bark, there is still much less paclitaxel being made in hairy roots than in cell cultures. Nevertheless, this *T. brevifolia* hairy root culture could be useful for biochemical pathway characterization and possibly biotransformation studies.

The Salvia species contain diterpenes with diverse structures, such as the labdane, abietane, clerodane, and pimarane types of diterpenes, which were reported to possess antioxidant, anti-inflammatory, antimicrobial, antitumor, and anticancer activities [41]. These diterpene compounds have been identified from hairy roots induced from various Salvia plants [41, 90, 91, 148]. Fraga et al. [41] also analyzed the biogenetic relationships of the abietane diterpenoids from *S. broussonetii* hairy roots and determined their antifeedant and toxic properties, thus providing new insights into this group of bioactive compounds. Hypotensive, positive inotropic, and chronotropic activities were shown for a labdane diterpenoid, forskolin, produced by the roots of *Coleus forskohlii* [27]. The hairy root culture of *C. forskohlii* yielded high concentrations of forskolin, and this production appeared to be affected by the *A. rhizogenes* strains and the culture media being used [143]. Diterpene lactones ginkgolides can be extracted from leaves and roots of *Ginkgo biloba*. Previous studies of differentiated *G. biloba* cell cultures produced ginkgolides [94], suggesting that the differentiated hairy roots could possibly be a source of these health-beneficial diterpenoids. Hairy roots of *G. biloba* were established; however, chemical analysis of hairy roots was not described in that study [7].

Triterpene saponins (C_{30}) are amphiphilic compounds with glycosidic chain(s) conjugated to the triterpene aglycone core structures. A group of triterpene saponin compounds, namely saikosaponins, contribute to the anti-inflammatory, antipyretic, and antitussive activities of the root extracts of the traditional Chinese herbal plant *Bupleurum falcatum* (Apiaceae) [2]. In consideration of the slow natural root growth, hairy root cultures of *B. falcatum* were established for speedy production of saikosaponins [2]. Sucrose and mineral contents in the growth medium showed differential impacts on hairy root growth and saikosaponin yield [2]. Further optimization of the culturing conditions is still needed for commercially viable production of this valuable metabolite. Another plant of the Apiaceae family, *Centella asiatica*, accumulates saponins, such as centellasaponin, asiaticoside, madecassoside, and sceffoleoside, mainly in leaves. These triterpene saponins are recognized for their antipyretic and anti-inflammatory properties [185]. Non-transformed roots and hairy roots of *C. asiatica* did not show significant accumulation of saponins. However, upon elicitation by MeJA, 7.12 mg g^{-1} asiaticoside accumulated in hairy roots [75]. This encouraging result suggested that saponin production could potentially be further enhanced in *C. asiatica* hairy roots using elicitors and additional manipulations.

The roots of *Panax ginseng* are rich in saponin ginsenosides, which are appreciated for their toning, immunomodulatory, adaptogenic, and antiaging activities. The lengthy cultivation time and various problems involved in culturing stimulated the development of Panax hairy root cultures. Interestingly, 2-fold more ginsenosides were achieved in *P. ginseng*-derived hairy roots than the wild-type roots [197]. Another important Panax species, *P. quinquefolium*, was subjected to hairy root induction, and the ginsenoside content reached 0.2 g g^{-1} dry weight at 10 weeks of hairy root growth [105]. The hybrid plant between *P. ginseng* and *P. quinquefolium* was more vigorous in ginsenoside production than either parent [83]. However, since the hybrid plant is sterile, hairy roots of the hybrid were developed to maintain this elevated biosynthetic capacity [188]. The promising finding from this work was that the 8-week-old hairy roots contained comparable amounts of ginsenosides to the roots of field-grown parental plants, suggesting that the biosynthetic potential of the parent plants was maintained in the hairy roots [188]. Structural analogs of ginsenosides, gypenosides, are produced by *Gynostemma pentaphyllum* (Cucurbitaceae) and are noted particularly for their anti-diabetic effects [119]. A maximum of 280 mg L^{-1} gypenoside accumulation was reached at day 49 of *G. pentaphyllum* hairy root growth, and kinetic evaluations revealed an intriguing concomitant production of gypenosides with the primary metabolites [26]. It will be interesting to understand how *G. pentaphyllum* hairy roots partition carbons between the primary and secondary metabolism.

The leaves and infusions of *Gymnema sylvestre* are valued in traditional Indian medicine for their antidiabetic properties. The active principles of *G. sylvestre*, triterpene saponins (e.g., gymnemic acid), were also reported to have additional health-promoting benefits. As with many other medicinal plants, overexploration of this herb caused concerns for a sustainable supply of these antidiabetic compounds. The *G. sylvestre* hairy roots generated almost 5-fold more gymnemic acid than the non-transformed roots [116]. The highly productive *G. sylvestre* hairy roots hold great potential for producing gymnemic acid. However, it remains to be determined whether the hairy root extracts exhibit comparable efficacy to that shown for the leaf and whole-plant extracts.

A special group of triterpene compounds are phytoecdysteroids, which are synthesized from the mevalonic acid pathway with cholesterol as a biosynthetic intermediate. These plant-produced ecdysteroid analogs are purposed to mimic the action of ecdysteroids (steroidal hormones) from insects and other anthropods to interfere with their molting processes (ecdysis). On the other hand, immunostimulatory effects of phytoecdysteroids have been described in humans [21]. High levels of phytoecdysteroids accumulated in the hairy roots of *Serratula tinctoria*, to an extent of 0.1–0.2 % of dry weight [31]. Biosynthesis of phytoecdysteroids in *S. tinctoria* hairy roots was supported by feeding of radiolabeled cholesterol and mevalonic acid and incorporation of the isotope into phytoecdysteroids [31]. The hydrolysis product of saponins, sapogenins (i.e., the aglycone, e.g., diosgenin), can be used for commercial synthesis of the human steroidal hormones. Fenugreek (*Trigonella foenum-graecum*) hairy roots were shown to produce diosgenin [110]. Cholesterol supplementation (i.e., precursor feeding) largely decreased the biomass of *T. foenum-graecum* hairy roots and only led to a moderately increased diosgenin content (5 %). However, addition of chitosan to the growth medium induced diosgenin production by 5-fold [110]. Further optimization of biomass growth and diosgenin yield could improve the prospect of using fenugreek hairy roots as human steroidal hormone "feedstocks".

2.2 Alkaloid and Glucosinolate Production in Hairy Roots

Alkaloids are nitrogen-containing (*N*-containing) compounds found in plants, microorganisms, and animals. The current classification of alkaloids is based on the *N*-containing structures, such as indole or quinoline alkaloids. The nitrogen atom in alkaloids is derived from an amino acid either directly or indirectly through a transamination reaction. Tropane alkaloids are likely synthesized from the monomethylated polyamine putrescine. Many solanaceous plants produce tropane alkaloids, particularly hyoscyamine and scopolamine, specifically in the root tissue. Tropane alkaloids can counteract the function of the neurotransmitter acetylcholine in the brain (i.e., anticholinergic) [100]. The pharmaceutical properties and the root-enriched accumulation of tropane alkaloids inspired hairy root studies in solanaceous plants (Table 1). A survey of Datura and Hyoscyamus species revealed large variations in the quantity of tropane alkaloids produced by hairy roots [86]. Most of the tropane alkaloids accumulated in *Datura candida* hairy roots, while only a small portion were released to the growth medium [28]. However, Sáenz-Carbonell and Loyola-Vargas [162] noticed that the use of ammonium as the nitrogen source for hairy root growth stimulated secretion of tropane alkaloids. It was found that UV B irradiation and acetylsalicylic acid elicitation could enhance tropane alkaloid production in *Anisodus luridus* hairy roots [132]. Interestingly, in addition to hyoscyamine and scopolamine, a novel polyamine cadaverine was detected in the hairy roots, but not the whole plant, of *Brugmansia candida* [22].

Indole alkaloids originate from the aromatic amino acid tryptophan. The terpene and indole alkaloid conjugates (TIAs) constitute one of the largest groups of alkaloids in plants. TIA biosynthesis presents an excellent example of highly complex and coordinated secondary metabolic pathways that encompass multiple distinct biosynthetic routes, diverse cell types, and different subcellular compartments. The rhizomes and roots of Rauvolfia species are rich sources of TIAs that have been used for treating nervous system disorders due to their anxiolytic activities. Hairy roots established from *Rauvolfia serpentine* produced ajmaline and serpentine in both solid and liquid cultures [14], while targeted analysis of ajamaline and ajamalicine in *Rauvolfia micrantha* hairy roots indicated that this hairy root system could synthesize both TIAs [160]. Development of hairy roots as sources of Rauvolfia TIAs will help alleviate overharvest of these endangered species.

C. roseus (Madagascar periwinkle) reportedly produces over 130 TIAs and is valued for its many health-beneficial activities. Two major alkaloids from Catharanthus, vinblastine and vincristine, are structurally very similar—each is composed of two alkaloid monomers, with vinblastine possessing an *N*-methyl group and vincristine possessing an *N*-formyl group at the corresponding position. In principle, both vinca alkaloids bind to tubulins, disrupt microtube functions, and terminate cell divisions. Yet, these dimeric TIAs exhibit distinct bioactivities and are components of cancer therapies. Since the yield of vincristine and vinblastine from whole plants is extremely low (0.0002 %), these chemicals are currently obtained from semisynthesis using intermediates isolated from *C. roseus*. Alternative methods are desirable for economic production of these valuable vinca alkaloids. Hairy root cultures of *C. roseus* were established, and various environmental, chemical, and biotic factors were applied to the hairy roots for yield increase [149]. In addition, biosynthetic and regulatory genes were also employed for manipulation of alkaloid content in hairy roots. To this end, the transcription factor octadecanoid-derivative-responsive catharanthus AP2 domain (*ORCA3*) and one of the biosynthetic genes geraniol 10-hydroxylase (*G10H*) were overexpressed in *C. roseus* hairy roots and resulted in enhanced accumulation of catharanthine, but not vindoline, vinblastine, and vincristine [187]. It remains to be explored how the enzyme organization (e.g., whether metabolons are involved and delineation of the regulons) and the subcellular localization of the biosynthetic enzymes affect the production of vinca alkaloids.

The TIA camptothecin (CPT) contains a quinoline-type alkaloid and is obtained from the Chinese happy tree *Camptotheca acuminata* [99]. CPT can bind to DNA and DNA topoisomerase I; the formation of such a complex blocks the activity of DNA topoisomerase I, which contributes to the antitumor, anticancer, and antiviral activities of CPT [85]. Synthetic analogs of CPT have already been approved by the FDA for treating cancers. Both *C. acuminata* and *Ophiorrhiza pumila* hairy roots produced quantities of CPT that were similar to the wild-type roots [99, 136]. The CPT produced in hairy roots was released to the liquid growth medium. It was shown that addition of polystyrene resins to the growth medium of *O. pumila* hairy roots absorbed CPT and further stimulated its secretion. These promising results bode well for hairy root production of this bioactive chemical. One may envision that high levels of CPT could be achieved by a combined elicitation, absorbent addition, and metabolic engineering of hairy roots.

Additional groups of alkaloids with piperidine or indolizidine structures were also identified from hairy root cultures. Piperidine alkaloids from *Lobelia inflata* (e.g., lobeline) have been used as antiasthmatic agents. Hairy roots of *L. inflata* produced similar or higher levels of lobeline when cultured in different growth media as compared to the soil-grown non-transformed roots [196]. Swainsonine, an indolizidine alkaloid derived from lysine, is produced from the hairy roots of the legume plant *Swainsona canescens* [113]. Its concentration in hairy roots was further increased by feeding biosynthetic precursors, lowering the pH of the growth medium, and supplementation of copper sulfate. These various treatments also stimulated the secretion of swainsonine to the growth medium [113]. Though release in the medium may facilitate downstream recovery of swainsonine, it is unclear whether swainsonine remains stable in the growth medium. One possible solution is adding absorbents to the growth medium to prevent degradation, assist in extraction, and stimulate further release of swainsonine.

Steroidal alkaloids are conjugates of a steroidal saponin and an alkaloid, such as solasodine found in solanaceous plants, and are often glycosylated to form steroidal glycoalkaloids. Steroidal alkaloids can impart toxicity to humans, particularly when accumulated at high concentrations in food crops. On the other hand, these steroidal alkaloids share similar structures to steroidal saponins and can serve as the biosynthetic precursors of steroid drugs and synthetic human hormones. Though these compounds are generally more concentrated in the photosynthetic and reproductive tissues of the plants, low concentrations of steroidal alkaloids were also detected in roots. By manipulating the composition of the growth medium, 6.2 mg g^{-1} solasodine was obtained in *Solanum aviculare* hairy roots and the solasodine-containing glycoalkaloids accounted for 0.3–1 % of the dry weight of *Solanum laciniatum* hairy roots [84, 122]. The yield of steroidal glycoalkaloids could be further improved in hairy roots by treatment with elicitors and feeding of biosynthetic precursors.

Besides alkaloids, another group of *N*-containing compounds in plants is known as glucosinolates, which are amino acid-derived thioglycosides. Hydrolysis of glucosinolates by myrosinase releases isothiocyanate that could be toxic to humans when ingested in large quantities. Conversely, cancer chemopreventive activities have been found in several glucosinolate compounds [58]. Hairy roots of *Arabis caucasica, Barbarea verna*, *Nasturtium officinale*, and *Tropaeolum majus* were obtained and produced gluconasturtiin, glucotropaeolin, and glucoiberverin [190, 191]. When amino acid precursors were added to the growth medium of the hairy roots, the yield of the corresponding glucosinolates was enhanced [190]. While the glucosinolate profiles varied in different plant species, these studies indicated that hairy roots can be used as sources of glucosinolates in addition to the parent plants.

2.3 Phenolic Production in Hairy Roots

Phenolic compounds contain at least one hydroxylated benzene ring in their chemical structures, including simple phenolics, such as coumarins, flavonoids, isoflavonoids, and anthocyanins, and more complex polymers, such as lignins and tannins. A majority of the plant phenolic metabolites are derived from the aromatic amino acids that are synthesized from the shikimate pathway. Phenolics are collectively valued for their wide variety of health-promoting activities. Some of the phenolic compounds, such as flavone glycosides from citrus, isoflavones from soy, and flavonoids and tannins from tea, were formerly referred to as vitamin P which are linked to antioxidant, cardioprotective, and cancer chemopreventive activities.

Coumarins originate from the orthohydroxylation of the cinnamic acid precursor and the subsequent lactone formation. Substitutions of the simple coumarin ring lead to the formation of furanocoumarins (a five-membered furan ring attached to the coumarin backbone) and pyranocoumarins (a six-membered pyran ring attached to the coumarin backbone). A plethora of pharmacological properties have been ascribed to coumarins, such as anti-inflammatory, antiviral, anticancer, and antihypertensive activities [180]. Coumarins are commonly found in the families of Umbelliferae and Rutaceae. Fruits of *Ammi majus* are rich in coumarins and furanocoumarins. Umbelliferone and furanocoumarins were identified from methanol extracts of hairy roots induced from *A. majus* stalk and leaf explants [88]. *Ruta graveolens* is also an abundant source of coumarins and furanocoumarins [155]. Unexpectedly, the hairy roots of *R. graveolens* synthesized novel coumarins, osthole and osthenol, which do not normally accumulate in the parent plants [155]. Another interesting observation was that the dark-grown hairy roots accumulated more coumarins than those grown under a light–dark regime. It was unclear why depletion of light may enhance the coumarin levels, which could result from an increase in coumarin biosynthesis, a decrease in its turnover, or a combination of these two processes.

The efficacy of flavonolignans in treating liver disorders is presumably due to their abilities to scavenge free radicals produced from the detoxification reactions in the liver. The flavonolignan silybins found in *Silybum marianum* are the oxidative coupling products of dihydroquercetin and coniferol alcohol. Hairy roots were induced from seeds of *S. marianum* in an attempt to produce silybins [3]. Although silybins and other variants of flavonolignans were identified from the non-transformed plant roots, isosilybin was the predominant flavonolignan in the hairy roots, where only trace amount of silychristin and silydianin and no silybin were identified. Though producing a different flavonolignan profile than the non-transformed roots, this hairy root system could be used for investigation of the flavonolignan biosynthetic pathway and for producing different types of flavonolignans.

The trihydroxyflavone jaceosidin acts as an inhibitor of neuroinflammation [117]. Jaceosidin has been isolated from several herbal plants. Hairy root clones of *Saussurea medusa* were obtained and able to form jaceosidin with a 37-fold increase in production within a 27-day growth period [202]. It was also found that the root tips were more active in biomass growth and jaecosidin production than the other sections of the hairy roots. Besides synthesizing flavonoids, the phenylalanine/tyrosine-derived ring structure in the common flavanonone precursor could rearrange to a different position (i.e., ring migration) and form isoflavonoids. In contrast to flavonoids that distribute ubiquitously in plants, isoflavonoids have more limited taxonomic distributions and are mostly found in Fabaceae. Isoflavonoids exhibit phytoestrogenic, anticancer, and anti-inflammatory activities [107]. With the exception of genistin (glycoside of genistein), isoflavonoids in *Lupinus mutabilis* hairy roots accumulated at a much higher level than the non-transformed roots [9]. The drop in genistin concentration could possibly be due to reduced glycosylation of genistein to genistin in *L. mutabilis* hairy roots. To improve isoflavonoid production, elicitors were applied to the hairy root culture and showed various effects. Elicitor treatments (chitosan, salicylic acid, yeast extract, polyamines) were more effective in enhancing daidzein and genistein accumulation than feeding the *Psoralea corylifolia* hairy roots with the phenylalanine precursor [153]. The biotic and abiotic elicitors stimulated isoflavonoid production, but did not affect the growth of *Pueraria candollei* hairy roots [177]. Of the five elicitors used (MeJA/MJ, chitosan, salicylic acid, yeast extract, lysate of *A. rhizogenes*), MJ was most effective and led to a 4.5-fold increase in total isoflavonoids than the non-elicited control roots [177]. These results suggest that, for optimal production of isoflavonoid metabolites, multiple elicitation strategies should be tested and evaluated on different hairy root cultures. It may be necessary to combine several elicitors to bring about the maximum metabolic potential of the hairy roots.

Anthocyanins are colorful pigments and excellent antioxidants produced largely in flowers, fruits, and tissues under stress. Anthocyanins and hydroxycinnamic acid esters were detected in the hairy roots of *Leontopodium alpinum* [61]. Anthocyanin production in both *L. alpinum* and *Campanula glomerata* hairy roots was increased upon addition of benzylaminopurine (i.e., benzyladenine), presumably due to the stimulated root growth [61, 166]. Light inhibited *C. glomerata* hairy root growth, but promoted anthocyanin accumulation [166]. The accumulation of polyacetylenes was not affected by light or benzylaminopurine, suggesting that these hormonal and environmental factors play distinct roles in producing different classes of secondary metabolites.

In addition to phenolic compounds with simple structures, phenolic polymers, such as condensed tannins (CTs) and hydrolyzable tannins (HTs), are also synthesized in hairy roots [67, 111, 123]. Though sharing the common name tannin, HTs and CTs have different biosynthetic origins; HTs are synthesized from an intermediate of the shikimate pathway, and CTs are derived from the end products of the shikimate pathway. Antioxidant and anticancer activities have been reported for both groups of tannins [62]. CTs are enriched in the *Lotus corniculatus* leaf, stem, root, flower, and seed tissues. Both *L. corniculatus* non-transformed and hairy roots produced CTs [111]. Interestingly, electron microscopy studies revealed that CTs distributed specifically in "tannin cells" of the hairy roots. Exogenously supplied auxin enhanced biomass growth, did not affect flavonoid accumulation, and inhibited CT production in *L. corniculatus* hairy roots. The underlying mechanism for a differential impact of auxin on flavonoid and CT accumulation is not known. Pomegranate fruit peel is highly abundant in HTs, particularly punicalagins. Hairy roots developed from radicle, cotyledon, and leaf explants also showed punicalagin production [123]. The pomegranate hairy root culture constitutes an excellent system for genetic characterization of HT biosynthetic genes. While HTs were the major tannins identified from hairy roots of *Geranium thunbergii*, catechin (precursor of CTs) also accumulated in this hairy root system [67], suggesting that *G.*

thunbergii hairy roots could potentially be used as a platform for understanding the biosynthetic relationships between the two groups of tannins.

2.4 Fatty Acid, Polyacetylene, Thiophene, and Polyketide Production in Hairy Roots

Fatty acids, polyacetylenes, thiophenes, and polyketides share the common biosynthetic scheme of condensing multiple acetate units and forming poly-β-keto chains and differ in the chain extension mechanisms. These various groups of secondary metabolites show diverse taxonomic distributions and exhibit distinct activities in plants and humans. Polyunsaturated fatty acids (PUFAs) are considered "good fats" as consumption of PUFAs has been linked to reduced levels of low-density lipoproteins (LDLs) and improvement on heart health [16]. Since extraction of PUFAs from fish contributes to a decline of fish stocks, plant oils have emerged as an alternative source of PUFAs. In addition to plant oils, hairy roots have also been explored for production of PUFAs and lipids. For example, hairy roots established from a PUFA-producing plant, *Echium acanthocarpum*, accumulated 18:2n-6 (linoleic acid) and 18:3n-6 (γ-linolenic acid) that accounted for 55 % of total fatty acids in the hairy roots [25]. Unlike oil seeds that are capable of amassing PUFAs to large quantities, hairy root production of PUFAs is limited by the competition of carbon precursors for formation of PUFAs or biomass. Nevertheless, hairy roots of PUFA-producing plants present an opportunity for gene discovery and functional characterization that could play a role in improving PUFA production in other plant organs.

Alkamides are amides of unsaturated fatty acids. Many plants that manufacture alkamides have been used for medicinal purposes. Herbal preparations of dried Echinacea roots or plant extracts are still being used nowadays as immunostimulants. Alkamides are active principles of Echinacea roots and comprise PUFAs as amides of 2-methylpropanamine or 2-methylbutanamine. Hairy root cultures of three Echinacea species were established and showed enhanced production of alkamides compared to the wild-type roots [133]. JA stimulated alkamide production in these hairy roots though hairy root growth was thwarted at high JA concentrations (10 μM or greater). The auxin indolebutyric acid (IBA) promoted hairy root branching and growth rate of biomass while having no impacts on alkamide production. This study suggested that a balanced growth and elicitation regime should be designed and implemented for optimal yield of alkamides in Echinacea hairy roots.

Polyacetylenes and thiophenes (heterocyclic sulfur-containing aromatic rings conjugated to acetylenes) share common biosynthetic pathways and are mainly found in roots of plants that belong to Asteraceae, Umbelliferae, and Campanulaceae. Antiviral, antimicrobial, and antitumor properties have been reported for these compounds [154]. Hairy root cultures have been established in multiple species of these families, and one interesting observation was that the hairy roots grew at least 2-fold faster than the corresponding non-transformed roots, while both root systems produced polyacetylenes and thiophenes [40]. The incorporation of radiolabeled oleic acid precursors to polyacetylenes and thiophenes, stable production of polyacetylenes and thiophenes over a 12-month period with multiple passages, and increased metabolite production upon elicitation by fungal culture filtrates collectively suggested that these compounds are synthesized locally in the root tissue, rather than remanents of the parent metabolites transferred to the roots [40]. One surprising observation was that the hairy roots turned green upon light exposure and the "green hairy roots" contain chloroplasts and perform photosynthesis. It is unclear whether carbon allocation to photosynthetic activities affects synthesis of secondary metabolites in these "green hairy roots".

Polyketides are synthesized in a similar fashion as fatty acids. Stilbenes are one class of polyketides that have the characteristic C6–C2–C6 backbone structures. A particular stilbene compound resveratrol has drawn much attention in recent years due to its cardioprotective, anticancer, antiangiogenic, and immunomodulatory activities [125]. To build a tissue culture-based platform for resveratrol production, hairy root lines of peanut (*Arachis hypogaea*) were established and resveratrol production was induced using five different elicitors that encompass biotic and abiotic stress factors, including copper sulfate, chitosan, cellulose, laminarin, and sodium acetate [108]. Sodium acetate was most effective in stimulating resveratrol formation in the *A. hypogaea* hairy roots, and 99 % of the resveratrol synthesized was released to the growth medium [108]. The mist bioreactor design used for *A. annua* hairy roots was also applied to *A. hypogaea* hairy roots and resulted in a slightly higher biomass gain as compared to the shake flask-grown hairy roots [157]. In addition to yield improvement, optimization of the extraction method could further increase the overall resveratrol production from *A. hypogaea* hairy roots.

2.5 Natural Dye and Biopesticide Production in Hairy Roots

Besides tackling various health-related problems, there are also growing interests from the food industry in using naturally occurring secondary metabolites, specifically pigments, for food coloration. The roots of red beet (*Beta vulgaris*) contain the water-soluble betalains, including the red/violet-colored betacyanins (aromatic indole pigments derived from tyrosine) and the yellow-colored betaxanthins. The betalain pigments mainly accumulated in vacuoles of the *B. vulgaris* hairy roots [169]. It was noted that temporary cessation of oxygen supply led to release of betalain pigments from the vacuolar compartments to the growth medium, thus facilitating downstream pigment extractions. More importantly, the oxygen starvation by cessation of shaking did not affect regeneration of hairy roots, and thus allowed regeneration and reuse of the root biomass for metabolite production. The betalain profiles in the growth medium resembled those of hairy roots and the parent roots, providing further support to the utility of this hairy root system for the production of natural betalain pigments.

Shikonin is a naphthoquinone (dicyclic quinone molecule) that is derived from the shikimate and methylerythritol phosphate (MEP) pathways. This reddish pigment has shown antimicrobial, anti-inflammatory, and anticancer activities [4] and has been used in food additives and cosmetics. Although shikonin is one of the first plant metabolites produced in cell cultures, it reportedly requires a complex two-step culture system for shikonin production. The cell culture-produced shikonin is secreted to the cell wall after fusion of ER-derived vesicles to the plasma membrane and trapped in between the plasma membrane and the cell wall. Upon optimization of culturing conditions, the hairy roots of *Lithospermum erythrorhizon* secreted shikonin to the growth medium [152]. Microscopic examination revealed that the mature, but not the newly grown, roots accumulated shikonin in aggregated granules [152]. Since about 25 % of shikonin produced was secreted to the growth medium, absorbents were added to the medium in an attempt to sequester shikonin and stimulate further secretion of this metabolite. Indeed, a 3-fold increase in shikonin yield was observed and over 85 % of total shikonin produced was sequestered in the XAD-2 absorbents. Overall, this is a promising example of producing a water-soluble pigment using the hairy root system.

Anthraquinone derivatives (tricyclic quinone molecules), such as alizarin and purpurin, are natural red dyes produced from the roots of the perennial herb *Rubia akane*. Hairy root cultures of various Rubia species have been established, and production of alizarin and purpurin correlated with hairy root growth [80, 126]. Maximum production of these compounds was reached after 20 days of culturing in liquid medium [126]. Upon oxygen starvation treatment, anthraquinones were secreted to the culture medium, similar to what was observed in *B. vulgaris* hairy roots with limited oxygen supply [169]. Like the *B. vulgaris* hairy roots, the usage of absorbents could also be examined for enhancing anthraquinone production and release to the liquid culture.

To protect crop plants from pathogens and insects, pesticides have been applied to agricultural fields in large quantities, causing environmental and human health concerns. To this end, pesticides produced from biosources are degradable and preferred by the general public. The principle insect repelling component of neem (*Azadirachta indica*) is azadirachtin. This tetranortriterpenoid structurally resembles the insect hormone ecdysones and works to disturb the hormone-controlled developmental processes of insect pests. Since neem trees are grown in more restricted climate and soil conditions, a hairy root culture system will ensure more sustainable production and broad utilization of this useful biopesticide. Crude extracts of *A. indica* hairy roots exhibited high antifeedant activities against the desert locust *Schistocerca gregaria*, which was previously demonstrated to be sensitive to azadirachtin [159]. It is worth noting that azadirachtin is photolabile and sensitive to the pH of the environment. Additional work still needs to be carried out for optimization of additives to this biopesticide to ensure its efficacy and shelf life. In this regard, the *A. indica* hairy roots provide a convenient system for carrying out these studies.

3 Preservation of Hairy Roots

As described in the previous sections, hairy roots produce many valuable secondary metabolites. It is conceivable that the established hairy roots should be maintained for long-term use. However, the constant efforts involved in maintaining hairy root cultures in solid or liquid media are cumbersome and the potential loss of hairy root cultures due to contamination or facility failures could be devastating. In addition, as cultured in solid or liquid media, shipping of hairy root specimens could be challenging. Taking these into consideration, it is highly desirable to develop methods for long-term storage of hairy roots and the capability for regeneration when needed. As with the induction of hairy roots, different plant species may have different optimal conditions for preservation and some may be recalcitrant for the preservation methods.

Cryopreservation is a process where tissues are frozen and preserved at subzero temperatures. Two major challenges encountered by cryopreservation of hairy roots are maintaining the integrity of cellular and subcellular structures and ensuring successful regeneration of the preserved hairy roots. To this end, multiple methods, including vitrification, encapsulation vitrification, and droplet vitrification, have been successfully applied to cryopreservation. Vitrification entails mixing root tips with chemical agents that reduce the cellular water content and suppress the formation of ice crystals. The vitrification method has traditionally been used for saving plant tissues and has been adapted to long-term storage of hairy roots. More recently, two new methods, encapsulation vitrification and droplet vitrification, build upon vitrification and encapsulate the explants in alginate beads or put them in individual droplets prior to freezing in liquid nitrogen for easy handling [138]; these improved methods are more suitable for processing a large quantity of samples. Successful cryopreservation of hairy roots has been reported in a number of plant species using either the traditional or the improved methods (Table 2). Rehydration and recovery of frozen hairy root tissues remain to be the challenging and often bottleneck steps in cryopreservation of hairy roots. In addition to ensuring viability of the cryopreserved tissues, it is also of importance for the recovered hairy roots to maintain the biochemical phenotype of the untreated controls, particularly for those that are used for valuable secondary metabolite production.

The perceivable limitations of cryopreservation inspired development of alternative methods for long-term storage of hairy roots. Previously, hairy roots were reported as a means to generate plantlets in the form of "artificial seeds" [178]. In these studies, the plantlets were induced from fragmented hairy roots, encapsulated and dehydrated prior to regeneration into whole seedlings. This artificial seed procedure was modified by manipulating the composition of the growth media, which was

Table 2 Examples of cryopreservation of hairy roots

Species	Common names	Preservation methods	References
Artemisia annua L.	Sweet wormwood	Slow freezing	[170]
Armoracia rusticana Gaertn. Mey. et Scherb.	Horseradish	Encapsulation dehydration	[59]
Beta vulgaris L.	Red beet	Slow and fast freezing	[15]
Nicotiana rustica L.	Mapacho	Slow and fast freezing	[15]
Panax ginseng C.A. Meyer	Asian ginseng	Vitrification	[198]
Atropa belladonna L.	Deadly nightshade	Vitrification	[174]
Maesa lanceolata Forssk.	False assegai	Encapsulation–dehydration	[93]
Medicago truncatula Gaertn.	Barrel medic	Encapsulation–dehydration	[93]
Rubia akane Nakai	Asian madder	Droplet vitrification	[77, 78, 139]

adapted for the preservation of hairy root clones without the induction of plantlets [179]. However, a caveat was that the rate of hairy root regeneration decreased as time lapsed. The longest period of viable storage for *Scutellaria baicalensis* (Baikal skullcap) hairy roots via the artificial seeds method was 4–5 months, after which the hairy roots needed to be regrown and recapsulated [179]. Though artificial seeds are still at the early stage of development and require further optimizations, they present a promising method for maintaining hairy roots.

4 Future Perspectives

Overall, the major groups of secondary metabolites have already been produced from hairy roots (Table 1). Novel compounds were also identified in hairy roots, and exogenously added metabolites could be biotransformed by the metabolic pathway or grid of hairy roots. Environmental factors, such as light, oxygen, and temperature, as well as abiotic and biotic stress factors, such as heavy metals and fungal elicitors, have all been applied to hairy roots for increased yield of secondary metabolites. In addition to these external stimuli, secondary metabolic pathways have also been modified for enhanced metabolite production, such as overexpression of biosynthetic genes and transcription factors, and suppression of catabolic or competing pathway genes. A better understanding of the biosynthetic pathway and regulation architecture of valuable secondary metabolites is crucial for genetic engineering and fully realizing the biosynthetic potential of hairy roots.

Progress has been made on commercialization of hairy root products. ROOTec Bioactives Ltd., founded in 2005 in Switzerland, currently produces phytochemicals from hairy roots induced from 17 plant species in their proprietary mist bioreactors. In the future, more investigations could be directed toward determining the efficacy of crude hairy root extracts or hairy root-produced chemicals. Plant biologists can work closely with engineers to tackle the challenges with scaling up hairy root cultures, such as optimal biomass growth and adaptation of the extraction methods to industrial-scale metabolite production. Looking forward, establishment of hairy roots guided by bioassays, augmented by elicitations and genetic manipulations, and coupled with efficient metabolite extractions will streamline the process and allow full exploitation of hairy roots as a production platform of valuable secondary metabolites.

References

1. Abhyankar G, Suprasanna P, Pandey BN, Mishra KP, Rao KV, Reddy VD (2010) Hairy root extract of *Phyllanthus amarus* induces apoptotic cell death in human breast cancer cells. Innov Food Sci Emerg Technol 11:526–532
2. Ahn J, Chong W, Kim Y, Hwang B (2006) Optimization of the sucrose and ion concentrations for saikosaponin production in hairy root culture of *Bupleurum falcatum*. Biotechnol Bioprocess Eng 11:121–126
3. Alikaridis F, Papadakis D, Pantelia K, Kephalas T (2000) Flavonolignan production from *Silybum marianum* transformed and untransformed root cultures. Fitoterapia 71:379–384
4. Andújar I, Ríos JL, Giner RM, Recio MC (2013) Pharmacological properties of shikonin—a review of literature since 2002. Planta Med 79:1685–1697
5. Arellano J, Vázquez F, Villegas T, Hernández G (1996) Establishment of transformed root cultures of *Perezia cuernavacana* producing the sesquiterpene quinone perezone. Plant Cell Rep 15:455–458
6. Asada Y, Li W, Yoshikawa T (1998) Isoprenylated flavonoids from hairy root cultures of *Glycyrrhiza glabra*. Phytochemistry 47:389–392
7. Ayadi R, Trémouillaux-Guiller J (2003) Root formation from transgenic calli of *Ginkgo biloba*. Tree Physiol 23:713–718
8. Babakov A, Bartova L, Dridze I, Maisuryan A, Margulis G, Oganian R, Voblikova V, Muromtsev G (1995) Culture of transformed horseradish roots as a source of fusicoccin-like ligands. J Plant Growth Regul 14:163–167
9. Babaoglu M, Davey MR, Power JB, Sporer F, Wink M (2004) Transformed roots of *Lupinus mutabilis*: induction, culture and isoflavone biosynthesis. Plant Cell Tissue Org Cult 78:29–36
10. Bais HP, Walker TS, Schweizer HP, Vivanco JM (2002) Root specific elicitation and antimicrobial activity of rosmarinic acid in hairy root cultures of *Ocimum basilicum*. Plant Physiol Biochem 40:983–995
11. Bakkali AT, Jaziri M, Foriers A, Vander Heyden Y, Vanhaelen M, Homès J (1997) Lawsone accumulation in normal and transformed cultures of henna, *Lawsonia inermis*. Plant Cell Tissue Org Cult 51:83–87
12. Banerjee S, Rahman L, Uniyal GC, Ahuja PS (1998) Enhanced production of valepotriates by *Agrobacterium rhizogenes* induced hairy root cultures of *Valeriana wallichii* DC. Plant Sci 131:203–208
13. Banerjee S, Singh S, Rahman LU (2012) Biotransformation studies using hairy root cultures—a review. Biotechnol Adv 30:461–468

14. Benjamin BD, Roja G, Heble MR (1994) Alkaloid synthesis by root cultures of *Rauwolfia serpentina* transformed by *Agrobacterium rhizogenes*. Phytochemistry 35:381–383
15. Benson E, Hamill J (1991) Cryopreservation and post freeze molecular and biosynthetic stability in transformed roots of *Beta vulgaris* and *Nicotiana rustica*. Plant Cell Tissue Org Cult 24:163–172
16. Beynen AC, Katan MB (1985) Why do polyunsaturated fatty acids lower serum cholesterol? Am J Clin Nutr 42:560–563
17. Bhattacharyya R, Bhattacharya S (2004) Development of a potent in vitro source of *Phylanthus amarus* roots with pronounced activity against surface antigen of the hepatitis B virus. In Vitro Cell Dev Biol Plant 40:504–508
18. Bonhomme V, Laurain-Mattar D, Lacoux J, Fliniaux MA, Jacquin-Dubreuil A (2000) Tropane alkaloid production by hairy roots of *Atropa belladonna* obtained after transformation with *Agrobacterium rhizogenes* 15834 and *Agrobacterium tumefaciens* containing rol A, B, C genes only. J Biotechnol 81:151–158
19. Bourgaud F, Bouque V, Guckert A (1999) Production of flavonoids by Psoralea hairy root cultures. Plant Cell Tissue Org Cult 56:96–103
20. Bourgaud F, Gravot A, Milesi S, Gontier E (2001) Production of plant secondary metabolites: a historical perspective. Plant Sci 161:839–851
21. Báthori M, Pongrácz Z (2005) Phytoecdysteroids—from isolation to their effects on humans. Curr Med Chem 12:153–172
22. Carrizo CN, Pitta-Alvarez SI, Kogan MJ, Giulietti AM, Tomaro MaL (2001) Occurrence of cadaverine in hairy roots of *Brugmansia candida*. Phytochemistry 57:759–763
23. Carron TR, Robbins MP, Morris P (1994) Genetic modification of condensed tannin biosynthesis in *Lotus corniculatus*. 1. Heterologous antisense dihydroflavonol reductase down-regulates tannin accumulation in "hairy root" cultures. Theor Appl Genet 87:1006–1015
24. Caspeta L, Quintero R, Villarreal ML (2005) Novel airlift reactor fitting for hairy root cultures: developmental and performance studies. Biotechnol Prog 21:735–740
25. Cequier-Sanchez E, Rodriguez C, Dorta-Guerra R, Ravelo A, Zarate R (2011) *Echium acanthocarpum* hairy root cultures, a suitable system for polyunsaturated fatty acid studies and production. BMC Biotechnol 11:42
26. Chang C, Chang K, Lin Y, Liu S, Chen C (2005) Hairy root cultures of *Gynostemma pentaphyllum* (Thunb.) Makino: a promising approach for the production of gypenosides as an alternative of ginseng saponins. Biotechnol Lett 27:1165–1169
27. Chinou I (2005) Labdanes of natural origin-biological activities (1981–2004). Curr Med Chem 12:1295–1317
28. Christen P, Roberts M, Phillipson JD, Evans W (1989) High-yield production of tropane alkaloids by hairy-root cultures of a *Datura candida* hybrid. Plant Cell Rep 8:75–77
29. Constabel CP, Towers GHN (1988) Thiarubrine accumulation in hairy root cultures of *Chaenactis douglasii*. J Plant Physiol 133:67–72
30. Davioud E, Kan C, Hamon J, Tempé J, Husson H-P (1989) Production of indole alkaloids by in vitro root cultures from *Catharanthus trichophyllus*. Phytochemistry 28:2675–2680
31. Delbeque J, Beydon P, Chapuis L (1995) In vitro incorporation of radio labelled cholesterol and mevalonic acid into ecdysteron by hairy root cultures of a plant *Serratula tinctoria*. Eur J Entomol 92:301–307
32. Deno H, Yamagata H, Emoto T, Yoshioka T, Yamada Y, Fujita Y (1987) Scopolamine production by root cultures of *Duboisia myoporoides*: II. Establishment of a hairy root culture by infection with *Agrobacterium rhizogenes*. J Plant Physiol 131:315–323
33. Dhakulkar S, Ganapathi TR, Bhargava S, Bapat VA (2005) Induction of hairy roots in *Gmelina arborea* Roxb. and production of verbascoside in hairy roots. Plant Sci 169:812–818
34. Djilani A, Dicko A (2012) The therapeutic benefits of essential oils. In: Bouayed J (ed) Nutrition, well-being and health. In Tech.

35. Dondorp AM, Nosten F, Yi P, Das D, Phyo AP, Tarning J, Lwin KM, Ariey F, Hanpithakpong W, Lee SJ, Ringwald P, Silamut K, Imwong M, Chotivanich K, Lim P, Herdman T, An SS, Yeung S, Singhasivanon P, Day NPJ, Lindegardh N, Socheat D, White NJ (2009) Artemisinin resistance in *Plasmodium falciparum* malaria. N Engl J Med 361:455–467
36. Von Eickstedt K, Rahman S (1969) Psychopharmacologic effect of valepotriates. Arzneim Forsch 19:316–319
37. Etsè KD, Aïdam AV, Melin C, Blanc N, Oudin A, Courdavault V, Creche J, Lanoue A (2014) Optimized genetic transformation of *Zanthoxylum zanthoxyloides* by *Agrobacterium rhizogenes* and the production of chelerythrine and skimmiamine in hairy root cultures. Eng Life Sci 14:95–99
38. Le Flem-Bonhomme V, Laurain-Mattar D, Fliniaux MA (2004) Hairy root induction of *Papaver somniferum* var. album, a difficult-to-transform plant, by *A. rhizogenes* LBA 9402. Planta 218:890–893
39. Flores HE, Vivanco JM, Loyola-Vargas VM (1999) 'Radicle' biochemistry: the biology of root-specific metabolism. Trends Plant Sci 4:220–226
40. Flores H, Pickard J, Hoy M (1988) Production of polyacetylenes and thiophenes in heterotrophic and photosynthetic root cultures of Asteraceae. In: Lam J, Breheler H, Arnason T, Hansen L (eds) Chemistry and biology of naturally occurring acetylenes and related compounds (NOARC) bioactive molecules, vol 7, pp 233–254
41. Fraga BM, Díaz CE, Guadaño A, González-Coloma A (2005) Diterpenes from *Salvia broussonetii* transformed roots and their insecticidal activity. J Agric Food Chem 53: 5200–5206
42. Fu C, Zhao D, Xue X, Jin Z, Ma F (2005) Transformation of *Saussurea involucrata* by *Agrobacterium rhizogenes*: hairy root induction and syringin production. Proc Biochem 40:3789–3794
43. Furmanowa M, Syklowska-Baranek K (2000) Hairy root cultures of Taxus × media var. Hicksii Rehd. as a new source of paclitaxel and 10-deacetylbaccatin III. Biotechnol Lett 22:683–686
44. Gangopadhyay M, Sircar D, Mitra A, Bhattacharya S (2008) Hairy root culture of *Plumbago indica* as a potential source for plumbagin. Biol Plant 52:533–537
45. Georgiev M, Pavlov A, Bley T (2007) Hairy root type plant in vitro systems as sources of bioactive substances. Appl Microbiol Biotechnol 74:1175–1185
46. Giri A, Banerjee S, Ahuja PS, Giri CC (1997) Production of hairy roots in *Aconitum heterophyllum* wall. using *Agrobacterium rhizogenes*. In Vitro Cell Dev Biol Plant 33:280–284
47. Giri A, Narasu M (2000) Transgenic hairy roots: recent trends and applications. Biotechnol Adv 18:1–22
48. Grzegorczyk I, Królicka A, Wysokińska H (2006) Establishment of *Salvia officinalis* L. hairy root cultures for the production of rosmarinic acid. Z Naturforsch C 61:351–356
49. Gränicher F, Christen P, Kapetanidis I (1992) High-yield production of valepotriates by hairy root cultures of *Valeriana officnalis* L. var. *sambucifolia* Mikan. Plant Cell Rep 11:339–342
50. Gränicher F, Christen P, Kapétanidis I (1995) Production of valepotriates by hairy root cultures of *Centranthus ruber* DC. Plant Cell Rep 14:294–298
51. Grąbkowska R, Królicka A, Mielicki W, Wielanek M, Wysokińska H (2010) Genetic transformation of *Harpagophytum procumbens* by *Agrobacterium rhizogenes*: iridoid and phenylethanoid glycoside accumulation in hairy root cultures. Acta Physiol Plant 32:665–673
52. Guillon S, Trémouillaux-Guiller J, Pati P, Rideau M, Gantet P (2006) Hairy root research: recent scenario and exciting prospects. Curr Opin Plant Biol 9:341–346
53. Guillon S, Trémouillaux-Guiller J, Pati P, Rideau M, Gantet P (2006) Harnessing the potential of hairy roots: dawn of a new era. Trends Biotechnol 24:403–409

54. Guillon S, Trémouillaux-Guiller J, Pati P, Gantet P (2008) Hairy roots: a powerful tool for plant biotechnological advances. In: Ramawat K, Merillon J (eds) Bioactive molecules and medicinal plants. Springer, Berlin Heidelberg, pp 271–283
55. Gunjan S, Lutz J, Bushong A, Rogers D, Littleton J (2013) Hairy root cultures and plant regeneration in *Solidago nemoralis* transformed with *Agrobacterium rhizogenes*. Am J Plant Sci 4:1675–1678
56. Głąb B, Furmanek T, Miklaszewska M, Banaś A, Królicka A (2013) Lipids in hairy roots and non-Agrobacterium induced roots of *Crambe abyssinica*. Acta Physiol Plant 35:2137–2145
57. Hamill J, Robins R, Rhodes M (1989) Alkaloid production by transformed root cultures of *Cinchona ledgeriana*. Planta Med 55:354–357
58. Hayes J, Kelleher M, Eggleston I (2008) The cancer chemopreventive actions of phytochemicals derived from glucosinolates. Eur J Nutr 47:73–88
59. Hirata K, Goda S, Phunchindawan M, Du D, Ishio M, Sakai A, Miyamoto K (1998) Cryopreservation of horseradish hairy root cultures by encapsulation-dehydration. J Ferment Bioeng 86:418–420
60. Hong S, Peebles CAM, Shanks JV, San K, Gibson SI (2006) Terpenoid indole alkaloid production by *Catharanthus roseus* hairy roots induced by *Agrobacterium tumefaciens* harboring rol ABC genes. Biotechnol Bioeng 93:386–390
61. Hook I (1994) Secondary metabolites in hairy root cultures of *Leontopodium alpinum* Cass. (Edelweiss). Plant Cell Tissue Org Cult 38:321–326
62. Huang W-Y, Cai Y-Z, Zhang Y (2009) Natural phenolic compounds from medicinal herbs and dietary plants: potential use for cancer prevention. Nutr Cancer 62:1–20
63. Huang SH, Vishwakarma RK, Lee TT, Chan HS, Tsay HS (2014) Establishment of hairy root lines and analysis of iridoids and secoiridoids in the medicinal plant *Gentiana scabra*. Bot Stud 55:17
64. Ikenaga T, Oyama T, Muranaka T (1995) Growth and steroidal saponin production in hairy root cultures of *Solanum aculeatissimum*. Plant Cell Rep 14:413–417
65. Inoguchi M, Ogawa S, Furukawa S, Kondo H (2003) Production of an Allelopathic Polyacetylene in Hairy Root Cultures of Goldenrod (*Solidago altissima* L.). Biosci Biotechnol Biochem 67:863–868
66. Ionkova I, Kartnig T, Alfermann W (1997) Cycloartane saponin production in hairy root cultures of *Astragalus mongholicus*. Phytochemistry 45:1597–1600
67. Ishimaru K, Shimomura K (1991) Tannin production in hairy root culture of *Geranium thunbergii*. Phytochemistry 30:825–828
68. Ishimaru K, Sudo H, Satake M, Matsunaga Y, Hasegawa Y, Takemoto S, Shimomura K (1990) Amarogentin, amaroswerin and four xanthones from hairy root cultures of *Swertia japonica*. Phytochemistry 29:1563–1565
69. Jaziri M, Legros M, Homes J, Vanhaelen M (1988) Tropine alkaloids production by hairy root cultures of *Datura stramonium* and *Hyoscyamus niger*. Phytochemistry 27:419–420
70. Jaziri M, Shimomura K, Yoshimatsu K, Fauconnier M-L, Marlier M, Homes J (1995) Establishment of normal and transformed root cultures of *Artemisia annua* L. for artemisinin production. J Plant Physiol 145:175–177
71. Jun Cheul A, Baik H, Tada H, Ishimaru K, Sasaki K, Shimomura K (1996) Polyacetylenes in hairy roots of *Platycodon grandiflorum*. Phytochemistry 42:69–72
72. Jung G, Tepfer D (1987) Use of genetic transformation by the Ri T-DNA of *Agrobacterium rhizogenes* to stimulate biomass and tropane alkaloid production in *Atropa belladonna* and *Calystegia sepium* roots grown in vitro. Plant Sci 50:145–151
73. Karwasara V, Dixit V (2009) *Agrobacterium rhizogenes* mediated genetic transformation of *Abrus precatorius* L. Pharmacognosy Mag 5:336–342
74. Kaur K, Jain M, Kaur T, Jain R (2009) Antimalarials from nature. Bioorg Med Chem 17:3229–3256

75. Kim O, Bang K, Shin Y, Lee M, Jung S, Hyun D, Kim Y, Seong N, Cha S, Hwang B (2007) Enhanced production of asiaticoside from hairy root cultures of *Centella asiatica* (L.) Urban elicited by methyl jasmonate. Plant Cell Rep 26:1941–1949
76. Kim S, Cha M, Lee E, Kim I, Kwon J, Kang S, Park T (2012) In vitro induction of hairy root from isoflavones-producing Korean wild arrowroot *Pueraria lobata*. J Plant Biotechnol 39:205–211
77. Kim H, Popova E, Shin D, Bae C, Baek H, Park S, Engelmann F (2012) Development of a droplet-vitrification protocol for cryopreservation of *Rubia akane* (nakai) hairy roots using a systematic approach. Cryo Lett 33:506–517
78. Kim H, Popova E, Yi J, Cho G, Park S, Lee S, Engelmann F (2010) Cryopreservation of hairy roots of *Rubia akane* (Nakai) using a droplet-vitrification procedure. Cryo Lett 31:473–484
79. Kim Y, Xu H, Park W, Park N, Lee S, Park S (2010) Genetic transformation of buckwheat (*Fagopyrum esculentum* M.) with *Agrobacterium rhizogenes* and production of rutin in transformed root cultures. Aust J Crop Sci 4:485–490
80. Kino-Oka M, Mine K, Taya M, Tone S, Ichi T (1994) Production and release of anthraquinone pigments by hairy roots of madder (*Rubia tinctorum* L.) under improved culture conditions. J Ferment Bioeng 77:103–106
81. Kisiel W, Stojakowska A (1997) A sesquiterpene coumarin ether from transformed roots of *Tanacetum parthenium*. Phytochemistry 46:515–516
82. Kisiel W, Stojakowska A, Malarz J, Kohlmüzer S (1995) Sesquiterpene lactones in *Agrobacterium rhizogenes*—transformed hairy root culture of *Lactuca virosa*. Phytochemistry 40:1139–1140
83. Kitanaka S, Kunishi M, Taira T, Takahashi S, Yazawa S, Miyazawa Y, Nakajima Y, Okuno K, Fukuoka S, Takido M (1997) Ginsenoside production, tissue culture and induction of regeneration of interspecific hybrid ginseng (*Panax ginseng* x *P. quinquefolium*). Nat Med 1–3
84. Kittipongpatana N, Hock R, Porter J (1998) Production of solasodine by hairy root, callus, and cell suspension cultures of *Solanum aviculare* Forst. Plant Cell Tissue Org Cult 52:133–143
85. Kjeldsen E, Svejstrup J, Gromova I, Alsner J, Westergaard O (1992) Camptothecin inhibits both the cleavage and religation reactions of eukaryotic DNA topoisomerase I. J Mol Biol 228:1025–1030
86. Knopp E, Strauss A, Wehrli W (1988) Root induction on several *Solanaceae* species by *Agrobacterium rhizogenes* and the determination of root tropane alkaloid content. Plant Cell Rep 7:590–593
87. Ko K, Ebizuka Y, Noguchi H, Sankawa U (1995) Production of polypeptide pigments in hairy root cultures of Cassia plants. Chem Pharm Bull (Tokyo) 43:274–278
88. Królicka A, Staniszewska I, Bielawski K, Maliński E, Szafranek J, Łojkowska E (2001) Establishment of hairy root cultures of *Ammi majus*. Plant Sci 160:259–264
89. Kursinszki L, Troilina J, Szõke É (1998) Determination of visnagin in *Ammi visnaga* hairy root cultures using solid-phase extraction and high-performance liquid chromatography. Microchem J 59:392–398
90. Kuźma Ł, Kisiel W, Królicka A, Wysokińska H (2011) Genetic transformation of *Salvia austriaca* by *Agrobacterium rhizogenes* and diterpenoid isolation. Pharmazie 66:904–907
91. Kuźma Ł, Skrzypek Z, Wysokinska H (2006) Diterpenoids and triterpenoids in hairy roots of *Salvia sclarea*. Plant Cell Tissue Org Cult 84:152–160
92. Kyo M, Miyauchi Y, Fujimoto T, Mayama S (1990) Production of nematocidal compounds by hairy root cultures of *Tagetes patula* L. Plant Cell Rep 9:393–397
93. Lambert E, Goossens A, Panis B, Van Labeke M, Geelen D (2009) Cryopreservation of hairy root cultures of *Maesa lanceolata* and *Medicago truncatula*. Plant Cell Tissue Org Cult 96:289–296
94. Laurain D, Trémouillaux-Guiller J, Chénieux JC, Van Beek T (1997) Production of ginkgolides and bilobalide in transformed and gametophyte derived cell cultures of *Ginkgo biloba*. Phytochemistry 46:127–130

95. Lee S, Xu H, Kim Y, Park S (2008) Rosmarinic acid production in hairy root cultures of *Agastache rugosa* Kuntze. World J Microbiol Biotechnol 24:969–972
96. Li F-X, Jin Z-P, Zhao D-X, Cheng L-Q, Fu C-X, Ma F (2006) Overexpression of the *Saussurea medusa* chalcone isomerase gene in *S. involucrata* hairy root cultures enhances their biosynthesis of apigenin. Phytochemistry 67:553–560
97. Lin HW, Kwok KH, Doran PM (2003) Development of *Linum flavum* hairy root cultures for production of coniferin. Biotechnol Lett 25:521–525
98. Lodhi AH, Bongaerts RJM, Verpoorte R, Coomber SA, Charlwood BV (1996) Expression of bacterial isochorismate synthase (EC 5.4.99.6) in transgenic root cultures of *Rubia peregrina*. Plant Cell Rep 16:54–57
99. Lorence A, Medina-Bolivar F, Nessler CL (2004) Camptothecin and 10-hydroxycamptothecin from *Camptotheca acuminata* hairy roots. Plant Cell Rep 22:437–441
100. De Luca V, St Pierre B (2000) The cell and developmental biology of alkaloid biosynthesis. Trends Plant Sci 5:168–173
101. Majumdar S, Garai S, Jha S (2011) Genetic transformation of *Bacopa monnieri* by wild type strains of *Agrobacterium rhizogenes* stimulates production of bacopa saponins in transformed calli and plants. Plant Cell Rep 30:941–954
102. Malarz J, Stojakowska A, Kisiel W (2002) Sesquiterpene lactones in a hairy root culture of *Cichorium intybus*. Z Naturforsch 57c:994–997
103. Mano Y, Nabeshima S, Matsui C, Ohkawa H (1987) Production of tropane alkaloids by hairy-root cultures of *Scopolia japonica*. Agric Biol Chem 50:2715–2722
104. Mano Y, Ohkawa H, Yamada Y (1989) Production of tropane alkaloids by hairy root cultures of *Duboisia leichhardtii* transformed by *Agrobacterium rhizogenes*. Plant Sci 59:191–201
105. Mathur A, Gangwar A, Mathur A, Verma P, Uniyal G, Lal R (2010) Growth kinetics and ginsenosides production in transformed hairy roots of American ginseng—*Panax quinquefolium* L. Biotechnol Lett 32:457–461
106. Matsumoto T, Tanaka N (1991) Production of phytoecdysteroids by hairy root cultures of *Ajuga reptans* var. atropurpurea. Agric Biol Chem 55:1019–1025
107. McCue P, Shetty K (2004) Health benefits of soy isoflavonoids and strategies for enhancement: a review. Crit Rev Food Sci Nutr 44:361–367
108. Medina-Bolivar F, Condori J, Rimando A, Hubstenberger J, Shelton K, O'Keefe S, Bennett S, Dolan M (2007) Production and secretion of resveratrol in hairy root cultures of peanut. Phytochemistry 68:1992–2003
109. Mencović N, Šavikin-Fodulović K, Vinterhalter B, Vinterhalter D, Jancović T, Krstić D (2000) Secoiridoid content in hairy roots of *Gentiana punctata*. Pharm Pharmacol Lett 10:73–75
110. Merkli A, Christen P, Kapetanidis I (1997) Production of diosgenin by hairy root cultures of *Trigonella foenum-graecum* L. Plant Cell Rep 16:632–636
111. Morris P, Robbins MP (1992) Condensed tannin formation by *Agrobacterium rhizogenes* transformed root and shoot organ cultures of *Lotus corniculatus*. J Exp Bot 43:221–231
112. Motomori Y, Shimomura K, Mori K, Kunitake H, Nakashima T, Tanaka M, Miyazaki S, Ishimaru K (1995) Polyphenol production in hairy root cultures of Fragaria x ananassa. Phytochemistry 40:1425–1428
113. Muji Ermayanti T, McComb JA, O'Brien PA (1994) Stimulation of synthesis and release of swainosonine from transformed roots of *Swainsona galegifolia*. Phytochemistry 36:313–317
114. Murakami Y, Shimomura K, Yoshihira K, Ishimaru K (1998) Polyacetylenes in hairy root cultures of *Trachelium caeruleum* L. J Plant Physiol 152:574–576
115. Murthy HN, Dijkstra C, Anthony P, White DA, Davey MR, Power JB, Hahn EJ, Paek KY (2008) Establishment of *Withania somnifera* hairy root cultures for the production of Withanolide A. J Integr Plant Biol 50:975–981
116. Nagella P, Thiruvengadam M, Jung SJ, Murthy HN, Chung IM (2013) Establishment of *Gymnema sylvestre* hairy root cultures for the production of gymnemic acid. Acta Physiol Plant 35:3067–3073

117. Nam Y, Choi M, Hwang H, Lee M-G, Kwon B-M, Lee W-H, Suk K (2013) Natural flavone jaceosidin is a neuroinflammation inhibitor. Phytother Res 27:404–411
118. Nin S, Bennici A, Roselli G, Mariotti D, Schiff S, Magherini R (1997) Agrobacterium-mediated transformation of *Artemisia absinthium* L. (wormwood) and production of secondary metabolites. Plant Cell Rep 16:725–730
119. Norberg A, Hoa N, Liepinsh E, Phan D, Thuan N, Jornvall H, Sillard R, Ostenson C (2004) A novel insulin-releasing substance, phanoside, from the plant *Gynostemma pentaphyllum*. J Biol Chem 279:41361–41367
120. Nussbaumer P, Kapétanidis I, Christen P (1998) Hairy roots of *Datura candida*×*D. aurea*: effect of culture medium composition on growth and alkaloid biosynthesis. Plant Cell Rep 17:405–409
121. Ogasawara T, Chiba K, Tada M (1993) Production in high-yield of a naphthoquinone by a hairy root culture of *Sesamum indicum*. Phytochemistry 33:1095–1098
122. Okršlar V, Štrukelj B, Kreft S, Bohanec B, Zel J (2002) Micropropagation and hairy root culture of *Solanum Laciniatum* Ait. In Vitro Cell Dev Biol Plant 38:352–357
123. Ono N, Bandaranayake PCG, Tian L (2012) Establishment of pomegranate (*Punica granatum*) hairy root cultures for genetic interrogation of the hydrolyzable tannin biosynthetic pathway. Planta 236:931–941
124. Ono N, Tian L (2011) The multiplicity of hairy root cultures: prolific possibilities. Plant Sci 180:439–446
125. Pangeni R, Sahni JK, Ali J, Sharma S, Baboota S (2014) Resveratrol: review on therapeutic potential and recent advances in drug delivery. Expert Opin Drug Deliv 11:1285–1298
126. Park S, Kim Y, Lee S (2009) Establishment of hairy root culture of *Rubia akane* Nakai for alizarin and purpurin production. Sci Res Essays 4:94–97
127. Parr A, Hamill J (1987) Relationship between *Agrobacterium rhizogenes* transformed hairy roots and intact, uninfected nicotiana plants. Phytochemistry 26:3241–3245
128. Patra N, Srivastava A (2014) Enhanced production of artemisinin by hairy root cultivation of *Artemisia annua* in a modified stirred tank reactor. Appl Biochem Biotechnol 1–14
129. Porter J (1991) Host range and implications of plant infection by *Agrobacterium rhizogenes*. Crit Rev Plant Sci 10:387–421
130. Pradel H, Dumke-Lehmann U, Diettrich B, Luckner M (1997) Hairy root cultures of *Digitalis lanata*. Secondary metabolism and plant regeneration. J Plant Physiol 151:209–215
131. Qin M, Li G, Yun Y, Ye H, Li G (1994) Induction of hairy root from *Artemisia annua* with *Agrobacterium Rhizogenes* and its culture in vitro. J Integr Plant Biol 36:165–170
132. Qin B, Ma L, Wang Y, Chen M, Lan X, Wu N, Liao Z (2014) Effects of acetylsalicylic acid and UV-B on gene expression and tropane alkaloid biosynthesis in hairy root cultures of *Anisodus luridus*. Plant Cell Tissue Organ Cult 117:483–490
133. Romero F, Delate K, Kraus G, Solco A, Murphy P, Hannapel D (2009) Alkamide production from hairy root cultures of Echinacea. In Vitro Cell Dev Biol Plant 45:599–609
134. Rostampour S, Sohi H, Jourabchi E, Ansari E (2009) Influence of *Agrobacterium rhizogenes* on induction of hairy roots and benzylisoquinoline alkaloids production in Persian poppy (*Papaver bracteatum* Lindl.): preliminary report. World J Microbiol Biotechnol 25:1807–1814
135. Ruiz L, Ruiz L, Maco M, Cobos M, Gutierrez-Choquevilca A-L, Roumy V (2011) Plants used by native Amazonian groups from the Nanay River (Peru) for the treatment of malaria. J Ethnopharm 133:917–921
136. Saito K, Sudo H, Yamazaki M, Koseki-Nakamura M, Kitajima M, Takayama H, Aimi N (2001) Feasible production of camptothecin by hairy root culture of *Ophiorrhiza pumila*. Plant Cell Rep 20:267–271
137. Saito K, Yamazaki M, Shimomura K, Yoshimatsu K, Murakoshi I (1990) Genetic transformation of foxglove (*Digitalis purpurea*) by chimeric foreign genes and production of cardioactive glycosides. Plant Cell Rep 9:121–124

138. Sakai A, Engelmann F (2007) Vitrification, encapsulation-vitrification and droplet-vitrification: a review. Cryo Lett 28:151–172
139. Salma M, Engelmann-Sylvestre I, Collin M, Escoute J, Lartaud M, Yi J-Y, Kim H, Verdeil J, Engelmann F (2014) Effect of the successive steps of a cryopreservation protocol on the structural integrity of *Rubia akane* Nakai hairy roots. Protoplasma 251:649–659
140. Santos PG, Figueiredo AC, Lourenço PL, Barroso J, Pedro L, Oliveira MM, Schripsema J, Deans S, Scheffer JC (2002) Hairy root cultures of *Anethum graveolens* (dill): establishment, growth, time-course study of their essential oil and its comparison with parent plant oils. Biotechnol Lett 24:1031–1036
141. Santos PAG, Figueiredo AC, Oliveira MM, Barroso JG, Pedro LG, Deans SG, Scheffer JJC (2005) Growth and essential oil composition of hairy root cultures of *Levisticum officinale* W. D.J. Koch (lovage). Plant Sci 168:1089–1096
142. Santos PM, Figueiredo AC, Oliveira MM, Barroso J, Pedro LG, Deans SG, Younus AKM, Scheffer JJC (1998) Essential oils from hairy root cultures and from fruits and roots of *Pimpinella anisum*. Phytochemistry 48:455–460
143. Sasaki K, Udagawa A, Ishimaru H, Hayashi T, Alfermann AW, Nakanishi F, Shimomura K (1998) High forskolin production in hairy roots of *Coleus forskohlii*. Plant Cell Rep 17:457–459
144. Sato K, Yamazaki T, Okuyama E, Yoshihira K, Shimomura K (1991) Anthraquinone production by transformed root cultures of *Rubia tinctorum*: influence of phytohormones and sucrose concentration. Phytochemistry 30:1507–1509
145. Sauerwein M, Ishimaru K, Shimomura K (1991) Indole alkaloids in hairy roots of *Amsonia elliptica*. Phytochemistry 30:1153–1155
146. Sauerwein M, Shimomura K (1991) Alkaloid production in hairy roots of *Hyoscyamus albus* transformed with *Agrobacterium rhizogenes*. Phytochemistry 30:3277–3280
147. Sauerwein M, Yamazaki T, Shimomura K (1991) Hernandulcin in hairy root cultures of *Lippia dulcis*. Plant Cell Rep 9:579–581
148. Savona M, Mascarello C, Bisio A, Romussi G, Profumo P, Warchol M, Bach A, Ruffoni B (2003) *S. cinnabarina* Martens et Galeotti: optimisation of the extraction of a new compound, tissue culture and hairy root transformation. Agr Med 133:28–35
149. Shanks JV, Bhadra R, Morgan J, Rijhwani S, Vani S (1998) Quantification of metabolites in the indole alkaloid pathways of *Catharanthus roseus*: implications for metabolic engineering. Biotechnol Bioeng 58:333–338
150. Shanks J, Morgan J (1999) Plant 'hairy root' culture. Curr Opin Biotech 10:151–155
151. Shi H, Kintzios S (2003) Genetic transformation of *Pueraria phaseoloides* with *Agrobacterium rhizogenes* and puerarin production in hairy roots. Plant Cell Rep 21:1103–1107
152. Shimomura K, Sudo H, Saga H, Kamada H (1991) Shikonin production and secretion by hairy root cultures of *Lithospermum erythrorhizon*. Plant Cell Rep 10:282–285
153. Shinde A, Malpathak N, Fulzele D (2009) Enhanced production of phytoestrogenic isoflavones from hairy root cultures of *Psoralea corylifolia* L. Using elicitation and precursor feeding. Biotechnol Bioprocess Eng 14:288–294
154. Siddiq A, Dembitsky V (2008) Acetylenic anticancer agents. Anticancer Agents Med Chem 8:132–170
155. Sidwa-Gorycka M, Krolicka A, Orlita A, Malinski E, Golebiowski M, Kumirska J, Chromik A, Biskup E, Stepnowski P, Lojkowska E (2009) Genetic transformation of *Ruta graveolens* L. by *Agrobacterium rhizogenes*: hairy root cultures a promising approach for production of coumarins and furanocoumarins. Plant Cell Tissue Organ Cult 97:59–69
156. Sircar D, Roychowdhury A, Mitra A (2007) Accumulation of p-hydroxybenzoic acid in hairy roots of *Daucus carota*. J Plant Physiol 164:1358–1366
157. Sivakumar G, Liu C, Towler MJ, Weathers PJ (2010) Biomass production of hairy roots of *Artemisia annua* and *Arachis hypogaea* in a scaled-up mist bioreactor. Biotechnol Bioeng 107:802–813

158. Spena A, Schmülling T, Koncz C, Schell J (1987) Independent and synergistic activity of *rol A*, *B*, and *C* loci in stimulating abnormal growth in plants. EMBO J 6:3891–3899
159. Srivastava S, Srivastava AK (2012) Azadirachtin production by hairy root cultivation of *Azadirachta indica* in a modified stirred tank reactor. Bioprocess Biosyst Eng 35:1549–1553
160. Sudha C, Obul Reddy B, Ravishankar G, Seeni S (2003) Production of ajmalicine and ajmaline in hairy root cultures of *Rauvolfia micrantha* Hook f., a rare and endemic medicinal plant. Biotechnol Lett 25:631–636
161. Swain S, Rout K, Chand P (2012) Production of triterpenoid anti-cancer compound taraxerol in Agrobacterium-transformed root cultures of Butterfly Pea (*Clitoria ternatea* L.). Appl Biochem Biotechnol 168:487–503
162. Sáenz-Carbonell L, Loyola-Vargas V (1997) *Datura stramonium* hairy roots tropane alkaloid content as a response to changes in Gamborg's B5 medium. Appl Biochem Biotechnol 61:321–337
163. Tada H, Nakashima T, Kunitake H, Mori K, Tanaka M, Ishimaru K (1996) Polyacetylenes in hairy root cultures of *Campanula medium* L. J Plant Physiol 147:617–619
164. Tada H, Shimomura K, Ishimaru K (1995) Polyacetylenes in Hairy Root Cultures of *Lobelia chinensis* Lour. J Plant Physiol 146:199–202
165. Takeda T, Kondo T, Mizukami H, Ogihara Y (1994) Bryonolic acid production in hairy roots of *Trichosanthes kirilowii* Max. var Japonica Kitam. Transformed with *Agrobacterium rhizogenes* and its cytotoxic activity. Chem Pharm Bull (Tokyo) 42:730–732
166. Tanaka N, Matsuura E, Terahara N, Ishimaru K (1999) Secondary metabolites in transformed root cultures of *Campanula glomerata*. J Plant Physiol 155:251–254
167. Tanaka N, Takao M, Matsumoto T (1994) *Agrobacterium rhizogenes* mediated transformation and regeneration of *Vinca minor* L. Plant Tissue Cult Lett 11:191–198
168. Tang W, Eisenbrand G (1992) Gentiana spp. In: Tang W, Eisenbrand G (eds) Chinese drugs of plant origin, chemistry, pharmacology, and use in traditional and modern medicine. Springer, Berlin, Heidelberg, pp 549–553
169. Taya M, Mine K, Kino-Oka M, Tone S, Ichi T (1992) Production and release of pigments by culture of transformed hairy root of red beet. J Ferment Bioeng 73:31–36
170. Teoh K, Weathers P, Cheetham R, Walcerz D (1996) Cryopreservation of transformed (hairy) roots of *Artemisia annua*. Cryobiol 33:106–117
171. Thiruvengadam M, Praveen N, Kim E-H, Kim S-H, Chung I-M (2014) Production of anthraquinones, phenolic compounds and biological activities from hairy root cultures of *Polygonum multiflorum* Thunb. Protoplasma 251:555–566
172. Thron U, Maresch L, Beiderbeck R, Reichling J (1989) Accumulation of unusual phenylpropanoids in transformed and non-transformed root cultures of *Coreopsis tinctoria*. Z Naturforsch 44:573–577
173. Tiwari RK, Trivedi M, Guang ZC, Guo GQ, Zheng GC (2007) Genetic transformation of *Gentiana macrophylla* with *Agrobacterium rhizogenes*: growth and production of secoiridoid glucoside gentiopicroside in transformed hairy root cultures. Plant Cell Rep 26:199–210
174. Touno K, Yoshimatsu K, Shimomura K (2006) Characteristics of *Atropa belladonna* hairy roots cryopreserved by vitrification method. Cryo Lett 27:65–72
175. Triplett B, Moss S, Bland J, Dowd M (2008) Induction of hairy root cultures from *Gossypium hirsutum* and *Gossypium barbadense* to produce gossypol and related compounds. In Vitro Cell Dev Biol Plant 44:508–517
176. Tusevski O, Stanoeva J, Stefova M, Kungulovski D, Pancevska N, Sekulovski N, Panov S, Simic S (2013) Hairy roots of *Hypericum perforatum* L.: a promising system for xanthone production. Cent Eur J Biol 8:1010–1022
177. Udomsuk L, Jarukamjorn K, Tanaka H, Putalun W (2011) Improved isoflavonoid production in *Pueraria candollei* hairy root cultures using elicitation. Biotechnol Lett 33:369–374
178. Uozumi N, Kobayashi T (1995) Artificial seed production through encapsulation of hairy root and shoot tips. In: Bajaj YPS (ed) Somatic embryogenesis and synthetic seed I. Springer, Berlin, Heidelberg, vol 30, pp 170–180

179. Vdovitchenko MY, Kuzovkina IN (2011) Artificial seed preparation as the efficient method for storage and production of healthy cultured roots of medicinal plants. Russ J Plant Physiol 58:524–530
180. Venugopala K, Rashmi V, Odhav B (2013) Review on natural coumarin lead compounds for their pharmacological activity. Biomed Res Int, 963248
181. Verma PC, Singh D, Rahman Lu, Gupta MM, Banerjee S (2002) In vitro-studies in *Plumbago zeylanica*: rapid micropropagation and establishment of higher plumbagin yielding hairy root cultures. J Plant Physiol 159:547–552
182. Verma P, ur Rahman L, Negi A, Jain D, Khanuja SPS, Banerjee S (2007) *Agrobacterium rhizogenes*-mediated transformation of *Picrorhiza kurroa* Royle ex Benth.: establishment and selection of superior hairy root clone. Plant Biotechnol Rep 1:169–174
183. Wagner H, Jurcic K (1979) On the spasmolytic activity of valeriana extracts. Planta Med 37:84–86
184. Wahby I, Arráez-Román D, Segura-Carretero A, Ligero F, Caba JM, Fernández-Gutiérrez A (2006) Analysis of choline and atropine in hairy root cultures of *Cannabis sativa* L. by capillary electrophoresis-electrospray mass spectrometry. Electrophoresis 27:2208–2215
185. Wan J, Gong X, Jiang R, Zhang Z, Zhang L (2013) Antipyretic and anti-inflammatory effects of Asiaticoside in lipopolysaccharide-treated rat through up-regulation of heme oxygenase-1. Phytother Res 27:1136–1142
186. Wang X, Beckham TH, Morris JC, Chen F, Gangemi JD (2008) Bioactivities of gossypol, 6-methoxygossypol, and 6,6′-dimethoxygossypol. J Agric Food Chem 56:4393–4398
187. Wang CT, Liu H, Gao XS, Zhang HX (2010) Overexpression of *G10H* and *ORCA3* in the hairy roots of *Catharanthus roseus* improves catharanthine production. Plant Cell Rep 29:887–894
188. Washida D, Shimomura K, Nakajima Y, Takido M, Kitanaka S (1998) Ginsenosides in hairy roots of a panax hybrid. Phytochemistry 49:2331–2335
189. Weathers PJ, Cheetham RD, Follansbee E, Teoh K (1994) Artemisinin production by transformed roots of *Artemisia annua*. Biotechnol Lett 16:1281–1286
190. Wielanek M, Królicka A, Bergier K, Gajewska E, Skłodowska M (2009) Transformation of *Nasturtium officinale*, *Barbarea verna* and *Arabis caucasica* for hairy roots and glucosinolate-myrosinase system production. Biotechnol Lett 31:917–921
191. Wielanek M, Urbanek H (1999) Glucotropaeolin and myrosinase production in hairy root cultures of *Tropaeolum majus*. Plant Cell Tissue Organ Cult 57:39–45
192. World Health Organization (2013) World malaria report. http://www.who.int/malaria/publications/world_malaria_report_2013/report/en/
193. Wysokiińska H, Rózga M (1998) Establishment of transformed root cultures of *Paulownia tomentosa* for verbascoside production. J Plant Physiol 152:78–83
194. Xu H, Park J, Kim Y, Park N, Lee S, Park S (2009) Optimization of growth and pyranocoumarins production in hairy root culture of *Angelica gigas* Nakai. J Med Plants Res 3:978–981
195. Yamanaka M, Ishibashi K, Shimomura K, Ishimaru K (1996) Polyacetylene glucosides in hairy root cultures of *Lobelia cardinalis*. Phytochemistry 41:183–185
196. Yonemitsu H, Shimomura K, Satake M, Mochida S, Tanaka M, Endo T, Kaji A (1990) Lobeline production by hairy root culture of *Lobelia inflata* L. Plant Cell Rep 9:307–310
197. Yoshikawa T, Furuya T (1987) Saponin production by cultures of *Panax ginseng* transformed with *Agrobacterium rhizogenes*. Plant Cell Rep 6:449–453
198. Yoshimatsu K, Yamaguchi H, Shimomura K (1996) Traits of Panax ginseng hairy roots after cold storage and cryopreservation. Plant Cell Rep 15:555–560
199. Young-Am C, Yu H, Song J, Chun H, Park S (2000) Indigo production in hairy root cultures of *Polygonum tinctorium* Lour. Biotechnol Lett 22:1527–1530
200. Zhai B, Clark J, Ling T, Connelly M, Medina-Bolivar F, Rivas F (2014) Antimalarial evaluation of the chemical constituents of hairy root culture of *Bixa orellana* L. Molecules 19:756–766

201. Zhang H, Liu J, Lu H, Gao S (2009) Enhanced flavonoid production in hairy root cultures of *Glycyrrhiza uralensis* Fisch by combining the over-expression of chalcone isomerase gene with the elicitation treatment. Plant Cell Rep 28:1205–1213
202. Zhao D, Fu C, Chen Y, Ma F (2004) Transformation of *Saussurea medusa* for hairy roots and jaceosidin production. Plant Cell Rep 23:468–474
203. Zhao J, Zou L, Zhang C, Li Y, Peng L, Xiang D, Zhao G (2014) Efficient production of flavonoids in *Fagopyrum tataricum* hairy root cultures with yeast polysaccharide elicitation and medium renewal process. Phcog Mag 10:234–240
204. Zhi BH, Alfermann AW (1993) Diterpenoid production in hairy root cultures of *Salvia miltiorrhiza*. Phytochemistry 32:699–703
205. Zhou Y, Hirotani M, Yoshikawa T, Furuya T (1997) Flavonoids and phenylethanoids from hairy root cultures of *Scutellaria baicalensis*. Phytochemistry 44:83–87

Adv Biochem Eng Biotechnol (2015) 149: 325–358
DOI 10.1007/10_2015_323

Published Online: 20 March 2015

Rheology of Lignocellulose Suspensions and Impact of Hydrolysis: A Review

Tien Cuong Nguyen, Dominique Anne-Archard and Luc Fillaudeau

Abstract White biotechnologies have several challenges to overcome in order to become a viable industrial process. Achieving highly concentrated lignocellulose materials and releasing fermentable substrates, with controlled kinetics in order to regulate micro-organism activity, present major technical and scientific bottlenecks. The degradation of the main polymeric fractions of lignocellulose into simpler molecules is a prerequisite for an integrated utilisation of this resource in a biorefinery concept. The characterisation methods and the observations developed for rheology, morphology, etc., that are reviewed here are strongly dependent on the fibrous nature of lignocellulose, are thus similar or constitute a good approach to filamentous culture broths. This review focuses on scientific works related to the study of the rheological behaviour of lignocellulose suspensions and their evolution during biocatalysis. In order to produce the targeted molecules (synthon), the lignocellulose substrates are converted by enzymatic degradation and are then metabolised by micro-organisms. The dynamics of the mechanisms is limited by coupled phenomena between flow, heat and mass transfers in regard to diffusion (within solid and liquid phases), convection (mixing, transfer coefficients, homogeneity) and specific inhibitors (concentration gradients). As lignocellulose suspensions consist of long entangled fibres for the matrix of industrial interest, they exhibit diverse and complex properties linked to this fibrous character (rheological, morphological, thermal, mechanical and biochemical parameters). Among the main variables to be studied, the rheological behaviour of such suspensions appears to be determinant for process efficiency. It is this behaviour that will determine the

T.C. Nguyen · L. Fillaudeau (✉)
Laboratoire d'Ingénierie des Systèmes Biologiques et des Procédés,
Université de Toulouse; INSA; INRA UMR792, CNRS UMR5504,
Toulouse, France
e-mail: lfillaud@insa-toulouse.fr

D. Anne-Archard
Université de Toulouse, IMFT (Institut de Mécanique des Fluides de Toulouse),
Allée Camille Soula, 31400 Toulouse, France

T.C. Nguyen · D. Anne-Archard · L. Fillaudeau
CNRS, Fédération de Recherche FERMAT (FR 3089),
2 Allée Camille Soula Toulouse, France

equipment to be used and the strategies applied (substrate and biocatalysis feed, mixing, etc.). This review provides an overview of (i) the rheological behaviour of fibrous materials in suspension, (ii) the methods and experimental conditions for their measurements, (iii) the main models used and (iv) their evolution during biocatalytic reactions with a focus on enzymatic hydrolysis.

Keywords Fibre suspension · Hydrolysis · Lignocellulose · Paper pulp · Rheology · Viscosity · Yield stress

Nomenclatures

Abbreviation

BAG	Sugarcane bagasse
CoSt	Corn stover
MCC	Microcrystalline cellulose
MFC	Microfibril cellulose
NCC	Nanocrystalline cellulose
PP	Paper pulp
RVA	Rapid viscosity analyser
WhSt	Wheat straw
WP	Whatman paper

Greek Letters

$\dot{\gamma}$	Shear rate (s^{-1})
$\dot{\gamma}^*$	Complex shear rate (s^{-1})
ϕ	Volume fraction (/)
ϕ_c	Critical volume fraction (/)
ϕ_m	Maximum volume fraction (/)
ϕ_{eff}	Effective volume fraction (/)
$[\mu]$	Intrinsic viscosity (/)
ρ	Density (kg m^{-3})
ρ_s	Substrate density (kg m^{-3})
μ	Viscosity (Pa s)
μ_0	Initial viscosity (Pa s)
μ_∞	Final viscosity (Pa s)
μ_s	Suspending viscosity (Pa s)
μ_{rel}	Relative viscosity (/)
τ	Shear stress (Pa)
τ_0	Yield stress (Pa)
ξ	Ionisation energy (mV)

Latin Letters

C*	Critical substrate concentration (gdm L^{-1})
Cm	Mass concentration (gdm L^{-1})

CrI Crystallinity (%)
d Particle diameter (m)
$d_{4,3}$ Volume mean diameter (m)
n Power-law index (/)
K_p Power constant (/)
K_s Metzner-Otto coefficient (/)
k Consistency index (Pa s^n)
k_c Cosson constant ($Pa^{1/2}$ $s^{1/2}$)
N_p Power number (/)
N_{po} Power number in turbulent flow (/)
Re Reynolds number (/)
Re_g Generalised Reynolds number (/)
Re* Rieger and Novak Reynolds (/)
G′ Elastic modulus (Pa)
G″ Viscous modulus (Pa)
G* Complex modulus (Pa)
N Mixing rate (round per second)
P Power consumption (W)

Contents

1 Introduction

For a number of years now, environmental preoccupations and energy control are at the heart of scientific and social debate. Behind the increase of energy consumption, and the depletion of fossil fuel resources, the developments of biorefineries and industrial biotechnology in general are of the greatest importance to make concrete the vision of an economy based on bioresources. The important role of white biotechnology is to replace petro-derived chemicals by those produced from biomass.

The main biomass resources include short rotation forestry (willow, poplar, *eucalyptus*), wood-based industries (pulp and paper industry, forest residues, sawmill and construction/industrial residues, etc.), sugar crops (sugar beet, sweet sorghum), starch crops (maize, wheat, Jerusalem artichoke, etc.), herbaceous lignocellulosic crops (*miscanthus*), oil crops (rapeseed, sunflower), agricultural waste (straw, slurry), municipal solid waste and refuse, and industrial waste (residues from food industry). These fractions can be used directly as desired biochemical or can be converted by thermal, mechanical, chemical, enzymatic and/or microbial approaches. Among these large users of lignocellulosic resources, the pulp and paper industry holds a strategic position. Pulp and paper industry [1] is able to provide a tried and tested industrial model for the processing of lignocellulosic biomass into pre-treated cellulose pulps. The pulp product of this industry is appropriate for modern biorefining, because it displays low lignin content, is free of inhibitory compounds that can perturb fermentations and is devoid of microbial contaminants. Cellulose biomass thus provides an abundant low-cost resource that has the potential to support large-scale production of fuels and chemicals via biotechnological routes [2].

The first hydrolysis methods used were chemical, but they are less competitive at the moment, because of the cost of reagents and the formation of numerous by-products and inhibitory compounds. They are now challenged by enzymatic methods, which allow more specific hydrolysis at higher yields and in less severe

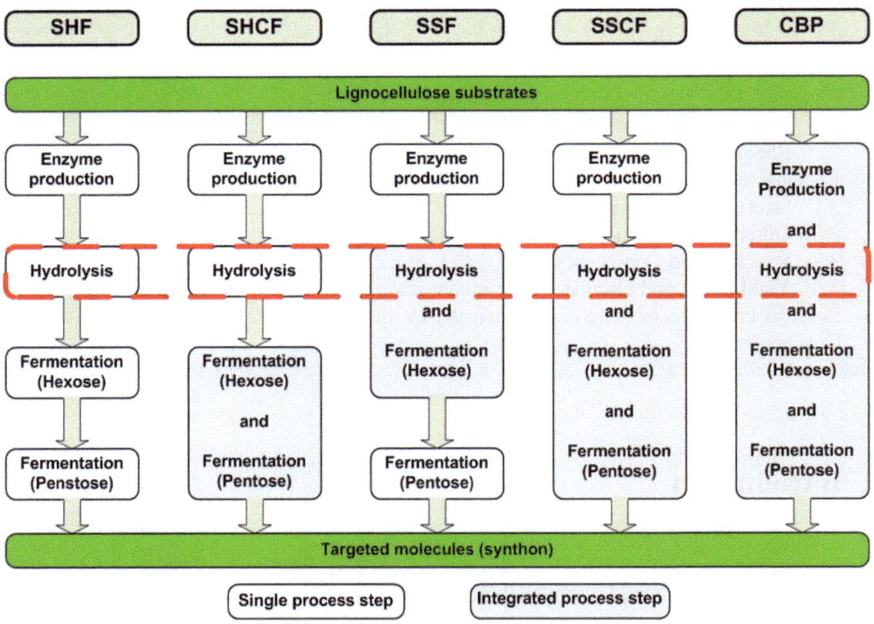

Fig. 1 Various configurations of biologically mediated processing steps during the biocatalytic conversion of lignocellulose and highlight on hydrolysis step with and without interaction with cell cultures. Abbreviations: separate hydrolysis and fermentation (*SHF*), separate hydrolysis and co-fermentation (*SHCF*), simultaneous saccharification and fermentation (*SSF*), simultaneous saccharification and co-fermentation (*SSCF*), and consolidated bioprocessing (*CBP*)

conditions [3]. An enzyme-based process can be divided into four principal steps (Fig. 1): (1) pre-treatment: due to the recalcitrant nature of native lignocelluloses, physical/chemical methods must be applied to generate an enzymatically convertible material; (2) enzymatic hydrolysis: where the cellulose and hemicelluloses are enzymatically degraded into sugar monomers; (3) fermentation: sugar monomers are converted into target molecules, often by yeast; and (4) distillation: to recover products [4–6]. In order to produce sugars, biomass is pre-treated with physical and/or thermo-chemical and possibly biological methods to reduce the size of the feedstock and to open up the plant structure. The cellulose and the hemicellulose portions are broken down (hydrolysed) by enzymes to yield simple sugars that are then fermented into target products by specific strains of micro-organisms. Currently, this biochemical route is the most commonly used [7].

To convert lignocellulose biomass into molecules of interest, the key steps focus on the pre-treatment technique and the conversion into fermentable sugars. The major challenges for cellulose-based products concern cost reduction for production, harvest, transportation and upfront processing in order to make cellulose-based products competitive with grain-based products [8]. Therefore, a better scientific understanding and ultimately technical control of these critical biocatalytic reactions, which involve complex matrices and high solids content, are currently the major challenges that must be won if biorefining operations are to become commonplace. Among the main parameters studied, the rheological behaviour of the hydrolysis suspension and the morphology and size of fibres stand out as major determinants of process efficiency and determine the equipment to be used and the strategies to be applied [9]. Indeed, suspensions exhibit a very wide range of rheological behaviour, and numerous examples can be found to illustrate shear-thinning/shear-thickening behaviour, viscoplasticity with observation of yield stress, elasticity or thixotropy. A lot of parameters influence the nature and the intensity of these non-Newtonian characteristics: the concentration, the particle size and the morphology of the objects, the nature and the magnitude of the particle interactions, etc. Many industries, especially when bioprocesses are involved, are confronted with these behaviours which can sometimes drastically affect hydrodynamics and transfer efficiency (mass or/and heat). It is then of crucial interest to explore the rheology of the suspension to ensure optimal process implementation and the right choice of equipment [10]. It requires a thorough knowledge of the rheological behaviour of the substrate suspensions.

These suspensions, after or simultaneously with hydrolysis, are converted into target molecules by various micro-organisms whether prokaryotes or eukaryotes. The physico-bio-chemical complexities of the broth resulting from these multiphase and dynamic media lead to additional complexity compounded by the fibrous morphology. The rheological behaviour is sensitive to multifactorial parameters (medium compositions, molecules produced, morphology and physiology of the micro-organisms present). Considering flow properties of suspensions, strong analogies between the growth of filamentous fungus cultures [11] and hydrolysis of complex lignocellulose matrices can be made concerning the rheological behaviour, hydrodynamics and transfers (definition of significant morphological criteria, impact of concentration and morphology, main rheological parameters: shear-thinning behaviour, yield stress). In this paper, the methods and the models related to the rheological behaviour of complex lignocellulosic suspensions are similar to or adapted from filamentous micro-organism suspensions. For simultaneous or consolidated bioprocesses, a good knowledge of the physico-bio-chemical properties of suspensions (lignocellulose materials, micro-organisms) is required to define accurate operating conditions (transfer efficiency) for substrate release kinetics (biocatalytic hydrolysis) and micro-organism metabolism (growth, production) kinetics.

A database interrogation was conducted to highlight scientific production focusing on "lignocellulosic materials, flow properties and biocatalytic degradation" for the last 40 years. In this aim, four specific profiles were defined:

- Profile 1: (Rheo* OR visco* OR newt*) AND (suspension* OR dissolution*) AND cellulose*
- Profile 2: (Rheol* OR visco*) AND fiber* AND cellulose*
- Profile 3: (Rheol* OR visco*) AND (paper pulp* OR pulp suspension*)
- Profile 4: Bioproce*

(* stands for truncation option, i.e.: visco* stands for viscosity, viscosimetry, viscous…).

The first and second profiles focus on the rheological behaviour of cellulose/fibre suspensions. The third one scrutinises specifically the pulp and paper industry, while the last one describes the global trends in biotechnological process. Figure 2 presents the chronological progression for these profiles with the number of publications per year. We observed a regular and remarkable increase since 1990 for all these profiles. This tendency is a medium-term consequence of the successive petroleum crises in the seventies. It is noticeable that the interest for the rheological behaviour of potential substrates/suspensions for second-generation biofuels has increased in the same way as research and developments in bioprocessing.

To carry out an overview of lignocellulose suspension rheology and its evolution during hydrolysis, the different published results reporting the methods of investigation and the results obtained (rheological behaviour and its relationship with different physical parameters: particle size, mechanical properties of fibre) are presented.

2 Lignocellulose: Nature and Characterisation

Considering rheological behaviour, the scientific publications simultaneously investigating physical properties and biochemical composition during suspension and/or hydrolysis have been extracted and are reported in Table 1. It is noticeable that this extraction leads to a restricted number of publications and limited information. Physical and biochemical properties are defined by particle diameter, aspect ratio, density, crystallinity, surface properties and biochemical compositions but are rarely integrally reported. Various substrates were studied and could be classified

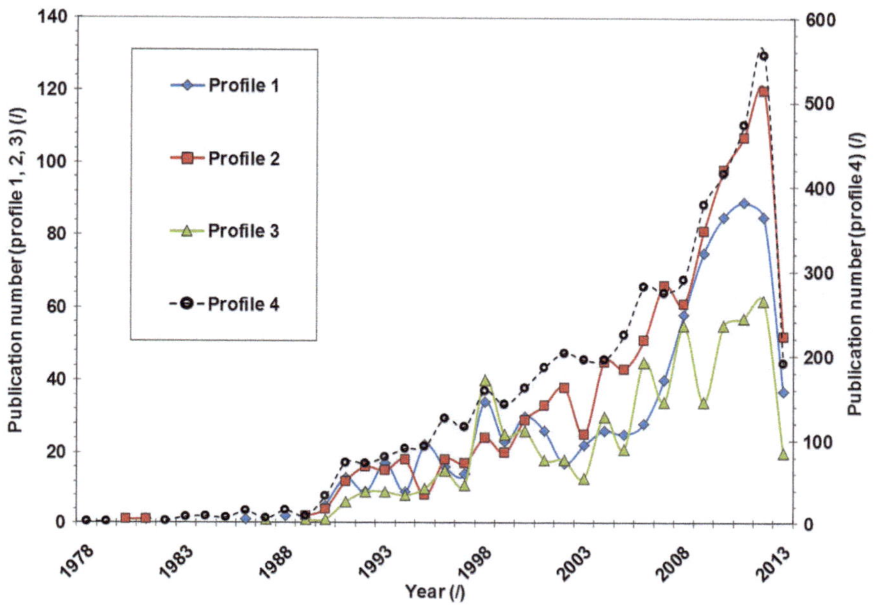

Fig. 2 Number of publications per year since 1978 (Profile 1: (Rheo* OR visco* OR newt*) AND (suspension* OR dissolution*) AND cellulose*; Profile 2: (Rheol* OR visco*) AND fiber* AND cellulose*; Profile 3: (Rheol* OR visco*) AND (paper pulp* OR pulp suspension*); Profile 4: Bioproce*)

into two groups: cellulose (simple matrices) and lignocellulose (complex matrices). Cellulose matrices are mainly nanocrystalline, microcrystalline and microfibril cellulose (NCC, MCC and MFC) with diameters ranging from 0.1 μm for NCC up to 100 μm for MCC. For complex matrices, a great variety of lignocellulosic substrates are derived from woody materials or by-products of agriculture (softwood pulp, hardwood pulp, wood powder, corn stover, bagasse sugarcane, paper). Such complex matrices have been processed with specific pre-treatments which confer various and singular characteristics with a strong influence of the fibrous morphology. These critical physical properties were unequally reported although they make a significant contribution to the rheological behaviour of the suspension.

2.1 Physical Characterisation of Fibres

Among the main parameters to be studied, the rheological behaviour of the hydrolysis suspension and the fibre size and morphology stand out as major determinants of the process efficiency. They are the key elements for the choice of the equipment and the feeding/hydrolysis strategy. However, numerous other parameters related to fibre affect bioprocess performances:

Table 1 Scientific publications linked to rheology, physico-chemical characterisations of lignocellulosic matrices and enzymatic hydrolysis

	Matrices	Publication	Study operations		Physical properties					Biochemical compositions				
			Suspension	Hydrolysis	d_m (µm)	A. ratio	ρ (kg/m^3)	ζ (mV)	CrI	% d_m	Cellu.	Hemi.	Lig.	Ash
Cellulose	NCC	Araki et al. [12]	x	–	0.18 ± 0.075	51 ± 21	–	–	–	–	–	–	–	–
	NCC	Boluk et al. [13]	x	–	0.18 ± 0.06	30 ± 14	–	−62.8	–	–	–	–	–	–
	NCC	Gonzalez-Labrada and Gray [14]	x	–	0.117	29	–	−47	–	–	–	–	–	–
	NCC	Lu et al. [15]	x	–	0.87	5.4	1560	−58	–	–	–	–	–	–
	MCC	Luukkonen et al. [16]	x	–	60	–	–	–	–	–	98	0	0	2
	MCC	Tatsumi et al. [17]	x	–	0.72–350	21–450	–	–	–	–	–	–	–	–
	MCC	Bayod et al. [18]	x	–	30–35	1.5–7	–	–	–	–	–	–	–	–
	MCC	Horvath and Lindstrom [19]	x	–	20–30	–	–	–	–	–	–	–	–	–
	MCC	Um and Hanley [20]	x	x	91	–	–	–	–	–	88	–	–	–
	MCC	Tatsumi et al. [21]	x	–	1.7	22	–	–	–	–	–	–	–	–
	MCC	Tozzi et al. [22]	x	–	24–27	–	–	–	–	90–92	68–72	7–7.8	4–5	0.15
	MFC	Lowys et al. [23]	x	–	–	–	–	−20	–	–	85	11	0	4
	MFC	Agoda-Tandjawa et al. [24]	x	–	10	80–500	–	–	11–13	–	93	–	–	3.6
	MFC	Saarikoski et al. [25]	x	–	–	–	–	–	–	–	–	–	–	–

(continued)

Table 1 (continued)

	Matrices	Publication	Study operations		Physical properties					Biochemical compositions				
			Suspension	Hydrolysis	d_m (μm)	A. ratio	ρ (kg/m^3)	ζ (mV)	CrI	% d_m	Cellu.	Hemi.	Lig.	Ash
Lignocellulose materials	BAG	Geddes et al. [26]	x	x	(25–50) × 10^3	–	–	–	–	–	38	28	–	–
	BAG	Pereira et al. [27]	–	x	<2000	–	1420	–	–	–	43.6	8.75	33.75	–
	CoSt	Pimenova and Hanley [28]	x	–	120	–	–	–	–	–	–	–	–	–
	CoSt	Stickel et al. [29]	x	–	100	1–20	–	–	–	–	–	–	–	–
	CoSt	Viamajala et al. [30]	x	–	–	–	–	–	–	93.5	33.2	21.6	–	–
	CoSt	Dunaway et al. [31]	x	x	–	–	–	–	–	–	60	5	32	3
	CoSt	Dibble et al. [32]	x	x	80–680	–	–	–	–	–	42–60	4.0–24	18–26	–
	WP	Samaniuk et al. [33]	x	x	600	–	–	–	–	100	99	–	–	–
	CoSt		x	x	3000	–	–	–	–	100	50	–	–	–
	PP	Damani et al. [34]	x	–	1180/1290	–	356/538	–	–	–	–	–	–	–
	PP	Chen et al. [35]	x	–	30/15	107/73	–	–	–	–	–	–	–	–
	PP	Blanco et al. [36]	x	–	1000	–	–	–	–	–	–	–	–	–
	PP	Derakhshandeh et al. [37]	x	–	670–2960	–	–	–	–	–	–	–	–	–
	PP	Le Moigne et al. [38]	–	x	–	–	–	–	46–49	–	90–94.7	2–3.5	1.1–1.8	–
	PP	Chaussy et al. [39]	x	–	1280	–	–	–	–	–	–	–	–	–
	PP	Wiman et al. [9]	x	x	188	–	–	–	–	13	48	3.2	45	–
	WhSt	Szijarto et al. [40]	–	x	–	–	–	–	–	34	58.9	–	–	–
	Wood	Dasari and Berson [41]	x	x	33–850	–	–	–	–	–	39.7	19.8	–	–
	Wood	Palmqvist and Liden [42]	–	x	<10,000	–	–	–	–	36.5/12.7	50.2/42.9	–	36.8/46	–

NCC nanocrystalline cellulose; *MCC* microcrystalline cellulose; *MFC* microfibril cellulose; *PP* paper pulp; *BAG* sugarcane bagasse; *CoSt* corn stover; *WhSt* wheat stover)

- Dimensions (equivalent diameter, thickness),
- Qualitative granulometry: sawdust and coarse particles (buchettes), short and long fibres, and flours,
- Morphology and specific surface areas (aspect ratio, complex shape factor: curvature, waviness and wrinkling),
- Mechanical properties (elastic modulus, tensile yield strength and ultimate elongation),
- Surface properties with electrostatic and non-electrostatic forces (Zeta potential, hydrophilicity/hydrophobicity, interfacial surface energy),
- Density (humid and intrinsic density) and porosity,
- Degree of lignification (correlated to the Kappa index in the pulp and paper industry) and crystallinity.
- Molecular mass, degree of polymerisation

There are various ways to define the size of a particle as a "diameter." Allen [43] listed thirteen possible ways to define an equivalent diameter for a given particle using the sphere as reference. Each definition of an "equivalent diameter" should give the same result when applied to spheres (i.e. diameter). Therefore, most of the particles are not spherical; the knowledge of more than one dimension is required to describe the shape of a particle and the associated distribution. The most useful diameter is the diameter of the volume equivalent sphere, d_v, which corresponds to the diameter of the sphere having the same volume. For irregular shapes, the magnitude of the equivalent diameter is affected by the measurement methods, which must then be selected with a particular care.

Because of the non-uniform size of particles, variations in a population are described by a size distribution. Frequency distributions or cumulative distributions are conventionally used. Many characterising techniques (microscopy, settling velocity, granulometry, morphometry, diffraction light scattering and focus beam reflectance) give distributions based on the number, length, surface area or volume of the particles. Considering the complexity of particle shapes and according to the properties highlighted, it is important to define a mean diameter (and standard deviation describing the width of the distribution around this average trend) for a given particle population. For instance, average diameters are defined as follows:

$$d_{p,q} = \left[\frac{\sum n_i \cdot d_i^p}{\sum n_i \cdot d_i^q}\right]^{1/p-q} \quad (1)$$

with n_i is the number of particles of diameter d_i, and p and q are the integers ($p = q + 1$ with $q = 0, 1, 2, 3$ for number weighted, length weighted, surface weighted and volume weighted, respectively) [44]. With these notations, $d_{1,0}$ is the number-average diameter, $d_{2,0}$ is the quadratic mean diameter, $d_{3,0}$ is the cube average diameter, $d_{4,3}$ is the mass or volume mean diameter, and $d_{3,2}$ is the area-average diameter or Sauter diameter.

2.2 *Rheological Characterisation of Suspensions*

Presenting an overview of the scientific literature, this chapter deals with (i) rheometry devices and set-ups for rheological characterisation of lignocellulose suspensions and (ii) rheological behaviour analysis.

2.2.1 Rheological Measurement Devices

Several approaches are suitable for rheological characterisation, and numerous examples of each can be found in the literature concerned with cellulose/lignocellulose suspensions. This is illustrated by studies listed in Table 2. These methods can be classified as a function of device complexity and associated methodologies. The cheapest and easiest-to-use viscometers, i.e. the capillary viscometer and the falling (or rolling) ball viscometer, are suitable for Newtonian fluids but are somewhat difficult to use with unknown non-Newtonian fluids. Their applicability is reduced for fibre suspensions. The works of Luukkonen et al. [16] and of Gonzalez-Labrada and Gray [14] on nano- and microcrystalline cellulose can be mentioned. Fine rheometry analysis was reported by Tozzi et al. [22] using magnetic resonance imaging (MRI) to determine the velocity field in a portion of a cylindrical duct. Completed by the measurement of the pressure drop, the velocity profile brings information on the nature and the characteristics of the fluid. However, this technique is of academic interest but hard to expand to industrial contexts.

The very classical and quite suitable way to obtain rheological characterisation (mostly viscosity, but not only) is the use of rheometers or viscometers equipped with cone and plate, narrow-gap coaxial cylinders or parallel plates. The flow generated is then a simple shearing flow with, except for the parallel plates, a constant shear rate all over the fluid. The shear rate $\dot{\gamma}$ and the shear stress τ are calculated from the rotation frequency and the torque, respectively, and from the characteristics of the geometry. The large-gap coaxial cylinders, which are frequently used in industry, can be considered separately among viscometers as, with a large gap, no assumptions as to the velocity profile can be made (the fluid's rheological characteristics must be known for that). The raw data cannot then be exploited as simply as in the Newtonian case [45, 46]. Coming back to the usual geometries for rheometers and viscometers, they usually require a small volume of fluid, but this advantage turns into a drawback when the size of the objects in suspension is not negligible when compared to a characteristic dimension of the geometry (for instance, a usual gap for parallel plates or concentric cylinders is around 1 mm and is lower for cone and plate geometries). To obtain values of the viscosity from the global measurements performed on a rheometer (torque and angular velocity), the fluid must indeed be considered as a homogeneous medium. It must be borne in mind that this assumption can become erroneous when suspensions are studied.

To overcome this difficulty, the mixing system (MS) of a process, when it exists, can be used as a rheometer, measuring the torque and the rotation rate (or

Table 2 Rheological characterisation of cellulose/lignocellulose suspensions, overview of methodologies used and quantities measured

Authors	Measurement system and mode	Quantities measured
Bennington et al. [45]	Rheometer/concentric cylinders	τ_0 (viscometry)
Damani et al. [34]	Rheometer/parallel plate	G′, G″
Araki et al. [12]	Viscometer/double-gap cylinders	μ
Swerin [46]	Rheometer/concentric cylinders	G′, G″
Wikström et al. [47]	Rheometer/Rushton-type turbine	τ_y (viscometry)
Tatsumi et al. [48]	Rheometer/cone plate	τ and G′, G″
Lowys et al. [23]	Rheometer/cone plate and parallel plates	μ, τ_0 and G′, G″
Luukkonen et al. [16]	Capillary viscometer	μ
Tatsumi et al. [17]	Rheometer/parallel plates	μ, τ_0 and G′, G″
Chen et al. [35]	Rheometer/parallel plates	μ
Pimenova and Hanley [28]	Viscometer/cone plate and double helical ribbon	μ, τ_0
Bayod et al. [18]	Rheometer/concentric cylinder	μ
Ein-Mozaffari et al. [49]	Viscometer/four-bladed vane	τ_0 (viscometry)
Dasari and Berson [41]	Rheometer/six-bladed vane	μ
Blanco et al. [36]	Mixing system/various mobiles	Torque
Horvath and Lindstrom [19]	Rheometer/parallel plates	τ_0 (oscillations), G′, G″
Stickel et al. [29]	Comparison of different methods	τ_0
Knutsen and Liberatore [50]	Rheometer/four-bladed vane (narrow gap)	τ_0 (viscometry), G′, G″
Agoda-Tandjawa et al. [24]	Rheometer/parallel plates	μ and G′, G″
Derakhshandeh et al. [37]	Viscometer/four-bladed vane + UD velocimetry	μ, τ_0 (viscometry + velocimetry)
Chaussy et al. [39]	Industrial disc refiner	μ
Samaniuk et al. [33]	Mixing system/counter-rotating screw elements	Torque
Wiman et al. [9]	Rheometer/four-bladed vane	μ and G′, G″
Gonzalez-Labrada and Gray [14]	Rolling ball viscometer	μ
Saarikoski et al. [25]	Rheometer/concentric cylinder	G′, G″
Nguyen et al. [51]	Rheometer/Serrated plates + Mixing system/four-pitched-blade turbine	μ and G′, G″, Torque
Mohtaschemi et al. [52]	Viscometer/four-bladed vane	μ, τ_0
Tozzi et al. [22]	Poiseuille flow + MRI velocimetry	μ

equivalently the power consumption and the rotation rate), but the difficulty here lies in the complex flow field generated [53]. Determination of viscosity then relies on the previous determination of the power number—Reynolds number characteristic curve of the MS, where the power number Np is a dimensionless number linked to power consumption. This single master curve depends only on impeller/reactor shape and geometry. In the laminar regime (Re < 10–100), the product $N_p \cdot \text{Re}$ is a constant, named *Kp,* and a deviation from $N_p \cdot \text{Re} = K_p$ indicates the end of the laminar regime. In fully turbulent flow (Re > 10^4–10^5) and for Newtonian fluids, the dimensionless power number N_p is assumed to be independent of mixing Reynolds number and equal to a constant, Np_0. Knowing N_p gives a value of the Reynolds number and an estimation of the viscosity if the fluid is a Newtonian one, and of an equivalent viscosity if the fluid is non-Newtonian. In this last case, interpretation of the rotation frequency and power consumption measurements requires an additional hypothesis that was first proposed by Metzner and Otto [54]. These authors showed that for a large range of shear-thinning fluids represented by a power-law model ($\tau = k\dot{\gamma}^n$), the equivalent shear rate corresponding to the equivalent viscosity is proportional to the rotation frequency via a K_s coefficient (known as the Metzner-Otto coefficient). In addition, the K_s coefficient depends only on the mixing system characteristics. It is thus sufficient to determine it in a preliminary stage by considering Metzner and Otto's concept [54] or Rieger and Novak's approach [55] based on the generalised Reynolds number, Re* assuming $N_p \cdot \text{Re}^* = K_p(n)$. In restricted cases, coefficient K_s can be extended to the transition region using a power equation [56]. This online viscosimetry method has been widely used [33, 51, 57, 58]. Note that such a device gives only the viscosity that is to say the shear-thinning/shear-thickening character of the fluid. Nevertheless, such rheometrical measurement systems are attractive as (i) they avoid suspension settling problems (with suitable geometry); (ii) because of the complex flow generated, they elude migration of particles/fibres which are responsible for heterogeneity in the fluid volume; (iii) it is not necessary to take samples (which, if processed, have to be representative and of a negligible volume); and (iv) they can be used online to follow a process.

One can also point out that geometries such as anchor or helical ribbon are proposed for rheometers which are then used as well-characterised mixing systems [40]. They have to be used with the same precautions as for mixing systems. More simply, but also less rich in information, such geometries can also be used in a relative way to follow the time course of global viscosity in some processes [26, 34]. A special geometry proposed for rheometers is the four-bladed (or six-bladed) vane. Never used as a mixing system, its use in rheometers is especially interesting for yield stress fluids to avoid the wall slip effects currently observed for suspensions (formation of a depletion layer caused by migration of fibres away from the wall). This geometry allows precise determination of the fluid behaviour for low shear rate and is thus of special interest for the determination of the yield stress τ_0. Its use for intermediate and high shear rates requires careful analysis when it is assimilated to the geometry of large-gap coaxial cylinders (see [59] for details). Such a geometry was used in a narrow-gap configuration by Knutsen and Liberatore [50] and in a large-gap configuration by Mohtaschemi et al. [52] for fibrillated cellulose suspensions and by Derakhshandeh et al. [37]. These last authors examined the validity of such geometry in more detail for fibre suspensions by analysing the velocity profile determined by ultrasonic Doppler velocimetry. Stickel et al. [29] proposed a comparison, for a corn stover suspension of a given concentration, of yield stress measurements using the different methods presented here.

The choice of rheometer type then depends on the nature of the fluid tested. Non-classical ways are commonly used to study fluids displaying complex characteristics including large particles, settling problems, slip and time dependency. Geometries that facilitate mixing, such as helical ribbons, anchors and paddles, can overcome settling problems and errors due to slip. Concerning lignocellulose suspensions, the main problems for measurement are heterogeneous suspension, floc formation [35] and particle decantation [51, 60, 61]. Wall slip is currently encountered with suspensions, and the use of surfaces with roughness equal to twofold or threefold the largest particle dimension is recommended [62]. Smooth surfaces, or inadequate roughness, lead to a depletion zone near the wall where the shear rate is higher, thus reducing the shear rate in the bulk fluid and distorting, sometimes significantly, the rheogram as illustrated in Fig. 3. It is thus important to take precautions when this kind of measurement must be carried out and some reports must be considered with caution when nothing is mentioned on this point. Aggregation of fibres is also a crucial point in suspension rheometry. It is caused by the constant and unidirectional shear rate applied to the suspension. An extreme case is presented as an illustration in Fig. 4. Formation of these structures depends on the value of the shear rate and the duration. Higher shear rates, as reported by Chen et al. [35] and Saarikoski et al. [25], can destroy the structures. These difficulties are not encountered in dynamic measurements (small-amplitude oscillations in the linear domain), and they are considerably reduced with online rheometry using a mixing system.

Focusing now on the data provided by these measurements, the main and best known characteristic is the viscosity. Suspensions are generally shear-thinning, and this behaviour is quantified by the power-law index n and the consistency k (fit of a

Fig. 3 Illustration of "wall slip" (depletion layer) in rheometry on a 4 % vol TiO_2 suspension

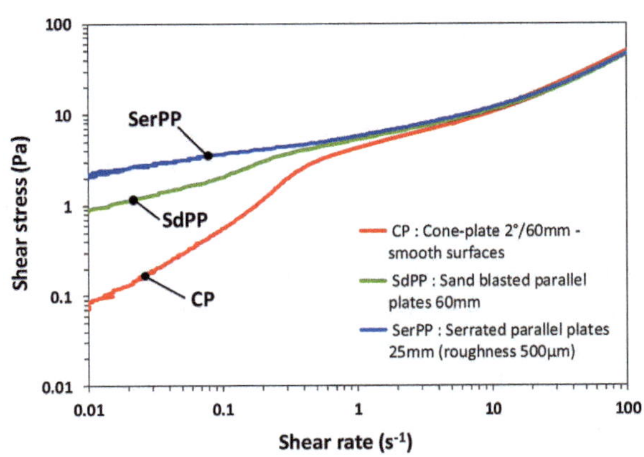

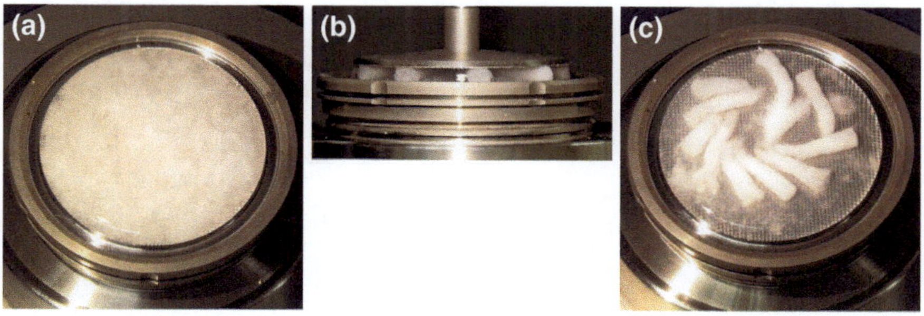

Fig. 4 An extreme case of aggregation caused by a constant unidirectional shear flow of Whatman paper suspension (1 % wdm/w), fibre length $d_{4,3}$: 310 μm, serrated parallel plates 60 mm diameter, gap: 4 mm, constant shear rate: 10 s^{-1}). **a** fluid before shearing, **b** during shearing, **c** after 15 s shearing. Note that the structures (visible on **b**) were somewhat disorganised when raising the upper plate (photo **c**), with no major changes in length or diameter

power law on the relevant region of the rheogram). This is, in practice, the only characteristic that can be obtained using a viscometer or a mixing system, together with the yield stress. The latter is evaluated from the low shear rate region of the rheogram, and several methods are used: shear stress limit when the shear rate tends to zero [17], adjustment of a Herschel-Bulkley-type model ($\tau = \tau_0 + k\dot{\gamma}^n$) [20, 23], peak torque (or shear stress) during transient flow experiments at a slow, steady rate [50] and stress ramp analysed in terms of deformation vs shear stress curve [45, 49].

With rheometers, additional rheological characterisations can be obtained with dynamic measurements. Oscillatory stress sweeps and oscillatory frequency sweeps are the most frequently used, particularly for lignocellulose suspensions [63]. Sinusoidal shear stress (or strain) gives information on elasticity (part of the strain which is phased with the shear stress) and viscosity (part of the shear rate which is phased with the shear stress). The complex modulus G* is defined as the ratio of shear stress to strain. Its real part G′ is the storage modulus (elastic contribution), and its imaginary out-phase component G″ is the loss modulus (viscous contribution). It is used to characterise the viscoelasticity or viscoplasticity of materials. Indeed, yield stress fluids behaved like solid elastic material in the linear domain (small-amplitude oscillatory), and the determination of the limit of this domain can be used to evaluate the yield stress [9, 19, 63].

2.2.2 Experimental Conditions

Lignocellulose suspensions were studied in various ranges of mass concentration varying between 0.1 and 600 gdm L^{-1} as indicated in Fig. 5. However, the majority of these studies focus on low and medium concentrations (1–200 gdm L^{-1}). Temperatures were usually between 20 and 30 °C. The pH of the suspension was generally neutral (around 7) excluding a few articles which investigated the effect of pH on the rheological behaviour (pH varies from 4 to 10) [24, 37].

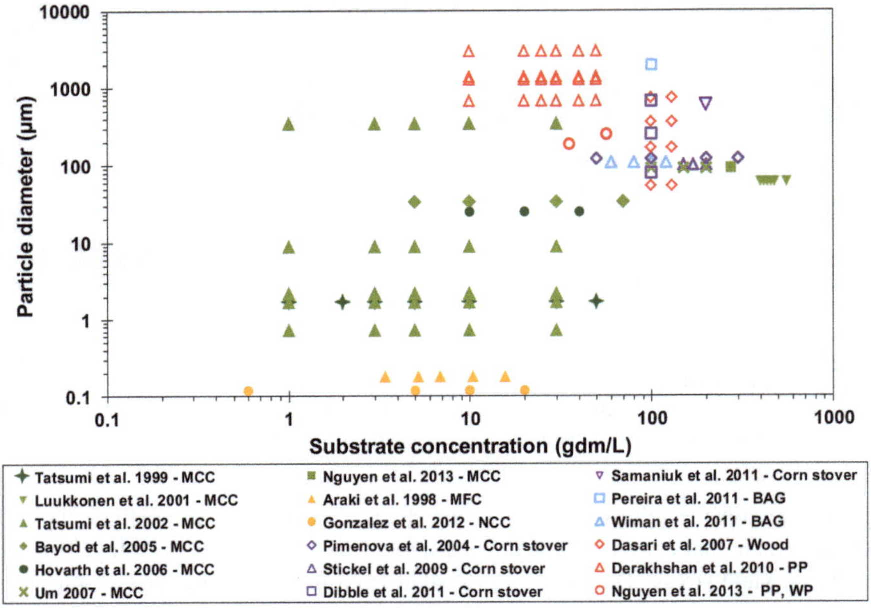

Fig. 5 Publication overview: particle diameters and substrate concentrations studied (*NCC* nanocrystalline cellulose, *MCC* microcrystalline cellulose, *MFC* microfibril cellulose, *WP* Whatman paper, *BAG* sugarcane bagasse, *PP* paper pulp)

Work scrutinising hydrolysis mechanisms was usually carried out in the range 40–55 °C for temperature, 4.8–5.0 for pH, 10–200 gdm L^{-1} for substrate concentration and 0.25–50 FPU/g cellulose for cellulase activity. Numerous studies explored two aspects: (i) hydrolysis kinetics and (ii) rheological behaviour [9, 20, 26, 27, 31–33, 38, 40–42, 51]. Items (i) and (ii) in addition to (iii) particle size evolution were investigated by Dibble et al. [32], Nguyen et al. [51] and Wiman et al. [9]. Studies of flow properties of suspensions together with hydrolysis kinetics represent half of all the articles published. This reveals the keen interest for applications concerning the valorisation of lignocellulose material by enzymatic methods, especially from 2010 on. However, simultaneous study of particle size and rheological behaviour during hydrolysis is a minor part of these studies despite the interest that this point presents.

3 Rheological Behaviour of Lignocellulose Suspensions

One critical physical parameter in lignocellulose materials is their morphology and the large aspect ratios that are encountered. They induce significant contacts among particles (fibres) even at low concentrations and bring about a strong effect on suspension rheology. An increase of aspect ratio implies increases in the suspension viscosity [64, 65]. Not only the aspect ratio but also the particle shape influences the viscosity, as reported by Barnes et al. [66]: for 20 % volume fraction suspensions of solid particles with a limited aspect ratio (23 for the highest), and compared to a quasi-monodisperse spherical particle suspension, viscosity is 1.5-fold larger for grain suspension, 4.3-fold for plate suspension and 7.7-fold for rods. Particles with more complex shapes show greater dependence of viscosity on concentration [67, 68]. This requires a detailed knowledge of this effect, and advanced models will have to introduce critical morphological parameters to describe these evolutions.

3.1 Overview of Classical Models

3.1.1 Basic Rheological Models

The two main parameters used in rheology are the shear rate $\dot{\gamma}$ (s^{-1}), which characterises the kinematics, and the shear stress, τ (Pa or N m^{-2}), which characterises the forces.

The flow curve $\tau(\dot{\gamma})$ or equivalently the rheogram $\mu(\dot{\gamma})$, with $\mu(\dot{\gamma}) = \tau/\dot{\gamma}$, provides a complete characterisation of the rheological behaviour if the fluid is neither elastic nor thixotropic (time-dependent viscosity). Table 3 shows the main rheological behaviour and associated models. The power-law index (or flow behaviour index) n and the consistency k of the power law have a direct physical interpretation for viscosity and are present in almost all other models. They are therefore largely used as information concerning non-Newtonian viscosity.

The shear-thinning fluids (also called pseudoplastic fluids) generally have, for low and high shear rates, a Newtonian plateau (with values μ_0 and μ_∞, respectively, for viscosity). This can be modelled by using three- or four-parameter models such as Sisko, Cross, Powell–Eyring or Carreau models [69].

3.1.2 Classical Rheological Model for Suspensions

Suspension viscosity is a function of numerous parameters linked to the solid phase (particle size, shape, concentration, separated fibres/aggregates/flocs as structural units, spatial orientation of the particles/units in the fluid, etc.) and to the suspending fluid (generally a Newtonian fluid).

The classical models describing suspension viscosity concern hard sphere suspensions (Table 4). These models mainly integrate the particle volume fraction effect, and after the Einstein equation for isolated monodisperse hard spheres, a physical complexity was progressively considered with hydrodynamic and particle–particle interactions up to percolation. Then, three concentration regimes are generally considered: dilute, semi-dilute and concentrated. The first one corresponds to very small values of the volume fraction, $\varphi \to 0^+$: the average distance between particles is large compared to their radius or length. The particles can move freely

Table 3 Rheological classification of fluids and usual associated models (quantities mentioned as constant are understood as "for given temperature and pressure")

Fluids		Rheological model(s)
Perfect fluid		$\mu = 0$
Newtonian fluids		$\tau = \mu \cdot \dot{\gamma}$ with μ = const.
Time-independent non-Newtonian viscous fluids	Shear thinning: $\mu(\dot{\gamma}) \downarrow (0<n<1)$	Power law (or Ostwald-de Waele): $\tau = k \cdot \dot{\gamma}^n$
	Shear thickening: $\mu(\dot{\gamma}) \uparrow (n > 1)$	
	Viscoplastic (Yield stress fluid: no flow if $\tau<\tau_s$)	Bingham: $\tau = \tau_y + \mu_p \cdot \dot{\gamma}$ Herschel–Bulkley: $\tau = \tau_y + k \cdot \dot{\gamma}^n$ Casson: $\tau^{1/2} = \tau_y^{1/2} + k_c \cdot \dot{\gamma}^{1/2}$
Time-dependent non-Newtonian fluids	Thixotropic	For $\dot{\gamma}$ = Const.: $\mu(t) \downarrow$
	Anti-thixotropic	For $\dot{\gamma}$ = Const.: $\mu(t) \uparrow$
Viscoelastic fluids		Complex rheological models linking the strain rate tensor and the stress tensor. For oscillatory shear flows: $\tau = G^* \cdot \dot{\gamma}$ with $G^* = G' + iG''$ accounting for the phase shift between τ and $\dot{\gamma}$

through the suspension under the action of Brownian forces without perturbation induced by neighbouring particles but with close fluid–particle interactions (local hydrodynamic interactions). In semi-dilute suspensions, the flow perturbations created by the presence of particles will influence the movement of close particles. The overall flow pattern is affected by interactions between particles and hydrodynamic interactions. The last regime corresponds to concentrated suspensions with a lot of contacts between the particles. The viscosity of the suspension increases rapidly with volume fraction. When ϕ reaches a critical value (glassy fraction, $\phi_G \approx 0.58$ for spherical monodisperse particles), each particle is confined in a cage formed by its nearest neighbours. For volume fractions above this value, only a vibration of the particles inside the cage remains possible, and this possibility completely disappears when ϕ reaches the value of random close packing (ϕ_{RCP} = 0.64 for monodisperse spheres) which is lower than the limit case ϕ_{FCC} = 0.74 corresponding to face-centred cubic arrangement.

The frontier between the different concentration domains strongly depends on the nature and the intensity of interaction between the particles. For instance, a 5 % vol. suspension of neutral monodispersed spheres behaves as a dilute and Newtonian suspension, while a 1 % vol or less clay suspension is viscoplastic with a very high yield stress together with a thixotropic behaviour.

The extension of the classical models of Table 4 had successively integrated particle complexity such as morphological aspects [66] and deformable particles [69]. A higher order of complexity was incorporated by considering the flocs or aggregates currently encountered in fibre suspensions. These complex associations

Table 4 Rheological models for hard sphere suspensions [69]

Dilute	Einstein	$\mu_{\text{rel}} = \frac{\mu}{\mu_s} = 1 + 2.5 \cdot \phi$
Dilute/semi-dilute	Batchelor	$\mu_{\text{rel}} = \frac{\mu}{\mu_s} = 1 + 2.5 \cdot \phi + 6.2 \cdot \phi^2$
Semi-dilute/ concentrated	Krieger-Dougherty	$\mu_{\text{rel}} = \frac{\mu}{\mu_s} = \left(1 - \frac{\phi}{\phi_m}\right)^{-[\mu] \cdot \phi_m}$ With $1 \leq [\mu] \cdot \phi_m \leq 2$ $[\mu]$: intrinsic viscosity ϕ_m: maximum volume fraction
Concentrated	Douglas-Garboczy	$\mu_{\text{rel}} = \frac{\mu}{\mu_s} = K\left(1 - \frac{\phi}{\phi^*}\right)$ ϕ^*: critical volume fraction (percolation threshold)

of particles can, for instance, be modelled by introducing structural units in an effective volume fraction ϕ_{eff} which includes the embedded fluid fraction through an average compactness [69].

3.2 Observed Rheological Behaviour

In dilute regime, Newtonian behaviour is generally observed for substrate concentrations less than 0.1 %. This critical concentration ($C^* = 0.1$ %) was proposed for both NCC [13] and MCC [17, 48]. Wu et al. [70] measured a slightly higher value for NCC ($C^* = 0.4$ %). No model of critical concentration was presented or proposed for complex matrices. This point can be explained by a lack of studies for very low concentrations which are not interesting from a lignocellulose valorisation point of view.

For substrate concentrations higher than 0.1 %, all studies report non-Newtonian behaviour, of shear-thinning or pseudoplastic type, as illustrated by Fig. 6. In this figure, these viscosities for MCC and BAG are well described by a power-law model on the studied shear rate range. Considering now the semi-dilute and concentrated regimes, a sharp increase of viscosity was observed by Nguyen et al. [51], who studied the rheology of different lignocellulose suspensions (microcrystalline cellulose, Whatman paper, paper pulp). They noticed a critical volume fraction, C** (based on humid matter), that signifies the regime transition. Depending on the material complexity, this value increased from Whatman paper ($C^{**} = 3$ %) to paper pulp ($C^{**} = 9$ %) and to microcrystalline cellulose ($C^{**} = 24$ %).

Table 5 illustrates the viscosity of corn stover suspensions for a given concentration and a given shear rate. These results indicate homogeneous values although measured by different authors using different methods. These similar values could originate from the same substrate type, the same pre-treatment method (diluted acid with H_2SO_4—standard protocol from NREL) and the same range of particle diameters (100–120 μm for volume mean diameter).

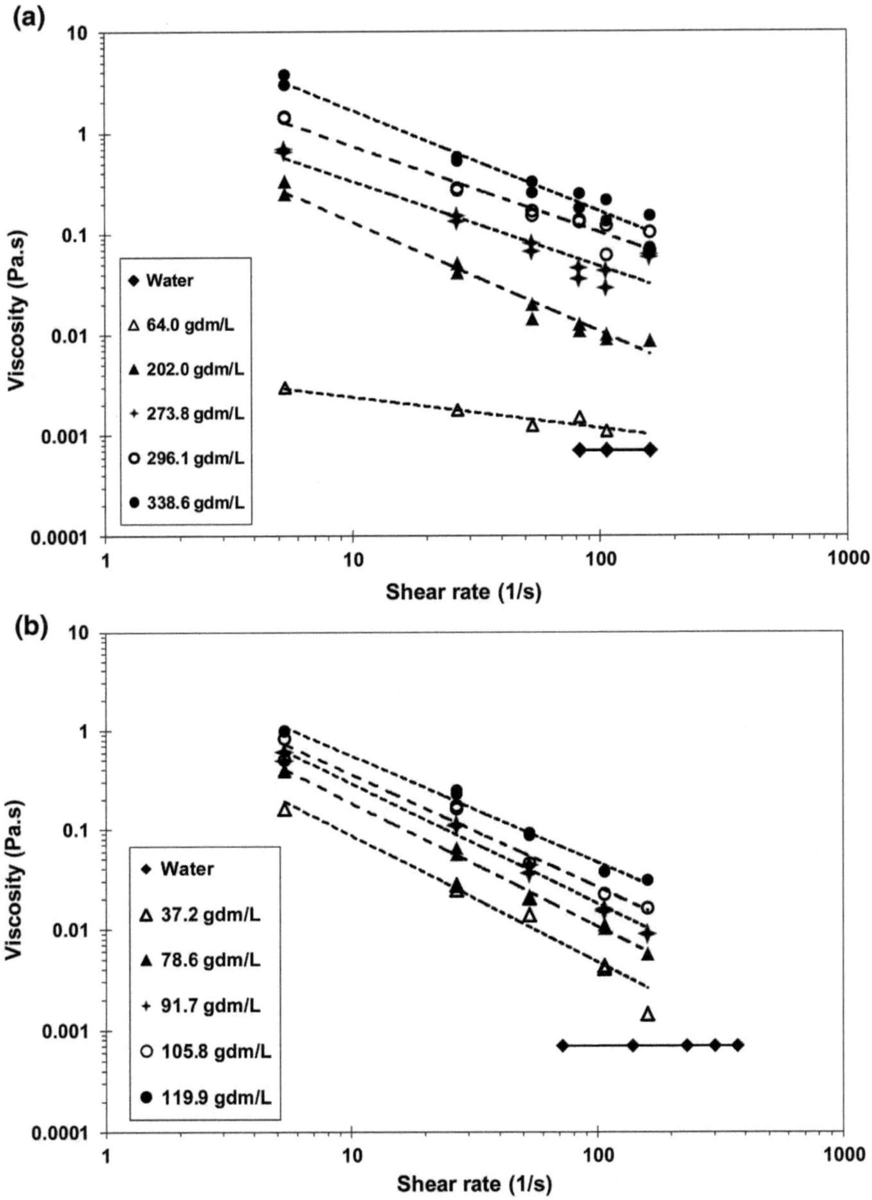

Fig. 6 Viscosity of MCC (90 μm (**a**)) and BAG (**b**) suspensions as a function of shear rate

Table 5 Comparison of viscosity for corn stover suspension at 15 % wdm/w

Authors	Viscosity (Pa s)
Pimenova and Hanley [28]	15
Stickel et al. [29]	25
Viamajala et al. [30]	20
Dunaway et al. [31]	10

Specific values are reported for a shear rate of 10 s^{-1}

3.3 Structure and Consistency Index

As mentioned before, the flow behaviour index *n* and the consistency *k* are good indicators of viscosity. Figures 7 and 8 present data extracted from the literature for the consistency (Fig. 7) and the flow behaviour index (Fig. 8) versus particle diameter and substrate concentration. Among all the publications cited, only 25 % present these two parameters. This point evokes the difficulty for data treatment and the observation of general tendencies. Furthermore, no results were found for particle diameters less than 30 μm or larger than 600 μm and only one result for substrate concentrations above 150 gdm L^{-1} [28]. Bayod et al. [18] and Pimenova and Hanley [28] who studied, respectively, MCC and corn stover suspensions gave totally different results compared to other authors. So their results are not presented in these two figures.

Concerning the consistency index *k* (Fig. 7), an increase is observed when the substrate concentration and/or the particle diameter increase. With PP, for example, *k* increased sixfold (from 3.3 to 17.1 Pa s^n) for concentrations varying from 28 to 42gdm L^{-1} [51]; at the same concentration of 50 gdm L^{-1}, *k* increased twofold (from 62 to 112 Pa s^n) for particle mean diameters varying from 670 to 2960 μm

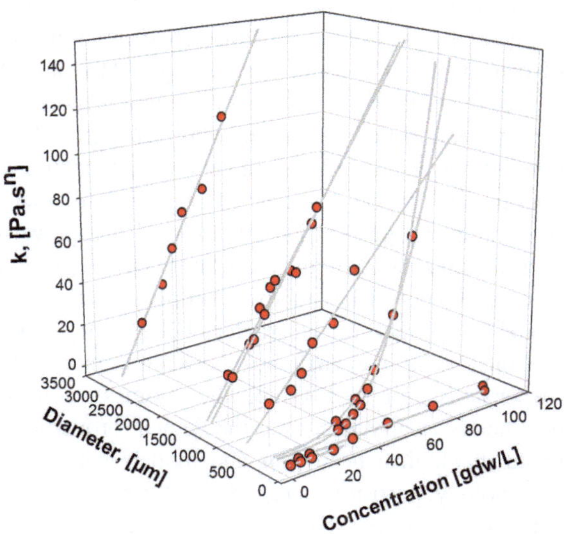

Fig. 7 Relationship between consistency index, particle diameter and substrate concentration

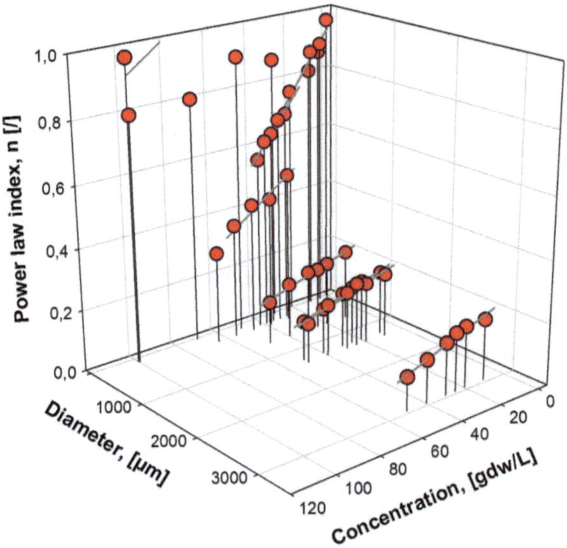

Fig. 8 Relationship between power-law index, particle diameter and substrate concentration

[37]. So, the effects of concentration and particle size on the consistency index are of the same order of magnitude. The relationship between these quantities can be modelled by a linear or a quadratic equation. This tendency is observed visually on graphics but also validated by Nguyen et al. [51] and Bayod et al. [18].

Now focusing on the power-law index n (Fig. 8), it decreases as the substrate concentration increases, thus underlining the fact that the non-Newtonian behaviour becomes increasingly pronounced as the concentration rises. With NCC, for example, n decreases from 0.962 to 0.75 for concentrations of 0.6 to 20 gdm L^{-1} [14], and for corn stover suspensions, n decreases from 0.91 to 0.5 for concentrations of 50 to 300 gdm L^{-1}. For particle diameters less than 1000 µm, n clearly tends to 1 as substrate concentration tends to 0, the behaviour of the suspension tends towards that of water [14, 20, 28, 51]. For particle diameters greater than 1000 µm, this tendency is not so clear on the graphs and n does not seem to tend to 1 as the concentration tends to 0. This point could be explained by a jump of power-law index for a very low concentration and could be specific to fibre morphology, aspect ratio, etc. For particle diameters greater than 1000 µm and concentrations greater than 10 gdm L^{-1}, the power-law index comes to 0.2 whatever the concentration and diameter [37]. This value can be considered as a critical value of the power-law index or simply regarded as resulting from the choice of the model (power-law model or Herschel–Bulkley model) and adjustment zone.

Considering the analogies between lignocellulose fibre and micro-organism suspensions, for cell broth, the biomass concentration and the morphology parameters clearly affect the rheological properties. Different authors tried to correlate rheological behaviour and models with various parameters (concentration, morphology, porosity) as reported in the review of Wucherpfennig et al. [11]. These authors gave a precise description of the morphological criteria of interest.

As reported by Wucherpfennig et al. [11], Tucker and Thomas [71] studied 75-h *Penicillium chrysogenum* samples and proposed that each rheological parameter (RP) could be expressed as a power-law function of the biomass concentration (C_m), the roughness and the compactness. In more detail, Riley et al. [72] used the same fungi and concluded that the average temporal value was 0.35 for a power-law index, n, with C_m ranging from 10 to 32 g L^{-1}. However, the data reported indicate that n evolves between 0.45 and 0.2. They proposed a correlation (Eq. 2) for the consistency index in relationship with biomass concentration C_m (g L^{-1}) and mean maximum dimension D_m (μm):

$$k = C_m^2 \cdot (5 \times 10^{-5} \cdot D_m - 10^{-3}) \tag{2}$$

Petersen et al. [73] proposed, using principal components analysis, a prediction of apparent viscosity, yield stress and consistency index from the size distribution and biomass concentration. Based on *Aspergillus orizae* cultures, they assumed a constant power-law index around 0.4 and reported accurate correlations between predicted and experimental parameters (k and τ_0). In a more detailed approach, Mohseni and Allen [74] investigated the influence of concentration (3–20 cdw/L) and particle morphology on the yield stress of filamentous broths of *Aspergilus niger* and *Streptomyces levoris*. They mainly focused on yield stress (ranging from 0.1 to 37 Pa) and proposed a power-law model for τ_0 as a function of biomass concentration (exponent between 2 and 3). In addition, Mohseni and Allen [74] examined different morphological parameters, reported empirical correlations and identified the most relevant morphological factors.

In view of the level of complexity of substrate suspension (cell culture broths, lignocellulose suspensions), the robustness of rheological models (consistency and power-law indexes, yield stress) has to be improved to reliably integrate the morphological parameters for filamentous and fibrous suspensions.

3.4 Yield Stress and Oscillatory Measurements

A central point in a bioprocess is to ensure correct homogenisation of the material throughout the reactor and the better control and an optimal efficiency of hydrolysis at each point. This becomes a challenge when complex rheological behaviours are encountered as is the case with lignocellulose suspensions, due to their fibrous nature, and with filamentous broth. In the dilute regime, increasing the dry matter content of a suspension is not a difficulty as it just corresponds to a reasonable increase in the viscosity which can be predicted. However, the limit for this regime depends on the nature and properties of the fibres, but whatever that may be, it is very low and extrapolation from low dry matter contents to high ones is not really feasible. But large-scale processes need higher dry matter contents to be considered as realistic from a production cost point of view. Increasing concentration is then an economical and environmental necessity but leads to complex rheological behaviour with the appearance of yield stress. Considering hydrolysis of ligno-cellulose matrices, a balance between feed strategy (viscosity increases due to substrate concentration) and bioreaction (viscosity decreases due to fibre degradation and solubilisation) can be proposed to reach high fermentable substrate concentrations (equivalent to high dry matter content).

Lowys et al. [23], working on MFC suspensions, observed that above a critical concentration $C^* \approx 0.3$ %w/w, the viscosity rapidly increases and suspensions exhibit a viscoplastic behaviour. This signifies, inter alia, that the shear stress has to be above this critical stress to initiate flow. Consequently, the regions where the stress is not high enough are dead zones or solid-like regions which are highly damageable for transfers and thus for process efficiency [49, 75]. This viscoplastic behaviour has been widely reported in fibre suspensions [63, 76] as well as in filamentous broth [74] and has to be considered in the design of process equipment [45].

As mentioned in Sect. 2.2, yield stress fluids behave as elastic/viscoelastic materials under the threshold and their behaviour is studied using various approaches throughout the literature: by shear viscosity measurements and/or by dynamic measurements (oscillatory strain sweeps and oscillatory frequency sweeps). The latter allow the determination of the elastic modulus G' and of the loss modulus G'' which can in turn be used to estimate a value for the yield stress. In the linear domain, the elastic modulus is generally found to be almost an order of magnitude higher than the loss modulus and independent or weakly dependent on frequency as expected for yield stress fluids.

Focusing now on pulp fibre suspensions in the domain of interest for industrial applications, increasing concentration leads from small flocs loosely connected by individual fibres to a network structure [34]. When these suspensions are submitted to small oscillatory strains, the linkages are deformed elastically. Increasing the strain amplitude and leaving the linear domain lead to inter-fibre or inter-floc failures and to a structural breakdown of the fibre network. The observed yield stress then indicates the appearance of a physical network and is associated with the strength of this network. So, yield stress and elastic modulus can also be regarded as sources of information on the fibre network.

The network properties and especially the yield stress depend on numerous parameters among which the concentration and the fibre characteristics: length, aspect ratio, pH, temperature, additives.

To illustrate the variety of these parameters, let us mention two studies. The first [23] concerns the dispersion method used to prepare the suspensions. The authors observed notably lower values for the yield stress and for the G' and G'' modulus for mechanical stirring when compared to ultrasound dispersion. Ultrasound achieves better dispersion and is linked to a greater strength of the network. Other interesting results are reported by Dibble et al. [32] who compared two reduction size methods on the same initial material. Their results show that mechanical dispersion does not reduce the yield stress nor the enzyme digestibility of the biomass, whereas dilute acid pre-treatment leading to a similar particle size distribution leads to a decrease in yield stress and to increased digestibility.

This illustrates the complex relationship existing between the fibre properties, their implementation and the rheological behaviour of the resulting suspension. Nevertheless, some interesting results are available in the literature. Most concern the variation of the yield stress and/or the elastic modulus with concentration. Bennington et al. [45] on the basis of previous works examined a power-law relation (Eq. 3) with the mass concentration C_m:

$$\tau_y = a \cdot C_m^b \tag{3}$$

A similar relation (Eq. 4) was proposed for the elastic modulus:

$$G' = \alpha \cdot C_m^\beta \tag{4}$$

Considering the general case of suspension rheological behaviour, the appropriate parameter is rather the volume fraction of the particles. For complex systems, such as cellulose/lignocellulose fibres, the volume of fibres/flocs is not easy to define and to evaluate because the pulp fibres swell up in the presence of water to constitute a porous medium with trapped water. Because the mass concentration can be accurately determined without ambiguity or having to choose between the objects to consider, it is preferred. The drawback for this choice is the dependence of the model parameters on the nature and properties of the fibres as the relationship between volume fraction and mass fraction is not a simple one. This point was raised by Bennington et al. [45]. The correlations proposed by different authors are presented in Table 6. A large dispersion is observed for parameters a and α and so they are not reported in this table.

The most striking result from Table 6 is the large dispersion observed for the values of β and b, and the fitted parameters appear to be specific to each pulp. Nevertheless, a tendency can be noted as the values observed for the smallest

Table 6 Correlation between suspension mass fraction and yield stress and/or elastic modulus

Authors	Matrices	% d_m (w/w) (%)	$G' \propto C_m^\beta$ β parameter	$\tau_y \propto C_m^b$ b parameter
Bennington et al. [45]	PP	0.5–35	–	2.31, 2.99 and 3.56
Tatsumi et al. [17]	MCC	0.05–1	2.25	2
Dalpke and Kerekes [77]	PP		–	2 to 4
Tatsumi et al. [21]	PP	1–10	3	–
Stickel et al. [29]	CoSt	5–25	–	5.7 ± 0.5
Agoda-Tandjawa et al. [24]	MFC	0.2–3	2.58	–
Ehrhardt et al. [78]	CoSt	20–35	–	3.7 to 4.2
Knutsen and Liberatore [50]	CoSt	5–17	–	6
Wiman et al. [9]	Pre-treated spruce	4–12	6.6 ± 0.3	5 ± 0.2

particles (MCC, MFC) are smaller than those for larger fibres (pulp fibres). To explain the dispersion observed in pulp fibres, the type of fibre processing is evoked [78]. Using Eq. 3 expressed with the volume fraction ϕ, Wiman et al. [9] calculated a fractal dimension of the network.

If the existence of a correlation between yield stress or elastic modulus and mass concentration seems to be generally acknowledged, no real correlation appears between these rheological parameters and the morphological properties of the fibres, namely length and aspect ratio. This point was examined by Bennington et al. [45] who were unable to conclude as to the existence of such a correlation although theoretical considerations had led the authors to expect a squared-aspect ratio dependence. Over a reduced range of aspect ratios, Wikström et al. [47] reported a value of around 1–1.3 for this exponent.

Some studies also explored the influence of pH and temperature and generally concluded that there was a low or null influence for MFC [23, 24] as well as for pulp [47] except Ehrhardt et al. [78] who observed a decreasing yield stress as the hydrolysis temperature was increased.

As increased yield stress will lower process efficiency and raise energy costs, one can expect that a good choice of additives will modify the rheological behaviour in the right direction. This was analysed by different authors. Thus, the yield stress and/or the G' G'' moduli are enhanced by additions of flocculent [46] or salt [23, 24]. Concerning polyelectrolyte additions, results are contrasted between Lowys et al. [23] on MFC and Horvath and Lindstrom [19] on bleached softwood kraft pulp. Testing a large range of surfactants on pre-treated corn stover, Knutsen and Liberatore [50] observed a decrease in the yield stress, while addition of protein had the opposite effect.

4 Physical Properties of Fibre and Suspension During Enzymatic Hydrolysis

Concerning the general evolution of viscosity, all studies conclude that the viscosity of suspensions decreases during enzymatic hydrolysis. This depends strongly on the nature of matrices, on the nature and the activity of the enzyme, and on the concentration of the substrates and the experimental conditions such as temperature or rotation speed. During hydrolysis, two phases can be observed for viscosity. First, a rapid decrease of viscosity is observed. Then, a steady value is maintained during a second phase. Dasari and Berson [41] studied the hydrolysis of red oak sawdust and demonstrated that viscosity decreased 10-fold after 24-h hydrolysis. For acid-pre-treated sugarcane bagasse, viscosity was reduced by 77 to 95 % after 6 h [26] and by 75 to 82 % within 10 h [27]. This decrease and the final plateau value depended on the enzyme loading [26]. For spruce pulp at 30 FPU/g cellulose, initial and final viscosities ($\mu_{initial}$/μ_{final}) were 0.24/0.028, 0.4/0.058 and 0.84/0.087 μm for concentrations of 10, 15 and 20 % (w/w), respectively [10]. This decrease of viscosity is due to mainly (i) substrate solubilisation and (ii) particle size diameter reduction. However, publications exploring the relationship between particle size and viscosity evolution are very few. Some publications can be cited like [9, 20]. Both of these authors concluded that a roughly twofold reduction occurred in particle mean diameter for 24-h or 48-h hydrolysis.

With the goal of enabling easier comparison of viscosity variations during enzymatic hydrolysis, authors propose the use of a time $t(\mu/\mu_0 = 0.1)$ defined as the duration necessary for a 90 % reduction in viscosity. A summary of past works with different experimental conditions is presented in Fig. 9. Observing only one data series (one publication), the relationship between $t(\mu/\mu_0 = 0.1)$ and cellulase activity is clearly demonstrated: the higher the quantity of enzyme, the shorter the time $t(\mu/\mu_0 = 0.1)$. However, the tendency for all lignocellulose substrates in all experimental conditions indicates a large dispersion of points. This clearly illustrates the complexity of rheology studies in hydrolysis conditions which depend not only on enzymatic activity but also on the nature of the substrate, the bio-chemical-physico properties of the substrate and of course on experimental conditions.

Szijarto et al. [40] explored hydrolysis experiments of hydrothermally pre-treated wheat straw with purified enzymes from *Trichoderma reesei*. The results obtained at 15 % (w/w) solids revealed that endoglucanases were the key enzymes to rapidly reduce the viscosity of the lignocellulose substrate. Cellobiohydrolases only played a minor role, and xylanase had practically no effect at all on the viscosity.

Palmqvist and Liden [42] monitored the impeller torque (and hence power input) in a stirred tank reactor throughout high solid enzymatic hydrolysis (<20 % w/w) of steam-pre-treated *Arundo donax* and spruce. The decrease in torque during spruce hydrolysis was much slower than *Arundo donax* hydrolysis because of a higher

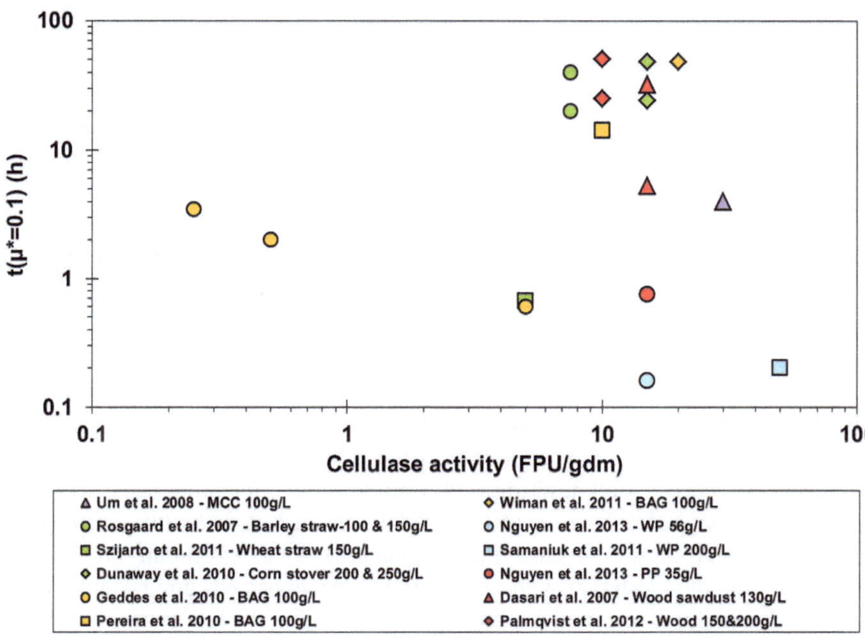

Fig. 9 Summary of $t(\mu/\mu_0 = 0.1)$ as a function of cellulase activity

amount of lignin compared to the arundo (46 and 37 %, respectively). The lignin structure is not broken down during the hydrolysis and might therefore contribute to maintaining high viscosities of the spruce material.

Concerning rheological behaviour, typical shear-thinning was confirmed during hydrolysis [27, 31, 79]. The consistency index decreases, and the power-law index increases as hydrolysis proceeds [9, 20, 31, 79]. Rosgaard et al. [79] studied the hydrolysis of steam-pre-treated barley straw and demonstrated that for a concentration of 15 % dry matter, the consistency index decreased from 16,500 to 185 Pa s^n, while the power-law index increased from 0.07 to 0.47 for 6- and 72-h hydrolysis, respectively. Dunaway et al. [31] surveyed the hydrolysis of pre-treated corn stover over a concentration range of 10–25 %. They concluded that the consistency index, k, decreases rapidly with time, with the largest rate of decrease (around 10-fold) occurring in the first 8 h.

One of the rheological parameters that were followed during enzymatic hydrolysis was the elasticity and/or the yield stress of the suspension through oscillatory measurements. However, only few articles were found on this point [9, 80, 81]. Wiman et al. [9] carried out a comprehensive rheological characterisation of dilute acid-pre-treated spruce during hydrolysis. Fillaudeau et al. [80] explored the enzymatic liquefaction and saccharification of paper pulp. Both authors confirmed that elastic modulus was always greater than viscous modulus in the initial step and during hydrolysis, confirming viscoplastic behaviour. These two modules decreased as a function of hydrolysis time around 100-fold for 48 and 100 h in spruce and paper pulp, respectively. Wiman et al. [9] demonstrated that the yield stress decreased dramatically with time. Roche et al. [81], for a 20 % w/w dilute acid-pre-treated corn stover, observed a low hydrolysis process (reduction of yield stress around 100-fold in 4 days) but obtained a single curve for yield stress versus total biomass conversion with enzyme concentrations of 5, 10 and 20 mg g^{-1}.

5 Conclusion

In a white biotechnology context, this review proposes a survey of the different methods and results concerning the rheological behaviour of lignocellulose fibre suspensions and their evolution during biocatalytic degradation. The research, initially centred on simple matrices (microcrystalline cellulose), was extended to complex substrates such as agriculture by-products and various woody substrates. The scientific literature with regard to the rheological behaviour of micro-organism suspensions (single cell up to pellet, ovoid up to mycelial morphology) constitutes a homogeneous corpus of knowledge. Nevertheless, a limited number of studies cover the rheological behaviour of lignocellulose matrices in highly concentrated suspensions and during biocatalytic degradation. A shear-thinning rheological behaviour was demonstrated with all substrates. As expected, this behaviour was conserved all along enzymatic attack, but its magnitude fell progressively. In addition, oscillatory measurements indicate a non-negligible yield stress for the initial suspensions of several complex matrices. This yield stress rapidly decreases and then disappears within the first ten hours of enzyme attack. During hydrolysis, the evolution of viscosity is dependent on the nature of the substrate and the enzyme concentration. Conventional or adapted models accurately describe rheological behaviour of suspensions as a function of shear rate and mass or volume concentration, but the fitted parameters strongly depend on the substrate. However, a major scientific bottleneck remains in the integration of various physical parameters to establish reliable and predictive engineering models. Beyond these models taking into account shear rate and volume/mass fraction, variables such as particle equivalent diameters are sometimes included, rarely the morphological criteria and their distribution. The establishment of structured rheological models integrating biocatalytic kinetics and using high dry matter contents remains a scientific goal and is a major condition for bio-industry development. In this paper, the methods and the models related to the rheological behaviour of complex lignocellulose suspensions are similar or adapted from filamentous micro-organism suspensions including concentration, mean diameter, hyphal propensity and shear rate. By analogy, the scientific study of the rheological behaviour of lignocellulose suspensions will aim to integrate additional physical parameters such as morphological parameters (equivalent diameter, mean length, aspect ratio, specific surface area, etc.) as well as the associated distribution functions.

Future prospects depend on this state of the art. Firstly, critical concentration and hydrolysis time can be used to define a "practical feed rate" which may enable suspension viscosity to be controlled and regulated along hydrolysis depending on the release kinetics of fermentable substrates and transfer limitations. Secondly, mixed (lignocellulosic and micro-organism) suspensions should be investigated to study their rheological behaviour, define the critical and limiting parameters and integrate consolidated rheological knowledge into the models applied.

References

1. Vallette P, De Choudens C (1992) Le bois, la pâte, le papier. Centre Technique de l'Industrie des Papiers, Cartons et Celluloses, Grenoble, France. ISBN 9782906579040
2. Gibbons WR, Hughes SR (2009) Integrated biorefineries with engineered microbes and high-value co-products for profitable biofuels production. Vitro Cell Dev Biol Plant 45(3):218–228. doi:10.1007/s11627-009-9202-1
3. Ogier JC, Ballerini D, Leygue JP, Rigal L, Pourquie J (1999) Ethanol production from lignocellulosic biomass. Oil Gas Sci Technol Revue IFP Energies Nouvelles 54(1):67–94. doi:10.2516/ogst:1999004
4. Bommarius AS, Katona A, Cheben SE, Patel AS, Ragauskas AJ, Knudson K, Pu Y (2008) Cellulase kinetics as a function of cellulose pretreatment. Metab Eng 10(6):370–381. doi:10.1016/j.ymben.2008.06.008
5. Lee J (1997) Biological conversion of lignocellulosic biomass to ethanol. J Biotechnol 56 (1):1–24
6. Zhang X, Qin W, Paice MG, Saddler JN (2009) High consistency enzymatic hydrolysis of hardwood substrates. Bioresour Technol 100(23):5890–5897

7. Karunanithy C, Muthukumarappan K, Gibbons WR (2013) Effect of extruder screw speed, temperature, and enzyme levels on sugar recovery from different biomasses. ISRN Biotechnol. doi:10.5402/2013/942810
8. Eggeman T, Elander RT (2005) Process and economic analysis of pretreatment technologies. Bioresour Technol 96(18):2019–2025. doi:10.1016/j.biortech.2005.01.017
9. Wiman M, Palmqvist B, Tornberg E, Liden G (2010) Rheological characterization of dilute acid pretreated softwood. Biotechnol Bioeng 108(5):1031–1041
10. Um BH (2007) Optimization of ethanol production from concentrated substrate. PhD thesis, Auburn University
11. Wucherpfennig T, Kiep KA, Driouch H, Wittmann C, Krull R (2010) Morphology and rheology in filamentous cultivations. Adv Appl Microbiol 72:89–136. doi:10.1016/S0065-2164(10)72004-9
12. Araki J, Wada M, Kuga S, Okano T (1998) Flow properties of microcrystalline cellulose suspension prepared by acid treatment of native cellulose. Colloids Surfaces A—Physicochemical Eng Aspects 142(1):75–82. doi:10.1016/s0927-7757(98)00404-x
13. Boluk Y, Lahiji R, Zhao L, McDermott MT (2011) Suspension viscosities and shape parameter of cellulose nanocrystals (CNC). Colloids and Surfaces A—Physicochem Eng Aspects 377(1–3):297–303. doi:10.1016/j.colsurfa.2011.01.003
14. Gonzalez-Labrada E, Gray DG (2012) Viscosity measurements of dilute aqueous suspensions of cellulose nanocrystals using a rolling ball viscometer. Cellulose 19(5):1557–1565. doi:10.1007/s10570-012-9746-9
15. Lu A, Hemraz U, Khalili Z, Boluk Y (2014) Unique viscoelastic behaviors of colloidal nanocrystalline cellulose aqueous suspensions. Cellulose 21(3):1239–1250. doi:10.1007/s10570-014-0173-y
16. Luukkonen P, Newton JM, Podczeck F, Yliruusi J (2001) Use of a capillary rheometer to evaluate the rheological properties of microcrystalline cellulose and silicified microcrystalline cellulose wet masses. Int J Pharm 216(1–2):147–157. doi:10.1016/s0378-5173(01)00585-3
17. Tatsumi D, Ishioka S, Matsumoto T (2001) Effect of fiber concentration and axial ratio on the rheological properties of cellulose fiber suspensions. J Soc Rheol Jpn 30(1):27–32
18. Bayod E, Bolmstedt U, Innings F, Tornberg E (2005) Rheological characterization of fiber suspensions prepared from vegetable pulp and dried fibers. A comparatible study. Ann Trans Nordic Rheol Soc 13:249–253
19. Horvath AE, Lindstrom T (2007) The influence of colloidal interactions on fiber network strength. J Colloid Interface Sci 309(2):511–517
20. Um BH, Hanley TR (2008) A comparison of simple rheological parameters and simulation data for Zymomonas mobilis fermentation broths with high substrate loading in a 3-L bioreactor. Appl Biochem Biotechnol 145(1/3):29–38
21. Tatsumi D, Hitoshi K, Chen B, Matsumoto T (2008) Effect of natural additives on the rheological properties of cellulose fiber disperse systems. Colloids Surf A: Physicochem Eng Aspects 316:151–158
22. Tozzi EJ, McCarthy MJ, Lavenson DM, Cardona M, Powell RL, Karuna N, Jeoh T (2014) Effect of fiber structure on yield stress during enzymatic conversion of cellulose. AIChE J 60 (5):1582–1590. doi:10.1002/aic.14374
23. Lowys M-P, Desbrières J, Rinaudo M (2000) Rheological characterization of cellulosic microfibril suspensions. Role of polymeric additives. Food Hydrocolloids 15:25–32
24. Agoda-Tandjawa G, Durand S, Berot S, Blassel C, Gaillard C, Garnier C, Doublier J-L (2010) Rheological characterization of microfibrillated cellulose suspensions after freezing. Carbohydr Polym 80:677–686
25. Saarikoski E, Saarinen T, Salmela J, Seppala J (2012) Flocculated flow of microfibrillated cellulose water suspensions: an imaging approach for characterisation of rheological behaviour. Cellulose 19(3):647–659. doi:10.1007/s10570-012-9661-0
26. Geddes CC, Peterson JJ, Mullinnix MT, Svoronos SA, Shanmugam KT, Ingram LO (2010) Optimizing cellulase usage for improved mixing and rheological properties of acid-pretreated

sugarcane bagasse. Bioresour Technol 101(23):9128–9136. doi:10.1016/j.biortech.2010.07.040
27. Pereira LTC, Pereira LTC, Teixeira RSS, Bon EPD, Freitas SP (2011) Sugarcane bagasse enzymatic hydrolysis: rheological data as criteria for impeller selection. J Ind Microbiol Biotechnol 38(8):901–907
28. Pimenova NV, Hanley AR (2004) Effect of corn stover concentration on rheological characteristics. Appl Biochem Biotechnol 113:347–360
29. Stickel JJ, Knutsen JS, Liberatore MW, Luu W, Bousfield DW, Klingenberg DJ, Tim Scott C, Root TW, Ehrhardt MR, Monz TO (2009) Rheology measurements of a biomass slurry: an inter-laboratory study. Rheol Acta 48(9):1005–1015. doi:10.1007/s00397-009-0382-8
30. Sridhar V, McMillan JD, Schell DJ, Elander RT (2009) Rheology of corn stover slurries at high solids concentrations—Effects of saccharification and particle size. Bioresour Technol 100(2):925–934. doi:10.1016/j.biortech.2008.06.070
31. Dunaway KW, Dasari RK, Bennett NG, Berson RE (2010) Characterization of changes in viscosity and insoluble solids content during enzymatic saccharification of pretreated corn stover slurries. Bioresour Technol 101(10):3575–3582. doi:10.1016/j.biortech.2009.12.071
32. Dibble CJ, Shatova TA, Jorgenson JL, Stickel JJ (2011) Particle morphology characterization and manipulation in biomass slurries and the effect on rheological properties and enzymatic conversion. Biotechnol Prog 27(6):1751–1759. doi:10.1002/btpr.669
33. Samaniuk JR, Scott CT, Root TW, Klingenberg DJ (2011) The effect of high intensity mixing on the enzymatic hydrolysis of concentrated cellulose fiber suspensions. Bioresour Technol 102(6):4489–4494. doi:10.1016/j.biortech.2010.11.117
34. Damani R, Powell RL, Hagen N (1993) Viscoelastic characterization of medium consistency pulp suspensions. Can J Chem Eng 71(5):676–684
35. Chen B, Tatsumi D, Matsumoto T (2003) Fiber orientation and flow properties of pulp fiber suspensions under shear flow conditions. Sen'I Gakkaishi 59(12):471–478
36. Blanco A, Negro C, Fuente E, Tijero J (2006) Rotor selection for a Searle-type device to study the rheology of paper pulp suspensions. Chem Eng Process 46:37–44
37. Derakhshandeh B, Hatzikiriakos SG, Bennington CPJ (2010) Rheology of pulp suspensions using ultrasonic Doppler velocimetry. Rheol Acta 49(11–12):1127–1140. doi:10.1007/s00397-010-0485-2
38. Le Moigne N, Jardeby K, Navard P (2010) Structural changes and alkaline solubility of wood cellulose fibers after enzymatic peeling treatment. Carbohydr Polym 79:325–332
39. Chaussy D, Martin C, Roux JC (2011) Rheological behavior of cellulose fiber suspensions: application to paper-making processing. Ind Eng Chem Res 50(6):3524–3533
40. Szijarto N, Siika-Aho M, Sontag-Strohm T, Viikari L (2011) Liquefaction of hydrothermally pretreated wheat straw at high-solids content by purified Trichoderma enzymes. Bioresour Technol 102(2):1968–1974. doi:10.1016/j.biortech.2010.09.012
41. Dasari RK, Berson RE (2007) The effect of particle size on hydrolysis reaction rates and rheological properties in cellulosic slurries. Appl Biochem Biotechnol 137:289–299. doi:10.1007/s12010-007-9059-x
42. Palmqvist B, Liden G (2012) Torque measurements reveal large process differences between materials during high solid enzymatic hydrolysis of pretreated lignocellulose. Biotechnol Biofuels 5:57. doi:10.1186/1754-6834-5-57
43. Allen T (1968) Particle size measurement. Chapman & Hall, London
44. Brittain HG (2001) Particle size distribution. Part I: representation of particle shape, size and distribution. Pharm Technol 25:38–45
45. Bennington CPJ, Kerekes RJ, Grace JR (1990) The yield stress of fiber suspensions. Can J Chem Eng 68(5):748–757
46. Swerin A (1998) Rheological properties of cellulosic fibre suspensions flocculated by cationic polyacrylamides. Colloids Surfaces A: Physicochem Eng Aspects 133 (3):279–294. doi:http://dx.doi.org/10.1016/S0927-7757(97)00212-4
47. Wikström T, Rasmuson A (1998) Yield stress of pulp suspensions. The influence of fibre properties and processing conditions. Nord Pulp Pap Res J 13(3):243–250

48. Tatsumi D, Ishioka S, Matsumoto T (1999) Effect of particle and salt concentrations on the rheological properties of cellulose fibrous suspensions. Nihon Reoroji Gakkaishi 27(4):243–248. doi:10.1678/rheology.27.243
49. Ein-Mozaffari F, Bennington CPJ, Dumont GA (2005) Suspension yield stress and the dynamic response of agitated pulp chests. Chem Eng Sci 60 (8–9):2399–2408. doi:http://dx.doi.org/10.1016/j.ces.2004.11.019
50. Knutsen JS, Liberatore MW (2010) Rheology modification and enzyme kinetics of high-solids cellulosic slurries: an economic analysis. Energy Fuels 24:6506–6512. doi:10.1021/ef100746q
51. Nguyen TC, Anne-Archard D, Coma V, Cameleyre X, Lombard E, Binet C, Nouhen A, To KA, Fillaudeau L (2013) In situ rheometry of concentrated cellulose fibre suspensions and relationships with enzymatic hydrolysis. Bioresour Technol 133:563–572. doi:10.1016/j.biortech.2013.01.110
52. Mohtaschemi M, Dimic-Misic K, Puisto A, Korhonen M, Maloney T, Paltakari J, Alava MJ (2014) Rheological characterization of fibrillated cellulose suspensions via bucket vane viscometer. Cellulose 21(3):1305–1312. doi:10.1007/s10570-014-0235-1
53. Cullen PJ, O'Donnell CP, Houska M (2003) Rotational rheometry using complex geometries—A review. J Texture Stud 34(1):1–20. doi:10.1111/j.1745-4603.2003.tb01052.x
54. Metzner AB, Otto RE (1957) Agitation of non-Newtonian fluids. AIChE J 3(1):3–10. doi:10.1002/aic.690030103
55. Rieger F, Novak V (1973) Power Consumption of Agitators in Highly Viscous non-Newtonian Liquids. Trans Inst Chem Eng 51:105–111
56. Jahangiri M, Golkar-Narenji MR, Montazerin N, Savarmand S (2001) Investigation of the viscoelastic effect on the Metzner and Otto coefficient through LDA velocity measurements. Chin J Chem Eng 9(1):77–83
57. Rao MA (1975) Measurement of flow properties of food suspensions with a mixer. J Texture Stud 6(4):533–539. doi:10.1111/j.1745-4603.1975.tb01426.x
58. Seyssiecq I, Marrot B, Djerroud D, Roche N (2008) In situ triphasic rheological characterisation of activated sludge in an aerated bioreactor. Chem Eng J 142(1):40–47. doi:10.1016/j.cej.2007.11.007
59. Rabia A, Yahiaoui S, Djabourov M, Feuillebois F, Lasuye T (2014) Optimization of the vane geometry. Rheol Acta 53(4):357–371. doi:10.1007/s00397-014-0759-1
60. Nguyen QD, Boger DV (1992) Measuring the flow properties of yield stress fluids. Annu Rev Fluid Mech 24:47–88
61. Barnes HA (1997) Thixotropy—A review. J Nonnewton Fluid Mech 70(1–2):1–33. doi:10.1016/s0377-0257(97)00004-9
62. Coussot P, Ancey C (1999) Rhéophysique des pâtes et suspensions. EDP Sciences, Paris
63. Derakhshandeh B, Kerekes RJ, Hatzikiriakos SG, Bennington CPJ (2011) Rheology of pulp fibre suspensions: a critical review. Chem Eng Sci 66(15):3460–3470. doi:10.1016/j.ces.2011.04.017
64. Marti I, Hofler O, Fischer P, Windhab EJ (2005) Rheology of concentrated suspensions containing mixtures of spheres and fibres. Rheol Acta 44(5):502–512. doi:10.1007/s00397-005-0432-9
65. Santamaria-Holek I, Mendoza CI (2010) The rheology of concentrated suspensions of arbitrarily-shaped particles. J Colloid Interface Sci 346(1):118–126. doi:10.1016/j.jcis.2010.02.033
66. Barnes HA, Hutton JF, Walters K (1989) An introduction to rheology. Elsevier
67. Larson RG (1999) The structure and rheology of complex fluids. Oxford University Press, New-York. ISBN 9780195121971
68. Tadros TF (2010) Rheology of dispersions. Wiley-VCH Verlag GmbH & Co. ISBN: 978-3-527-32003-5
69. Quemada D (2006) Modélisation Rhéologique Structurelle. Dispersions Concentrées et Fluides Complexes. Edition Tec & Doc Lavoisier, Paris

70. Wu Q, Meng,YJ, Wang SQ, Li YJ, Fu SY, Ma LF, Harper D (2014) Rheological behavior of cellulose nanocrystal suspension: influence of concentration and aspect ratio. J Appl Polym Sci 131(15). doi:10.1002/app.40525
71. Tucker KG, Thomas CR (1993) Effect of biomass concentration and morphology on the rheological parameters of Penicillium chryso-genurin fermentation broths. Trans Inst Chem Eng 71:111–117
72. Riley GL, Tucker KG, Paul GC, Thomas C (2000) Effect of biomass concentration and mycelial morphology on fermentation broth rheology. Biotechnol Bioeng 68(2):160–172
73. Petersen N, Stocks S, Gernaey K (2008) Multivariate models for prediction of rheological characteristics of filamentous fermentation broth from the size distribution. Biotechnol Bioeng 100(1):61–71
74. Mohseni M, Allen DG (1995) The effect of particle morphology and concentration on the directly measured yield stress in filamentous suspensions. Biotechnol Bioeng 48:257–265
75. Hirata Y, Aoshima Y (1996) Formation and growth of cavern in yield stress fluids agitated under baffled and non-baffled conditions. Chem Eng Res Des 74(4):438–444
76. Buscall R, Mills PDA, Stewart RF, Sutton D, White LR, Yates GE (1987) The rheology of strongly-flocculated suspensions. J Nonnewton Fluid Mech 24(2):183–202. doi:10.1016/0377-0257(87)85009-7
77. Dalpke B, Kerekes RJ (2005) The influence of fibre properties on the apparent yield stress of flocculated pulp suspensions. J Pulp Pap Sci 31(1):39–43
78. Ehrhardt MR, Monz TO, Root TW, Connelly RK, Scott CT, Klingenberg DJ (2010) Rheology of dilute acid hydrolyzed corn stover at high solids concentration. Appl Biochem Biotechnol 160(4):1102–1115. doi:10.1007/s12010-009-8606-z
79. Rosgaard L, Andric P, Dam-Johansen K, Pedersen S, Meyer AS (2007) Effects of substrate loading on enzymatic hydrolysis and viscosity of pretreated barley straw. Appl Biochem Biotechnol 143(1):27–40. doi:10.1007/s12010-007-0028-1
80. Fillaudeau L, Babau M, Cameleyre X, Lombard E, Anne-Archard D (2011) Libération de substrats fermentescibles à partir de matrices lignocellulosiques issues de l'industrie papetière. *Récents Progrès en Génie des Procédés* 101, Lavoisier Ed., Paris
81. Roche CM, Dibble CJ, Knutsen JS, Stickel JJ, Liberatore MW (2009) Particle concentration and yield stress of biomass slurries during enzymatic hydrolysis at high-solids loadings. Biotechnol Bioeng 104:290–300

Adv Biochem Eng Biotechnol (2015) 149: 359–364
DOI 10.1007/978-3-319-20511-3

Index

V

W

X

Y

MIX
Papier aus verantwortungsvollen Quellen
Paper from responsible sources
FSC® C105338

If you have any concerns about our products,
you can contact us on
ProductSafety@springernature.com

In case Publisher is established outside the EU,
the EU authorized representative is:
Springer Nature Customer Service Center GmbH
Europaplatz 3, 69115 Heidelberg, Germany

Printed by Libri Plureos GmbH
in Hamburg, Germany